Multivariable Calculus

Multivariable Calculus

Ivan Dimov
Aleksandar Petkov

Kruger Brentt
Publishers

2025

Kruger Brentt Publishers UK. LTD.
Company Number 9728962

Regd. Office: 68 St Margarets Road, Edgware, Middlesex HA8 9UU

© 2025 AUTHORS

ISBN: 978-1-78715-377-6

For information on all our publications visit our website at http://krugerbrentt.com/

PREFACE

Multivariable calculus serves as a cornerstone of advanced mathematics, providing the framework for understanding and analyzing functions of multiple variables in both theoretical and applied contexts. From modeling physical phenomena in engineering and physics to optimizing processes in economics and biology, multivariable calculus plays a fundamental role in numerous scientific disciplines and real-world applications. Understanding the principles, techniques, and applications of multivariable calculus is essential for students, researchers, and practitioners seeking to tackle complex problems and explore the interconnectedness of variables in multidimensional spaces.

This handbook, titled "Multivariable Calculus," is a comprehensive exploration of the principles, methodologies, and applications of multivariable calculus. Drawing upon the expertise of mathematicians, educators, and researchers, this volume offers theoretical insights, practical guidance, and computational examples for understanding and applying multivariable calculus concepts.

From the fundamentals of vector calculus and partial derivatives to the applications of multivariable optimization and integration techniques, the "Multivariable Calculus" handbook covers a wide range of topics essential for mastering the field. Through clear explanations, mathematical derivations, and real-world examples, readers will gain a deep understanding of the principles and methods used to analyze functions of multiple variables and their geometric interpretations.

Moreover, this handbook explores the applications of multivariable calculus across different scientific and engineering disciplines, including physics, engineering, economics, and computer science. By examining case studies, research findings, and computational methods, we highlight the diverse ways in which multivariable calculus is used to model complex systems, solve optimization problems, and analyze data in multidimensional spaces.

As we continue to push the boundaries of scientific inquiry and technological innovation, the importance of multivariable calculus in understanding complex systems and phenomena has never been greater. It is our hope that the "Multivariable Calculus" handbook will serve as a valuable resource for students, educators, and practitioners in the field of mathematics and its applications. Whether you are studying mathematical theory, conducting research in applied fields, or solving practical problems in engineering and science, this handbook invites you to explore the rich and diverse world of multivariable calculus and its myriad applications.

Ivan Dimov

Aleksandar Petkov

CONTENTS

Contents

PART-I
PRELIMINARIES

1
Chapter

$\mathbb{R}^N$

Let R denote the set of real numbers. Its elements are also called scalars. If n is a positive integer, then $\mathbb{R}^n$ is defined to be the set of all sequences x of n real numbers:

$$\mathbf{x} = (x_1, x_2, \ldots, x_n). \tag{1.1}$$

The elements of $\mathbb{R}^n$ are called **points**, vectors, or **n-tuples**. We follow the convention of indicating vectors in boldface and scalars in plainface. For a vector x, the individual scalar entries x_i for i = 1, 2, ..., n are called **coordinates** or **components**.

Multivariable calculus studies functions between these sets, that is, functions of the form $f: \mathbb{R}^n \to Rm$, or, more accurately, of the form $f: A \to Rm$, where A is a subset of $\mathbb{R}^n$. In this context, if x represents a typical point of $\mathbb{R}^n$, the coordinates $x_1, x_2, \ldots, x_n$ are referred to as **variables**. For example, first-year calculus studies real-valued functions of one variable, functions of the form $f: R \to R$.

This chapter collects some of the background information about $\mathbb{R}^n$ that we use throughout the book. The presentation is meant to be self-contained, though readers who have studied linear algebra are likely to have a greater perspective on how the pieces fit together as part of a bigger picture.

1.1 VECTOR ARITHMETIC

There are two basic algebraic operations in $\mathbb{R}^n$. Let $x = (x_1, x_2, \ldots, x_n)$ and $y = (y_1, y_2, \ldots, y_n)$ be elements of $\mathbb{R}^n$, and let c be a real number. The aforementioned operations are defined as follows.

- **Addition:** $x + y = (x_1 + y_1, x_2 + y_2, \ldots, x_n + y_n)$.
- **Scalar multiplication:** $cx = (cx_1, cx_2, \ldots, cx_n)$.

Because of our familiarity with $\mathbb{R}^2$, we illustrate these concepts there in some detail. An element of $\mathbb{R}^2$ is an ordered pair $x = (x_1, x_2)$. Geometrically, x is a point in the plane plotted in the usual way. In particular, the origin (0, 0) is called the **zero vector** and is denoted by **0**. Alternatively, we may visualize x by drawing the arrow starting at (0, 0) and ending at (x_1, x_2). We'll go back and forth freely between the point/arrow viewpoints.

Given two vectors $x = (x_1, x_2)$ and $y = (y_1, y_2)$ in $\mathbb{R}^2$, the sum $x + y$ as defined above is the point that results from adding the displacements in each of the horizontal and vertical directions, respectively. For instance, if $x = (1, 2)$ and $y = (3, 4)$, then $x + y = (4, 6)$. Thinking of x and y as arrows emanating from 0, this places $x + y$ at the vertex opposite the origin in the parallelogram determined by x and y, as on the left of Figure 1.1. If we think of $x + y$ as an arrow as well, it is one of the diagonals of the parallelogram, as shown on the right.

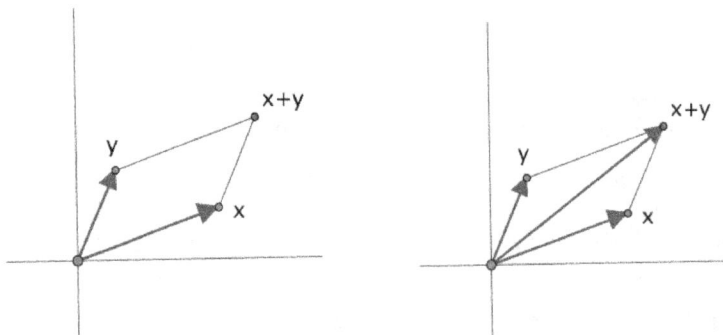

Figure 1.1: Vector addition

Another way to reach $x + y$ is to move the arrow representing y so that it retains the same length and direction but begins at the endpoint of x. This is called a "translation" of the original vector y. Then $x + y$ is the destination if you go along x followed by the translated version of y, as illustrated in Figure 1.2, like following two displacements x and y in succession. It is often convenient to translate vectors, especially when they represent quantities for which length and direction are the most relevant characteristics. Nevertheless, it's important to remember that the translations are only copies: the real vector starts at the origin.

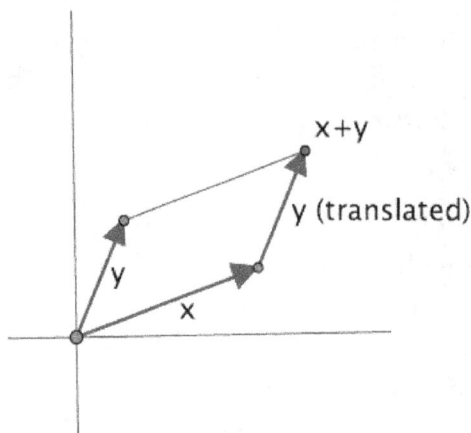

Figure 1.2: Using the translation of a vector

Similarly, if c is a real number, then cx results from multiplying each of the coordinate dis- placements by a factor of c. For instance, if x = (1, 2), then 3x = (3, 6). In general, cx is an arrow |c| times as long as x and in the same direction as x if c > 0, the opposite direction if c < 0. See Figure 1.3. In particular, (–1)x is a copy of x rotated by 180° to reverse the direction. It is usually denoted by –x since it satisfies x + (–1)x = 0.

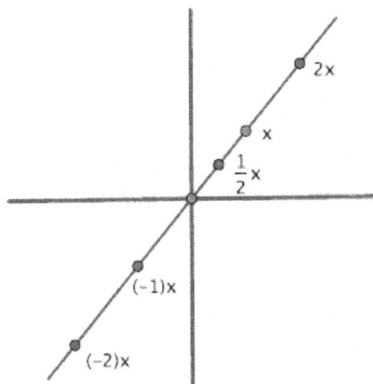

Figure 1.3: Scalar multiplication

Going back to the parallelogram used to visualize x+y, we could also look at the other diagonal, say drawn as an arrow from y to x, as indicated in Figure 1.4. We sometimes denote this arrow by $\overrightarrow{yx}$. It is what you would add to y to get to x. In other words, it is the difference x - y:

$$\overrightarrow{yx} = x - y.$$

Again, the real x - y should be returned so that it starts at 0.

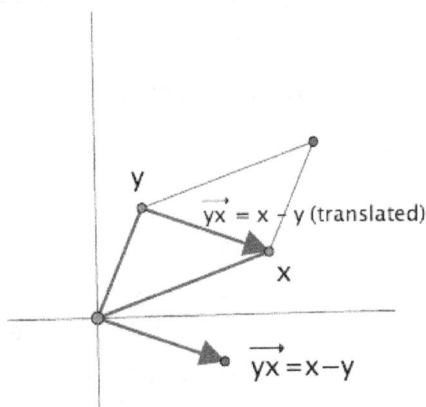

Figure 1.4: The difference of two vectors

In the case of $\mathbb{R}^2$, the coordinates are usually denoted by x, y, rather than x_1, x_2. Then $\mathbb{R}^2$ is the usual xy-plane. The notation is potentially confusing since x is also often used to denote the generic vector in R^n, as in equation (1.1) above. Hopefully, the context and the use of boldface will clarify whether a coordinate or vector is intended. Similarly, R^3 represents three-dimensional space. Its coordinates are often denoted by x, y, z.

Returning to the general case, every vector x = (x_1, x_2, . . . , x_n) in $\mathbb{R}^n$ can be decomposed as a sum along the coordinate directions:

$$x = (x_1, 0, \ldots, 0) + (0, x_2, \ldots, 0) + \cdots + (0, 0, \ldots, x_n)$$
$$= x_1(1, 0, \ldots, 0) + x_2(0, 1, \ldots, 0) + \cdots + x_n(0, 0, \ldots, 1).$$

The vectors e_1 = (1, 0, . . . , 0), e_2 = (0, 1, . . . , 0), . . . , e_n = (0, 0, , 1), with a 1 in a single component and zeros everywhere else, are called the standard basis vectors. Thus:

$$x = x_1 e_1 + x_2 e_2 + \cdots + x_n e_n. \tag{1.2}$$

The scalar coefficients x_i are the coordinates of x.

In general, a set of n vectors $\{v_1, v_2, \ldots, v_n\}$ in $\mathbb{R}^n$ is called a basis if every x in $\mathbb{R}^n$ can be written in a unique way as a combination x = $c_1 v_1$ +$c_2 v_2$ +$\cdots$+$c_n v_n$ for some scalars $c_1, c_2, \ldots\ldots, c_n$.

Any basis can be used to define its own coordinate system in $\mathbb{R}^n$, and $\mathbb{R}^n$ has many different bases.

Apart from a few occasions, however, we'll stick with the standard basis.

In $\mathbb{R}^2$, the standard basis vectors are usually denoted by i = (1, 0) and j = (0, 1). In R^3, the corresponding names are i = (1, 0, 0), j = (0, 1, 0), and k = (0, 0, 1).

1.2 LINEAR TRANSFORMATIONS

Linear transformations are functions that respect vector addition and scalar multiplication. More precisely:

Definition. A function T : $\mathbb{R}^n$ → Rm is called a linear transformation if:

⊙ T (x + y) = T (x) + T (y) and

⊙ T (cx) = c T (x).

The conditions must be satisfied for all x, y in $\mathbb{R}^n$ and all c in R.

This definition may seem austere, but many familiar functions are linear transformations. We just may not be used to thinking of them that way.

Example 1.1. Let T : $\mathbb{R}^2$ → $\mathbb{R}^2$ be counterclockwise rotation by $\frac{\pi}{3}$ about the origin. That is, if x ∈ $\mathbb{R}^2$, then T (x) is the point that is reached after x is rotated counterclockwise by $\frac{\pi}{3}$ about 0. We give a geometric argument that T satisfies the two requirements for being a linear transformation.

First, consider the parallelogram determined by vectors x and y in $\mathbb{R}^2$. Then T rotates this parallelogram to another parallelogram, the one determined by T (x) and T (y). In particular, the vertex x + y is rotated to the vertex T (x) + T (y). This says exactly that T (x + y) = T (x) + T (y). This is illustrated at the left of Figure 1.5.

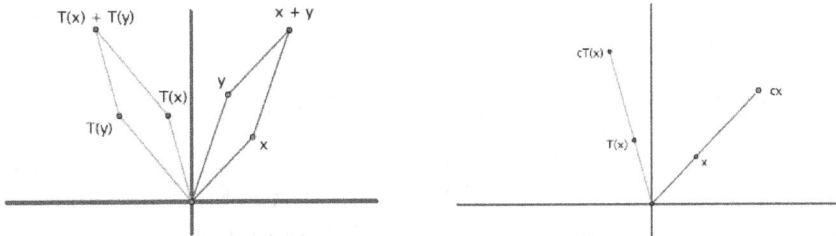

Figure 1.5: Rotations are linear transformations.

This geometric argument is so simple that it may not be clear that there is any actual reasoning behind it. The reader is encouraged to go through it carefully to pin down how it proves what we want. Also, the case where x and y are collinear, so that they don't determine an honest parallelogram, requires a separate argument. We won't give it, though see the reasoning in the next paragraph.

Similarly, T rotates the line through the origin and x to the line through the origin and T (x). The point on the first line c times as far from the origin as x is rotated to the point c times as far from the origin as T (x), as indicated on the right of the figure. In other words, T (cx) = c T (x).

Example 1.2. Likewise, the function $T : \mathbb{R}^2 \to \mathbb{R}^2$ that reflects a point $x = (x_1, x_2)$ in the x_1-axis is a linear transformation. The appropriate supporting diagrams are shown in Figure 1.6.

Example 1.3. Let $T : R^3 \to \mathbb{R}^2$ be the function that projects a point x = (x1, x2, x3) in R^3 onto the $x_1 x_2$-plane, that is, $T (x_1, x_2, x_3) = (x_1, x_2)$. Rather than use pictures, this time we show that T is linear by calculating.

For instance:

$$T(\mathbf{x} + \mathbf{y}) = T(x_1 + y_1, x_2 + y_2, x_3 + y_3) = (x_1 + y_1, x_2 + y_2).$$

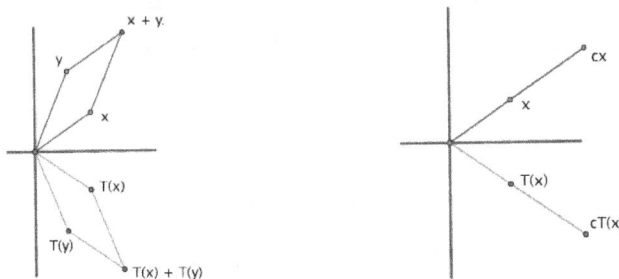

Figure 1.6: So are reflections.

On the other hand:

$$T(\mathbf{x}) + T(\mathbf{y}) = (x_1, x_2) + (y_1, y_2) = (x_1 + y_1, x_2 + y_2).$$

Both expressions equal the same thing, so T (x + y) = T (x) + T (y).

Similarly, T (cx) = T (cx1, cx2, cx3) = (cx1, cx2) while c T (x) = c(x1, x2) = (cx1, cx2). Thus T (cx) = c T (x) as well.

1.3 THE MATRIX OF A LINEAR TRANSFORMATION

Let T : $\mathbb{R}^n$ → Rm be a linear transformation. One of the things that make such a function easy to work with is that, although T (x) is defined for all x in $\mathbb{R}^n$, the transformation is actually determined by a finite amount of input. To make sense of this, recall from equation (1.2) that every x in $\mathbb{R}^n$ can be expressed in terms of the standard basis vectors: x = $x_1e_1 + x_2e_2 + \cdots + x_ne_n$. Thus, by the defining properties of a linear transformation:

$$\begin{aligned}
T(\mathbf{x}) &= T(x_1\mathbf{e}_1 + x_2\mathbf{e}_2 + \cdots + x_n\mathbf{e}_n) \\
&= T(x_1\mathbf{e}_1) + T(x_2\mathbf{e}_2) + \cdots + T(x_n\mathbf{e}_n) \\
&= x_1T(\mathbf{e}_1) + x_2T(\mathbf{e}_2) + \cdots + x_nT(\mathbf{e}_n) \\
&= x_1\mathbf{a}_1 + x_2\mathbf{a}_2 + \cdots + x_n\mathbf{a}_n,
\end{aligned} \tag{1.3}$$

where a_j = T (e_j) for j = 1, 2, . . . , n. Conversely, given any n vectors $a_1, a_2, \ldots, a_n$ in Rm, it's straightforward to check by calculation that the formula T (x) = $x_1a_1 + x_2a_2 + \cdots + x_na_n$ satisfies the two requirements for being a linear transformation. This is Exercise 2.3 at the end of the chapter. Moreover, T (e_j) = T (0, 0, . . . , 1, . . . , 0) = 0 · a_1 + 0 · a_2 + · · · + 1 · a_j + · · · + 0 · a_n = a_j for each j. In other words, a linear transformation T : $\mathbb{R}^n$ → Rm is completely determined, as in (1.3), once you know the values T (e_1) = a_1, T (e_2) = a_2, . . . , T (e_n) = a_n, and there are no restrictions on what these values can be.

To understand how this might be used, we consider first the case of a linear transformation T : $\mathbb{R}^n$ → R whose values are real numbers. Then every T (e_j) is a scalar, say T (e_j) = a_j and equation (1.3) becomes:

$$T(\mathbf{x}) = x_1a_1 + x_2a_2 + \cdots + x_na_n. \tag{1.4}$$

Sums of this type are an indispensable part of vector algebra.

Definition. Given x = ($x_1, x_2, \ldots, x_n$) and y = ($y_1, y_2, \ldots, y_n$) in $\mathbb{R}^n$, the **dot product**, denoted by **x · y**, is defined by:

$$\mathbf{x} \cdot \mathbf{y} = x_1y_1 + x_2y_2 + \cdots + x_ny_n.$$

For example, in R^3, if x = (1, 2, 3) and y = (4, 5, 6), then x·y = 1·4+2·5+3·6 = 4+10+18 = 32.

The dot product satisfies a variety of elementary properties, such as x · y = y · x. The ones we shall use are pretty obvious, so we won't bother listing them out, though please see Exercises 5.5–5.10 if you'd like to see some of them collected together.

Returning to the main point, we have shown in equation (1.4) that every real-valued linear transformation $T : \mathbb{R}^n \to \mathbb{R}$ has the form:

$$T(\mathbf{x}) = \mathbf{a} \cdot \mathbf{x}$$

for some vector a = $(a_1, a_2, \ldots, a_n)$ in $\mathbb{R}^n$.

The analysis for the general case of a linear transformation $T : \mathbb{R}^n \to \mathbb{R}m$ follows the same pattern except that now the values $a_j = T(e_j)$ are vectors in $\mathbb{R}m$. We record these values by putting them in the columns of a rectangular table. That is, for each $j = 1, 2, \ldots, n$, say a_j is the vector $(a_{1j}, a_{2j}, \ldots, a_{mj})$ in $\mathbb{R}m$, and let A be the table:

$$A = \begin{bmatrix} a_{11} & a_{12} & \cdots & a_{1j} & \cdots & a_{1n} \\ a_{21} & a_{22} & \cdots & a_{2j} & \cdots & a_{2n} \\ \vdots & \vdots & & \vdots & & \vdots \\ a_{m1} & a_{m2} & \cdots & a_{mj} & \cdots & a_{mn} \end{bmatrix},$$

where a_j is highlighted in red in the jth column. Such a table is called a **matrix**. In fact, A is called an m by n matrix, which means that it has m rows and n columns. The subscripting has been chosen so that a_{ij} is the entry in row i, column j, where the rows are numbered starting from the top and the columns starting from the left.

The matrix obtained in this way is called the **matrix of T** with respect to the standard bases. Since, by equation (1.3), the transformation T is completely determined by the columns a_j, the matrix contains all the data we need to find T (x) for all x in $\mathbb{R}^n$.

We illustrate this with the three examples of linear transformations considered earlier.

Example 1.4. Let $T : \mathbb{R}^2 \to \mathbb{R}^2$ be the counterclockwise rotation by $\frac{\pi}{3}$ about the origin. Then T rotates the vector $e_1 = (1, 0)$ to the vector on the unit circle that makes an angle of $\frac{\pi}{3}$ with the x_1-axis. That is, $T(e_1) = (\cos\frac{\pi}{3}, \sin\frac{\pi}{3}) = (\frac{1}{2}, \frac{\sqrt{3}}{2})$. Similarly, $T(e_2) = (\cos\frac{5\pi}{6}, \sin\frac{5\pi}{6}) = (-\frac{\sqrt{3}}{2}, \frac{1}{2})$.

Hence the matrix of T with respect to the standard bases is:

$$A = \begin{bmatrix} \frac{1}{2} & -\frac{\sqrt{3}}{2} \\ \frac{\sqrt{3}}{2} & \frac{1}{2} \end{bmatrix}. \tag{1.5}$$

Example 1.5. If $T : \mathbb{R}^2 \to \mathbb{R}^2$ is the reflection in the x1-axis, then T (e_1) = T (1, 0) = (1, 0) and T (e_2) = T (0, 1) = (0, –1). Hence:

$$A = \begin{bmatrix} 1 & 0 \\ 0 & -1 \end{bmatrix}. \tag{1.6}$$

Example 1.6. Lastly, if $T : R^3 \to \mathbb{R}^2$ is the projection of $x_1x_2x_3$-space onto the x_1x_2-plane, then $T(e_1) = T(1, 0, 0) = (1, 0)$, $T(e_2) = T(0, 1, 0) = (0, 1)$, and $T(e_3) = T(0, 0, 1) = (0, 0)$, so:

$$A = \begin{bmatrix} 1 & 0 & 0 \\ 0 & 1 & 0 \end{bmatrix}. \tag{1.7}$$

To use the matrix A to compute T (x) in a systematic way, we observe the convention that vectors are identified with matrices having a single column. Thus:

$$\mathbf{x} = \begin{bmatrix} x_1 \\ x_2 \\ \vdots \\ x_n \end{bmatrix} \quad \text{and} \quad \mathbf{a}_j = \begin{bmatrix} a_{1j} \\ a_{2j} \\ \vdots \\ a_{mj} \end{bmatrix}$$

represent vectors in $\mathbb{R}^n$ and Rm, respectively. Sometimes they're referred to as column vectors. With this notation, equation (1.3) becomes:

$$T(\mathbf{x}) = x_1 \begin{bmatrix} a_{11} \\ a_{21} \\ \vdots \\ a_{m1} \end{bmatrix} + x_2 \begin{bmatrix} a_{12} \\ a_{22} \\ \vdots \\ a_{m2} \end{bmatrix} + \cdots + x_n \begin{bmatrix} a_{1n} \\ a_{2n} \\ \vdots \\ a_{mn} \end{bmatrix}$$

$$= \begin{bmatrix} a_{11}x_1 + a_{12}x_2 + \cdots + a_{1n}x_n \\ a_{21}x_1 + a_{22}x_2 + \cdots + a_{2n}x_n \\ \vdots \\ a_{m1}x_1 + a_{m2}x_2 + \cdots + a_{mn}x_n \end{bmatrix}. \tag{1.8}$$

This is similar to the real-valued case (1.4) in that each component is a dot product. Specifically, the ith component is the dot product of the ith row of A and x. This expression is given a name.

Definition (Preliminary version of matrix multiplication). Let A be an m by n matrix, and let x be a column vector in $\mathbb{R}^n$. Then the **product** Ax is defined to be the column vector y in Rm whose ith component is the dot product of the ith row of A and x:

$$y_i = a_{i1}x_1 + a_{i2}x_2 + \cdots + a_{in}x_n.$$

In other words, Ax is the column vector given by equation (1.8) above.

We've labeled this as a "preliminary" version, because we define a more general matrix multi- plication shortly that includes this as a special case.

Using this definition, equation (1.8) can be written in the simple form T (x) = Ax. The preceding discussion is summarized in the following result.

Proposition 1.7.

1. Given any linear transformation T : $\mathbb{R}^n$ → Rm, there is an m by n matrix A such that:

⊙ the jth column of A is T (e$_j$) for j = 1, 2, . . . , n, and

⊙ T (x) = Ax for all x in $\mathbb{R}^n$.

2. Conversely, given any m by n matrix A, the formula T (x) = Ax defines a linear transformation T : $\mathbb{R}^n$ → Rm.

We apply this to the examples previously considered. For instance, if T is the counterclockwise rotation of $\mathbb{R}^2$ by $\frac{\pi}{3}$ about the origin, then by (1.5):

$$T(\mathbf{x}) = \begin{bmatrix} \frac{1}{2} & -\frac{\sqrt{3}}{2} \\ \frac{\sqrt{3}}{2} & \frac{1}{2} \end{bmatrix} \begin{bmatrix} x_1 \\ x_2 \end{bmatrix} = \begin{bmatrix} \frac{1}{2}x_1 - \frac{\sqrt{3}}{2}x_2 \\ \frac{\sqrt{3}}{2}x_1 + \frac{1}{2}x_2 \end{bmatrix} = \left(\frac{1}{2}x_1 - \frac{\sqrt{3}}{2}x_2, \frac{\sqrt{3}}{2}x_1 + \frac{1}{2}x_2 \right).$$

Once you get the hang of it, this may be the simplest way to find a formula for a rotation.

For the reflection in the x$_1$-axis, (1.6) gives:

$$T(\mathbf{x}) = \begin{bmatrix} 1 & 0 \\ 0 & -1 \end{bmatrix} \begin{bmatrix} x_1 \\ x_2 \end{bmatrix} = \begin{bmatrix} 1 \cdot x_1 + 0 \cdot x_2 \\ 0 \cdot x_1 - 1 \cdot x_2 \end{bmatrix} = \begin{bmatrix} x_1 \\ -x_2 \end{bmatrix} = (x_1, -x_2).$$

We didn't need matrix methods to come up with this formula, but at least it's correct.

Lastly, for the projection of R³ onto the x$_1$x$_2$-plane, we find from (1.7) that:

$$T(\mathbf{x}) = \begin{bmatrix} 1 & 0 & 0 \\ 0 & 1 & 0 \end{bmatrix} \begin{bmatrix} x_1 \\ x_2 \\ x_3 \end{bmatrix} = \begin{bmatrix} x_1 + 0 + 0 \\ 0 + x_2 + 0 \end{bmatrix} = \begin{bmatrix} x_1 \\ x_2 \end{bmatrix} = (x_1, x_2).$$

as expected.

1.4 MATRIX MULTIPLICATION

Let T : $\mathbb{R}^n$ → Rm and S : Rm → R^p be functions, where, as indicated, the codomain of T equals the domain of S. The composition S ° T : $\mathbb{R}^n$ → R^p is defined to be the function given by (S ° T)(x) = S(T (x)) for all x in $\mathbb{R}^n$.

If S and T are linear transformations, it's easy to check that S ° T is a linear transformation, too:

⊙ S(T (x + y)) = S(T (x) + T (y)) = S(T (x)) + S(T (y)),

⊙ $S(T(cx)) = S(c\,T(x)) = c\,S(T(x))$.

In each case, the first equation follows from the definition of linear transformation applied to T and the second applied to S. As a result, S ° T is represented by a matrix C with respect to the standard bases. Let A and B be the matrices of S and T, respectively, with respect to the standard bases. We shall describe C in terms of A and B.

By Proposition 1.7, the jth column c_j of C is $S(T(e_j))$. For the same reason, $T(e_j)$ is the jth column of B. Let's call it b_j, so $cj = S(b_j)$. On the other hand, S is represented by the matrix A, so c_j is the matrix product $c_j = Ab_j$. Its ith component is the ith row of A dotted with b_j, that is, the dot product of the ith row of A and the jth column of B. Doing this for all j = 1, 2, . . . , n, fills out the matrix C. The matrix obtained in this way is called the product of A and B.

Definition (Final version of matrix multiplication). Let A be a p by m matrix and B an m by n matrix. Their **product**, denoted by AB, is the p by n matrix C whose (i, j) th entry is the dot product of the ith row of A and the jth column of B:

$$c_{ij} = a_{i1}b_{1j} + a_{i2}b_{2j} + \cdots + a_{im}b_{mj} = \sum_{k=1}^{m} a_{ik}b_{kj}.$$

Visually, the action is something like this:

$$\begin{bmatrix} & \vdots & \\ \cdots & c_{ij} & \cdots \\ & \vdots & \end{bmatrix} = \begin{bmatrix} \vdots & \vdots & & \vdots \\ a_{i1} & a_{i2} & \cdots & a_{im} \\ \vdots & \vdots & & \vdots \end{bmatrix} \begin{bmatrix} \cdots & b_{1j} & \cdots \\ \cdots & b_{2j} & \cdots \\ & \vdots & \\ \cdots & b_{mj} & \cdots \end{bmatrix}.$$

This only makes sense when the number of columns of A equals the number of rows of B.

The definitions have been rigged so that the following statement is an immediate consequence.

Proposition 1.8. *Composition of linear transformations corresponds to matrix multiplication.* That is, let S : Rm → R^p and T : $\mathbb{R}^n$ → Rm be linear transformations with matrices A and B, respectively, with respect to the standard bases. Then S ° T is a linear transformation, and its matrix with respect to the standard bases is the product AB.

Example 1.9. Let T : $\mathbb{R}^2$ → $\mathbb{R}^2$ be the counterclockwise rotation by $\frac{\pi}{3}$ about the origin, S : $\mathbb{R}^2$ → $\mathbb{R}^2$ the reflection in the x_1-axis, and consider the composition S°T (first rotate, then reflect). Using the matrices from (1.5) and (1.6), the matrix of S °

$$\begin{bmatrix} 1 & 0 \\ 0 & -1 \end{bmatrix} \begin{bmatrix} \frac{1}{2} & -\frac{\sqrt{3}}{2} \\ \frac{\sqrt{3}}{2} & \frac{1}{2} \end{bmatrix} = \begin{bmatrix} 1 \cdot \frac{1}{2} + 0 \cdot \frac{\sqrt{3}}{2} & 1 \cdot (-\frac{\sqrt{3}}{2}) + 0 \cdot \frac{1}{2} \\ 0 \cdot \frac{1}{2} + (-1) \cdot \frac{\sqrt{3}}{2} & 0 \cdot (-\frac{\sqrt{3}}{2}) + (-1) \cdot \frac{1}{2} \end{bmatrix} = \begin{bmatrix} \frac{1}{2} & -\frac{\sqrt{3}}{2} \\ -\frac{\sqrt{3}}{2} & -\frac{1}{2} \end{bmatrix}.$$

Working backwards and looking at the columns of this matrix, this tells us that $(S \circ T)$ $(e_1) = (\frac{1}{2}, -\frac{\sqrt{3}}{2}) = (\cos(-\frac{\pi}{3}), \sin(-\frac{\pi}{3}))$ and $(S \circ T)(e_2) = (-\frac{\sqrt{3}}{2}, -\frac{1}{2}) = (\cos(\frac{7\pi}{6}), \sin(\frac{7\pi}{6}))$. These points are plotted in Figure 1.7. A linear transformation that has the same effect on e_1 and e_2 is the reflection in the line ℓ that passes through the origin and makes an angle of $-\frac{\pi}{6}$ with the positive x_1-axis.

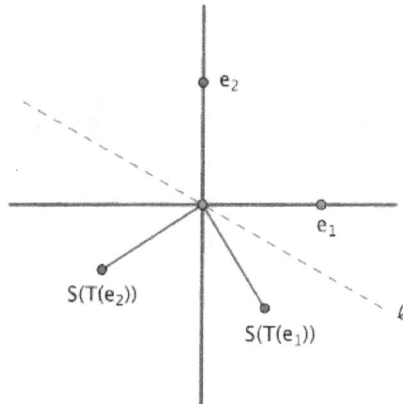

Figure 1.7: The composition of a rotation and a reflection

But linear transformations are completely determined by what they do to the standard basis: two transformations that do the same thing must be the same transformation. Thus we conclude that $S \circ T$ is the reflection in ℓ. This can be verified with geometric reasoning as well.

By comparison, the composition $T \circ S$ (first reflect, then rotate) is represented by the same matrices multiplied in the opposite order:

$$\begin{bmatrix} \frac{1}{2} & -\frac{\sqrt{3}}{2} \\ \frac{\sqrt{3}}{2} & \frac{1}{2} \end{bmatrix} \begin{bmatrix} 1 & 0 \\ 0 & -1 \end{bmatrix} = \begin{bmatrix} \frac{1}{2} & \frac{\sqrt{3}}{2} \\ \frac{\sqrt{3}}{2} & -\frac{1}{2} \end{bmatrix}. \tag{1.9}$$

Note that this is different from the matrix of $S \circ T$. Matrix multiplication need not obey the commutative law AB = BA. (Can you describe geometrically the linear transformation that (1.9) represents?)

1.5 THE GEOMETRY OF THE DOT PRODUCT

We now discuss some geometric properties of the dot product, or, perhaps more accurately, how the dot product can be used to develop geometric intuition about $\mathbb{R}^n$ when n is large enough to be outside direct experience.

Definition. The norm, or magnitude, of a vector $x = (x_1, x_2, \ldots, x_n)$ in $\mathbb{R}^n$, denoted by $\|x\|$, is defined to be:

$$\|x\| = \sqrt{x_1^2 + x_2^2 + \cdots + x_n^2}.$$

For instance, in $\mathbb{R}^3$, if x = (1, 2, 3), then $\|x\| = \sqrt{1+4+9} = \sqrt{14}$.

The following simple property gets used a lot.

Proposition 1.10. If $x \in \mathbb{R}^n$, then $x \cdot x = \|x\|^2$.

Proof. Both sides equal $x_1^2 + x_2^2 + \cdots + x_n^2$.

We return to the familiar setting of the plane and examine these notions there. For instance, if x = (x_1, x_2), then $\|x\| = \sqrt{x_1^2 + x_2^2}$. By the Pythagorean theorem, this is the length of the hypotenuse of a right triangle with legs |x1| and |x2|. If we think of x as an arrow emanating from the origin, then $\|x\|$ is the length of the arrow. If we think of x as a point, then $\|x\|$ is the distance from x to the origin.

Given two points x and y in $\mathbb{R}^2$, the distance between them is the length of the arrow that connects them, $\vec{yx} = x - y$. Hence:

Distance between x and y = $\|x - y\|$.

Next, let x = (x_1, x_2) and y = (y_1, y_2) be nonzero elements of $\mathbb{R}^2$, regarded as arrows emanating from the origin. Suppose that the arrows are perpendicular. If x and y do not form a horizontal/vertical pair, then the slopes of the lines through the origin that contain them are defined and are negative reciprocals. The slope is the ratio of vertical displacement to horizontal displacement, so this gives $\frac{x_2}{x_1} = -\frac{y_1}{y_2}$. See Figure 1.8. This is easily rearranged to become $x_1 y_1 + x_2 y_2 = 0$, or x · y = 0. If x and y do form a horizontal/vertical pair, say x = $(x_1, 0)$ and y = $(0, y_2)$, then x · y = 0 once again. Conversely, if x · y = 0, the preceding reasoning can be reversed to conclude that x and y are perpendicular. Thus:

In $\mathbb{R}^2$, x and y are perpendicular if and only if x · y = 0.

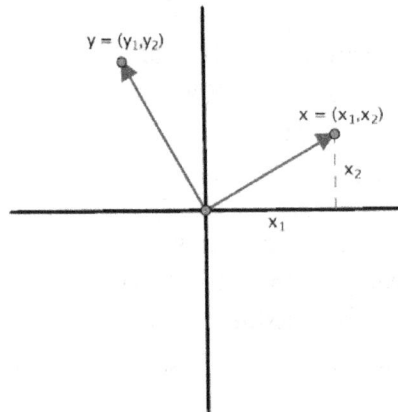

Figure 1.8: Perpendicular vectors in the plane: for instance, note that x has slope x_2/x_1.

For vectors in the plane in general, let θ denote the angle between two given vectors x and y, where $0 \le \theta \le \pi$. To study the relationship between the dot product and θ, assume for the moment that neither x nor y is a scalar multiple of the other, so $\theta \ne 0, \pi$, and consider the triangle whose vertices are 0, x, and y. Two of the sides of this triangle have lengths $\|x\|$ and $\|y\|$, and the length of the third side is the length of the arrow $\overrightarrow{yx} = x - y$. See Figure 1.9. Thus, by the law of cosines, $\|x - y\|^2 = \|x\|^2 + \|y\|^2 - 2\|x\|\,\|y\| \cos\theta$. By Proposition 1.10, this is the same as:

$$(x - y) \cdot (x - y) = x \cdot x + y \cdot y - 2\|x\|\,\|y\| \cos\theta. \tag{1.10}$$

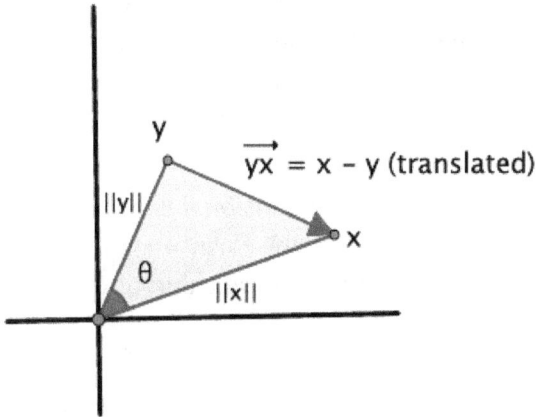

Figure 1.9: Vectors x and y in $\mathbb{R}^2$ and the angle θ between them

Multiplying out the left-hand side gives $(x-y) \cdot (x-y) = x \cdot x - x \cdot y - y \cdot x + y \cdot y = x \cdot x - 2x \cdot y + y \cdot y$. This expansion uses some of the elementary, but obvious, algebraic properties of the dot product that we declined to list. After substitution into (1.10), we obtain:

$$x \cdot x - 2x \cdot y + y \cdot y = x \cdot x + y \cdot y - 2\|x\|\,\|y\| \cos\theta.$$

Hence, after some cancellation:

$$x \cdot y = \|x\|\,\|y\| \cos\theta. \tag{1.11}$$

In the case that x or y is a scalar multiple of the other, then $\theta = 0$ or π. Say, for instance, that $y = cx$. Then $x \cdot y = x \cdot (cx) = c\|x\|^2$. Meanwhile, $\|y\| = |c|\,\|x\|$, so $\|x\|\,\|y\| \cos\theta = |c|\,\|x\|^2 \cos\theta = \pm c\|x\|^2 \cos\theta$. The plus sign occurs when $|c| = c$, i.e., the scalar multiple c is nonnegative, in which case $\theta = 0$ and $\cos\theta = 1$. The minus sign occurs when $c < 0$, whence $\cos\theta = \cos\pi = -1$. Either way, $\|x\|\,\|y\| \cos\theta = c\|x\|^2$, the same as $x \cdot y$. Thus (1.11) is valid for all x, y in $\mathbb{R}^2$.

We use our knowledge of the plane to try to create some intuition about $\mathbb{R}^n$ when $n \ge 3$. This is especially important when $n > 3$, since visualization in those spaces is almost completely an act of imagination. For instance, if v is a nonzero vector in

$\mathbb{R}^n$, we think of the set of all scalar multiples cv as a "line." If w is a second vector, not a scalar multiple of v, and c and d are scalars, the parallelogram law of addition suggests that the combination cv+ dw should lie in a "plane" determined by v and w and that, as c and d range over all possible scalars, cv + dw should sweep out this plane. As a result, the set of all vectors cv + dw, where c, d $\in$ R, is called the plane spanned by v and w. We denote it by P. The only reason that it's a plane is because that's what we have chosen to call it.

We would like to make P into a replica of the familiar plane $\mathbb{R}^2$. We sketch one approach for doing this, without filling in many technical details. We begin by choosing two vectors u_1 and u_2 in P such that $u_1 \cdot u_1 = u_2 \cdot u_2 = 1$ and $u_1 \cdot u_2 = 0$. These vectors play the role of the standard basis vectors e_1 and e_2, which of course satisfy the same relations. (It's not hard to show that such vectors u_1 and u_2 exist. In fact, in terms of the spanning vectors v and w, one can check that $u_1 = \frac{1}{\|v\|}v$ and $u_2 = \frac{1}{\|z\|}z$, where $z = w - \frac{v \cdot w}{v \cdot v}v$, is one possibility, though there are many others.)

We use u_1 and u_2 to establish a two-dimensional coordinate system internal to P. Every element of P can be written in the form $x = a_1 u_1 + a_2 u_2$ for some scalars a_1 and a_2. In terms of our budding intuition, we think of $\sqrt{a_1^2 + a_2^2}$ as representing the distance from the origin, or the length, of x.

If $y = b_1 u_2 + b_2 u_2$ is another element of P, then:

$$\begin{aligned}
x \cdot y &= (a_1 u_1 + a_2 u_2) \cdot (b_1 u_1 + b_2 u_2) \\
&= a_1 b_1 u_1 \cdot u_1 + (a_1 b_2 + a_2 b_1) u_1 \cdot u_2 + a_2 b_2 u_2 \cdot u_2 \\
&= a_1 b_1 + a_2 b_2.
\end{aligned}$$

Thus the dot product in $\mathbb{R}^n$ agrees with the result we would expect in terms of the newly created internal coordinates in P. In particular, $\|x\|^2 = x \cdot x = a_1^2 + a_2^2$. so the norm in $\mathbb{R}^n$ represents our intuitive notion of length in P. This is true even though the coordinates of $x = (x_1, x_2, \ldots, x_n)$ in $\mathbb{R}^n$ don't really have anything to do with P.

Continuing in this way, we can build up P as a copy of $\mathbb{R}^2$ and transfer over the familiar concepts of plane geometry, such as distance and angle. The following terminology reflects this intuition.

Definition. A vector x in $\mathbb{R}^n$ is called a **unit vector** if $\|x\| = 1$.

For example, the standard basis vectors $e_i = (0, 0, \ldots, 0, 1, 0, \ldots, 0)$ are unit vectors.

Corollary 1.11. If x is a nonzero vector in $\mathbb{R}^n$, then $u = \frac{1}{\|x\|}x$.is a unit vector. It is called the **unit vector in the direction of x.**

Proof. We calculate that $u \cdot u = (\frac{1}{\|x\|}x) \cdot (\frac{1}{\|x\|}x) = \frac{1}{\|x\|^2}(x \cdot x) = \frac{1}{\|x\|^2}\|x\|^2 = 1.$ Thus, by Proposition 1.10, $\|u\| = 1$.

Our main result in this section is that the relation between the dot product and angles that we derived for the plane in equation (1.11) is true in $\mathbb{R}^n$ for all n.

Theorem 1.12. If x and y are vectors in $\mathbb{R}^n$, then:

$$\mathbf{x} \cdot \mathbf{y} = \|\mathbf{x}\| \, \|\mathbf{y}\| \cos \theta,$$

where θ is the angle between x and y in the plane that they span.

Proof. The case that x or y is a scalar multiple of the other is proved in the same way as in $\mathbb{R}^2$. Otherwise, x and y span a plane P, and the points 0, x, and y are the vertices of a "triangle" in P. Since we can make P into a geometric replica of $\mathbb{R}^2$, the law of cosines remains true, and, since the norm in $\mathbb{R}^n$ represents length in P, this takes the form:

$$\|\mathbf{x} - \mathbf{y}\|^2 = \|\mathbf{x}\|^2 + \|\mathbf{y}\|^2 - 2\|\mathbf{x}\| \, \|\mathbf{y}\| \cos \theta.$$

From here, the argument is identical to the one for $\mathbb{R}^2$.

Lastly, we introduce the standard terminology for the case of perpendicular vectors, that is, when $\theta = \frac{\pi}{2}$, so cos θ = 0.

Definition. Two vectors x and y in $\mathbb{R}^n$ are called **orthogonal** if x · y = 0.

For instance, the standard basis vectors in $\mathbb{R}^n$ are orthogonal: $e_i \cdot e_j$ = 0 whenever i≠ j.

1.6 DETERMINANTS

The determinant a function that assigns a real number to an n by n matrix. There's a separate function for each n. We shall focus almost exclusively on the cases n = 2 and n = 3, since those are the cases we really need later. The determinant not defined for matrices in which the numbers of rows and columns are unequal.

The determinant of a 2 by 2 matrix is defined to be:

$$\det \begin{bmatrix} a_{11} & a_{12} \\ a_{21} & a_{22} \end{bmatrix} = a_{11}a_{22} - a_{12}a_{21}.$$

It's the product of the diagonal entries minus the product of the off-diagonal entries. For instance, $\det \begin{bmatrix} 1 & 2 \\ 3 & 4 \end{bmatrix} = 1 \cdot 4 - 2 \cdot 3 = 4 - 6 = -2$.

One often thinks of the determinant as a function of the rows of the matrix. If x = (x_1, x_2) and y = (y_1, y_2), let x y denote the matrix whose rows are x and y:

$$\det \begin{bmatrix} \mathbf{x} & \mathbf{y} \end{bmatrix} = \det \begin{bmatrix} x_1 & x_2 \\ y_1 & y_2 \end{bmatrix} = x_1 y_2 - x_2 y_1.$$

The main geometric property of 2 by 2 determinants is the following.

Proposition 1.13. Let x and y be vectors in $\mathbb{R}^2$, neither a scalar multiple of the other. Then:

|det [x y]| = Area of the parallelogram determined by x and y.

Proof. We show the equivalent result that $(\det [x \ y])^2 = (\text{Area})^2$. The area of the parallelogram is (base)(height). As base, we use the arrow x, which has length $\|x\|$. For the height h, we drop a perpendicular from the point y to the base. See Figure 1.10. Let θ be the angle between x and y, where as usual $0 \le \theta \le \pi$. Then the height is given by $h = \|y\| \sin \theta$, so Area $= \|x\| \ \|y\| \sin \theta$ and:

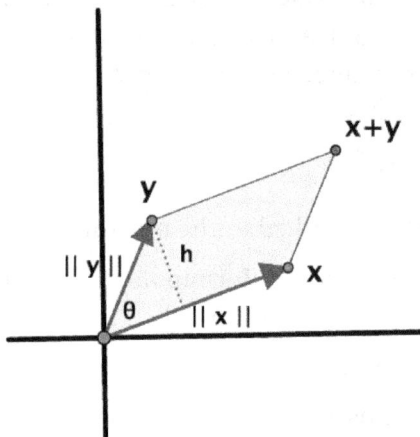

Figure 1.10: Area of a parallelogram

$$(\text{Area})^2 = \|\mathbf{x}\|^2 \|\mathbf{y}\|^2 \sin^2 \theta. \tag{1.12}$$

Meanwhile:

$$
\begin{aligned}
\left(\det \begin{bmatrix} \mathbf{x} & \mathbf{y} \end{bmatrix}\right)^2 &= (x_1 y_2 - x_2 y_1)^2 \\
&= x_1^2 y_2^2 - 2 x_1 x_2 y_1 y_2 + x_2^2 y_1^2 \\
&= (x_1^2 + x_2^2)(y_1^2 + y_2^2) - x_1^2 y_1^2 - x_2^2 y_2^2 - 2 x_1 x_2 y_1 y_2 \\
&= (x_1^2 + x_2^2)(y_1^2 + y_2^2) - (x_1 y_1 + x_2 y_2)^2 \\
&= \|\mathbf{x}\|^2 \|\mathbf{y}\|^2 - (\mathbf{x} \cdot \mathbf{y})^2 \\
&= \|\mathbf{x}\|^2 \|\mathbf{y}\|^2 - \|\mathbf{x}\|^2 \|\mathbf{y}\|^2 \cos^2 \theta \\
&= \|\mathbf{x}\|^2 \|\mathbf{y}\|^2 (1 - \cos^2 \theta) \\
&= \|\mathbf{x}\|^2 \|\mathbf{y}\|^2 \sin^2 \theta. \tag{1.13}
\end{aligned}
$$

Comparing equations (1.12) and (1.13) gives the result.

If x or y is a scalar multiple of the other, for instance, if y = cx for some scalar c, then the "parallelogram" they determine degenerates into a line segment, which has area 0. At the same time, $\det \begin{bmatrix} x & y \end{bmatrix} = \begin{bmatrix} x_1 & x_2 \\ cx_1 & cx_2 \end{bmatrix} = cx_1 x_2 - cx_1 x_2 = 0$. so the proposition remains valid when things degenerate, too.

Determinants satisfy a great many algebraic properties. We list only the ones that we shall use, which barely begins to scratch the surface. In part (c) of the following proposition, the **transpose** of a matrix A, denoted by A^t, refers to the matrix obtained by turning the rows into columns and the columns into rows. So the (i, j) th entry of At is the (j, i)th entry of A. For instance:

$$\begin{bmatrix} 1 & 2 \\ 3 & 4 \end{bmatrix}^t = \begin{bmatrix} 1 & 3 \\ 2 & 4 \end{bmatrix}. \qquad \begin{bmatrix} 1 & 2 & 3 \\ 4 & 5 & 6 \end{bmatrix}^t = \begin{bmatrix} 1 & 4 \\ 2 & 5 \\ 3 & 6 \end{bmatrix}. \qquad \begin{bmatrix} 1 \\ 2 \\ 3 \end{bmatrix}^t = \begin{bmatrix} 1 & 2 & 3 \end{bmatrix}.$$, and so on.

If A is an m by n matrix, then A^t is n by m.

Proposition 1.14.

If two of the rows are equal, the determinant 0.

Interchanging two rows flips the sign of the determinant.

$\det(A^t) = \det A$.

$\det(AB) = (\det A)(\det B)$.

Proof. For 2 by 2 determinants, the proofs are easy.

$$\det \begin{bmatrix} x & x \end{bmatrix} = \det \begin{bmatrix} x_1 & x_2 \\ x_1 & x_2 \end{bmatrix} = x_1 x_2 - x_2 x_1 = 0.$$

$$\det \begin{bmatrix} y & x \end{bmatrix} = \det \begin{bmatrix} y_1 & y_2 \\ x_1 & x_2 \end{bmatrix} = y_1 x_2 - y_2 x_1 = -(x_1 y_2 - x_2 y_1) = -\det \begin{bmatrix} x & y \end{bmatrix}.$$

$$\det(A^t) = \det \begin{bmatrix} a_{11} & a_{21} \\ a_{12} & a_{22} \end{bmatrix} = a_{11} a_{22} - a_{21} a_{12} = \det \begin{bmatrix} a_{11} & a_{12} \\ a_{21} & a_{22} \end{bmatrix} = \det A.$$

$$\det(AB) = \det\left(\begin{bmatrix} a_{11} & a_{12} \\ a_{21} & a_{22} \end{bmatrix} \begin{bmatrix} b_{11} & b_{12} \\ b_{21} & b_{22} \end{bmatrix} \right) = \det \begin{bmatrix} a_{11}b_{11}+a_{12}b_{21} & a_{11}b_{12}+a_{12}b_{22} \\ a_{21}b_{11}+a_{22}b_{21} & a_{21}b_{12}+a_{22}b_{22} \end{bmatrix} = (a_{11}b_{11} + a_{12}b_{21})(a_{21}b_{12}$$

$+a_{22}b_{22}) - (a_{11}b_{12}+a_{12}b_{22})(a_{21}b_{11}+a_{22}b_{21})$. After multiplying out, the terms involving $a_{11}a_{21}b_{11}b_{12}$ and $a_{12}a_{22}b_{21}b_{22}$ cancel in pairs, leaving:

$$\det(AB) = a_{11}b_{11}a_{22}b_{22} + a_{12}b_{21}a_{21}b_{12} - a_{11}b_{12}a_{22}b_{21} - a_{12}b_{22}a_{21}b_{11}.$$

One can check that $(\det A)(\det B) = a_{11}a_{22} - a_{12}a_{21})(\ b_{11}b_{22} - b_{12}b_{21})$ expands to the same thing.

For 3 by 3 matrices, the determinant defined in terms of the 2 by 2 case:

$$\det \begin{bmatrix} a_{11} & a_{12} & a_{13} \\ a_{21} & a_{22} & a_{23} \\ a_{31} & a_{32} & a_{33} \end{bmatrix} = a_{11} \cdot \det \begin{bmatrix} a_{22} & a_{23} \\ a_{32} & a_{33} \end{bmatrix} - a_{12} \cdot \det \begin{bmatrix} a_{21} & a_{23} \\ a_{31} & a_{33} \end{bmatrix} + a_{13} \cdot \det \begin{bmatrix} a_{21} & a_{22} \\ a_{31} & a_{32} \end{bmatrix}. \qquad (1.14)$$

The signs in the sum alternate, and the pattern is that the terms run along the entries of the first row, a1j, each multiplied by the 2 by 2 determinant obtained by deleting the first row and jth column of the original matrix. The process is known as **expansion along the first row**. For instance:

$$\det \begin{bmatrix} 1 & 2 & 3 \\ 4 & 5 & 6 \\ 7 & 8 & 9 \end{bmatrix} = 1 \cdot \det \begin{bmatrix} 5 & 6 \\ 8 & 9 \end{bmatrix} - 2 \cdot \det \begin{bmatrix} 4 & 6 \\ 7 & 9 \end{bmatrix} + 3 \cdot \det \begin{bmatrix} 4 & 5 \\ 7 & 8 \end{bmatrix}$$

$$= 1 \cdot (45 - 48) - 2 \cdot (36 - 12) + 3 \cdot (32 - 35) = -3 + 12 - 9 = 0.$$

One can show that 3 by 3 determinants satisfy all four of the algebraic properties of Proposition 1.14. The calculations are longer than in the 2 by 2 case but still straightforward, except for the product formula det(AB) = (det A)(det B) which calls for a new approach.

One consequence of the properties is that there's a formula for expanding the determinant along any row. To expand along the ith row, interchange row i with the row above it repeatedly until it reaches the top, then expand using equation (1.14). The result has the same form as (1.14), except that the leading scalar factors come from the ith row and sometimes the signs might alternate beginning with a minus sign depending on how many sign flips were introduced in getting the ith row to the top.

In addition, since det(A^t) = det A, any general statement about rows applies to columns as well, so there are formulas for expanding along any column, too. We won't write down those formulas precisely.

There is also a geometric interpretation of 3 by 3 determinants in terms of three-dimensional volume. This is discussed in the next chapter.

For larger matrices, the same pattern continues, that is, n by n determinants can be defined in terms of (n - 1) by (n - 1) determinants using expansion along the first row. For the 4 by 4 case, the formula is:

$$
\det \begin{bmatrix} a_{11} & a_{12} & a_{13} & a_{14} \\ a_{21} & a_{22} & a_{23} & a_{24} \\ a_{31} & a_{32} & a_{33} & a_{34} \\ a_{41} & a_{42} & a_{43} & a_{44} \end{bmatrix} = a_{11} \cdot \det \begin{bmatrix} a_{22} & a_{23} & a_{24} \\ a_{32} & a_{33} & a_{34} \\ a_{42} & a_{43} & a_{44} \end{bmatrix} - a_{12} \cdot \det \begin{bmatrix} a_{21} & a_{23} & a_{24} \\ a_{31} & a_{33} & a_{34} \\ a_{41} & a_{43} & a_{44} \end{bmatrix}
$$

$$
+ a_{13} \cdot \det \begin{bmatrix} a_{21} & a_{22} & a_{24} \\ a_{31} & a_{32} & a_{34} \\ a_{41} & a_{42} & a_{44} \end{bmatrix} - a_{14} \cdot \det \begin{bmatrix} a_{21} & a_{22} & a_{23} \\ a_{31} & a_{32} & a_{33} \\ a_{41} & a_{42} & a_{43} \end{bmatrix} .
$$

The algebraic properties of Proposition 1.14 remain true for n by n determinants whatever the value of n, though, to prove this, one should really develop the algebraic structure of determinants in a systematic way rather than hope for success with brute force calculation. We leave this level of generality for a course in linear algebra. In what follows, we work for the most part with the cases n = 2 and n = 3.

1.7 EXERCISES FOR CHAPTER 1

Section 1 Vector arithmetic

In Exercises 1.1–1.4, let x and y be the vectors x = (1, 2, 3) and y = (4, −5, 6) in R^3. Also, 0 denotes the zero vector, 0 = (0, 0, 0).

1.1. Find x + y, 2x, and 2x - 3y.

1.2. Find $\overrightarrow{yx}$ and $y + \overrightarrow{yx}$.

1.3. If x + y + z = 0, find z.

1.4. If x - 2y + 3z = 0, find z.

1.5. Let x and y be points in $\mathbb{R}^n$.

    a. Show that the midpoint of line segment xy is given by $m = \frac{1}{2}x + \frac{1}{2}y$. (Hint: What do you add to x to get to the midpoint?)

    b. Find an analogous expression in terms of x and y for the point p that is 2/3 of the way from x to y.

    c. Let x = (1, 1, 0) and y = (0, 1, 1). Find the point z in R^3 such that y is the midpoint of line segment xz.

Section 2 Linear transformations

2.1. Let $T : R^3 \to \mathbb{R}^2$ be the projection of $x_1x_2x_3$-space onto the x_1x_2-plane, T $(x_1, x_2, x_3) = (x_1, x_2)$. Draw pictures that show that T satisfies the requirements for being a linear trans-formation.

2.2. Let $T : \mathbb{R}^2 \to R^3$ be the function defined by T $(x_1, x_2) = (x_1, x_2, 0)$. Show that T is a linear transformation.

2.3. Let $a_1, a_2, \ldots, a_n$ be vectors in Rm. Verify that the function $T : \mathbb{R}^n \to$ Rm given by is a linear transformation.

$$T(x_1, x_2, \ldots, x_n) = x_1 a_1 + x_2 a_2 + \cdots + x_n a_n$$

Section 3 The matrix of a linear transformation

3.1. Let $T : \mathbb{R}^2 \to \mathbb{R}^2$ be the linear transformation such that T $(e_1) = (1, 2)$ and T (e2) = (3, 4).

    a. Find the matrix of T with respect to the standard bases.

    b. Find T (5, 6).

    c. Find a general formula for T (x_1, x_2).

3.2. Let $T : R^3 \to R^3$ be the linear transformation such that T $(e_1) = (1, 0, -1)$, T $(e_2) = (-1, 1, 0)$, and T $(e_3) = (0, -1, 1)$.

    a. Find the matrix of T with respect to the standard bases.

    b. Find T (1, -1, 1).

    c. Find the set of all points x = (x1, x2, x3) in R3 such that T (x) = 0.

In Exercises 3.3–3.6, find the matrix of the given linear transformation $T : \mathbb{R}^2 \to \mathbb{R}^2$ with respect to the standard bases.

3.3. T is the counterclockwise rotation by $\frac{\pi}{2}$ about the origin.

3.4. T is the reflection in the x_2-axis.

3.5. T is the identity function, T (x) = x.

3.6. T is the dilation about the origin by a factor of 2, T (x) = 2x.

3.7. Let $\rho_\theta : \mathbb{R}^2 \to \mathbb{R}^2$ be the rotation about the origin counterclockwise by an angle θ. Show that the matrix of ρ_θ with respect to the standard bases is:

$$R_\theta = \begin{bmatrix} \cos\theta & -\sin\theta \\ \sin\theta & \cos\theta \end{bmatrix}.$$ (1.15)

The matrix Rθ is called a **rotation matrix**.

3.8. Let $T : \mathbb{R}^2 \to \mathbb{R}^2$ be the linear transformation whose matrix with respect to the standard bases is $A = \begin{bmatrix} 0 & 1 \\ 1 & 0 \end{bmatrix}$. Describe T geometrically.

3.9. Let $T : \mathbb{R}^2 \to \mathbb{R}^2$ be the linear transformation whose matrix with respect to the standard bases is $A = \begin{bmatrix} 1 & 2 \\ 2 & 4 \end{bmatrix}$.

    a. Find T (e_1) and T (e_2).

    b. Describe the image of T , that is, the set of all y in $\mathbb{R}^2$ such that y = T (x) for some x in $\mathbb{R}^2$.

3.10. Let v_1 = (1, 1) and v_2 = (-1, 1).

    a. Find scalars c_1, c_2 such that $c_1 v_1 + c_2 v_2 = e_1$.

    b. Find scalars c_1, c_2 such that $c_1 v_1 + c_2 v_2 = e_2$.

    c. Let $T : \mathbb{R}^2 \to \mathbb{R}^2$ be the linear transformation such that T (v_1) = (-1, -1) and T (v_2) = (-2, 2). Find the matrix of T with respect to the standard bases.

    d. Let $S : \mathbb{R}^2 \to \mathbb{R}^2$ be the linear transformation such that $S(v_1)$ = e_1 and $S(v_2)$ = e_2. Find the matrix of S with respect to the standard bases.

3.11. Let $T : \mathbb{R}^2 \to \mathbb{R}^3$ be the linear transformation given by T (x_1, x_2) = $(x_1, x_2, 0)$. Find the matrix of T with respect to the standard bases.

3.12. Let $T : \mathbb{R}^3 \to \mathbb{R}^3$ be the rotation about the x3-axis by $\frac{\pi}{2}$ counterclockwise as viewed looking down from the positive x3-axis. Find the matrix of T with respect to the standard bases.

3.13. Let $T : \mathbb{R}^3 \to \mathbb{R}^3$ be the rotation by π about the line x_1 = x_2, x_3 = 0, in the $x_1 x_2$-plane. Find the matrix of T with respect to the standard bases.

3.14. Let A and B be m by n matrices. If Ax = Bx for all column vectors x in $\mathbb{R}^n$, show that

Section 4 Matrix multiplication

In Exercises 4.1–4.6, find the indicated matrix products.

4.1. AB and BA, where $A = \begin{bmatrix} 1 & -1 \\ 1 & 1 \end{bmatrix}$ and $B = \begin{bmatrix} 2 & 4 \\ -1 & 3 \end{bmatrix}$

4.2. AB and BA, where $A = \begin{bmatrix} 1 & 2 \\ 3 & 4 \end{bmatrix}$ and $B = \begin{bmatrix} 1 & 0 \\ 0 & 1 \end{bmatrix}$

4.3. AB and BA, where $A = \begin{bmatrix} 0 & 0 & 1 \\ 1 & 0 & 0 \\ 0 & 1 & 0 \end{bmatrix}$ and $B = \begin{bmatrix} 0 & 1 & 0 \\ 0 & 0 & 1 \\ 1 & 0 & 0 \end{bmatrix}$

4.4. AB and BA, where $A = \begin{bmatrix} 1 & 2 & 3 \\ 0 & 4 & 5 \\ 0 & 0 & 6 \end{bmatrix}$ and $B = \begin{bmatrix} 1 & 1 & 1 \\ 0 & 1 & 1 \\ 0 & 0 & 1 \end{bmatrix}$

4.5. AB and BA, where $A = \begin{bmatrix} 2 & -1 & -4 \\ 3 & 2 & 1 \end{bmatrix}$ and $B = \begin{bmatrix} 2 & 3 \\ -1 & 2 \\ -4 & 1 \end{bmatrix}$

4.6. AB and BA, where $A = \begin{bmatrix} 1 & 2 & 3 \end{bmatrix}$ and $B = \begin{bmatrix} 4 \\ 5 \\ 6 \end{bmatrix}$

4.7. We have seen that the commutative law AB = BA does not hold in general for matrix multiplication. In fact, the situation is worse than that: it's rare for AB and BA even to both be defined. In that sense, the preceding half dozen exercises are misleading.

    a. Find an example of matrices A and B such that neither AB nor BA is defined.

    b. Find an example where AB is defined but BA is not.

4.8.   a. Let $T : \mathbb{R}^n \to \mathbb{R}m$, $S : \mathbb{R}m \to \mathbb{R}^p$ and $R : \mathbb{R}^p \to \mathbb{R}^q$ be functions. Show that:

$$((R \circ S) \circ T)(\mathbf{x}) = (R \circ (S \circ T))(\mathbf{x}) \quad \text{for all } \mathbf{x} \text{ in } \mathbb{R}^n.$$

    b. Show that matrix multiplication is associative, that is, if A is a q by p matrix, B a p by m matrix, and C an m by n matrix, show that:

$$(AB)C = A(BC).$$

Since it does not matter how the terms are grouped, we shall write the product henceforth simply as ABC, without parentheses. (Hint: See Exercise 3.14.)

In Exercises 4.9–4.10:

◉   let $R_\theta = \begin{bmatrix} \cos\theta & -\sin\theta \\ \sin\theta & \cos\theta \end{bmatrix}$ be the matrix (1.15) that represents the rotation ρ_θ

    $: \mathbb{R}^2 \to \mathbb{R}^2$ about the origin counterclockwise by angle θ, and

◉   let $S = \begin{bmatrix} 1 & 0 \\ 0 & -1 \end{bmatrix}$ be the matrix (1.6) that represents the reflection $r : \mathbb{R}^2 \to \mathbb{R}^2$

    in the x1-axis.

4.9. Use matrices to prove that $r \circ \rho_\theta \circ r = \rho\text{-}\theta$.

4.10.   a. Compute the product $A = R_\theta S R_{-\theta}$.

    b. By thinking about the corresponding composition of linear transformations, give a geometric description of the linear transformation $T : \mathbb{R}^2 \to \mathbb{R}^2$ that is represented with respect to the standard bases by A.

Section 5 The geometry of the dot product

5.1. Let x = (-1, 1, -2) and y = (4, -1, -1).

    a. Find x · y.

    b. Find ‖x‖ and ‖y‖.

    c. Find the angle between x and y.

5.2. Find the unit vector in the direction of x = (2, -1, -2).

5.3. Find all unit vectors in $\mathbb{R}^2$ that are orthogonal to x = (1, 2).

5.4. What does the sign of x · y tell you about the angle between x and y?

In Exercises 5.5–5.10, show that the dot product satisfies the given property. The properties are true for vectors in $\mathbb{R}^n$, though you may assume in your arguments that the vectors are in $\mathbb{R}^2$, i.e., x = (x_1, x_2), y = (y_1, y_2), and so on. The proofs for $\mathbb{R}^n$ in general are similar.

5.5. x · y = y · x

5.6. (cx) · y = c(x · y) for any scalar c

5.7. ‖cx‖ = |c| ‖x‖ for any scalar c

5.8. w · (x + y) = w · x + w · y

5.9. (v + w) · (x + y) = v · x + v · y + w · x + w · y (Hint: Make use of the preceding exercises.)

5.10. (x + y) · (x - y) = ‖x‖² - ‖y‖²

5.11. Recall that a rhombus is a planar quadrilateral whose sides all have the same length. Use the dot product to show that the diagonals of a rhombus are perpendicular to each other.

5.12. The Pythagorean theorem states that, for a right triangle in the plane, a2 + b2 = c2, where a and b are the lengths of the legs and c is the length of the hypotenuse. Use vector algebra and the dot product to verify that the theorem remains true for right triangles in $\mathbb{R}^n$. (Hint: The hypotenuse is a diagonal of a rectangle.)

Section 6 Determinants

In Exercises 6.1–6.4, find the given determinant.

6.1. $\det \begin{bmatrix} 2 & 5 \\ -3 & 4 \end{bmatrix}$

6.2. $\det \begin{bmatrix} 0 & 5 \\ -3 & 4 \end{bmatrix}$

6.3. $\det \begin{bmatrix} 1 & -2 & 3 \\ -4 & 5 & -6 \\ 7 & -8 & 9 \end{bmatrix}$

$$\det \begin{bmatrix} 1 & 0 & 0 \\ 2 & 3 & 0 \\ 4 & 5 & 6 \end{bmatrix}$$

6.4.

6.5. Find the area of the parallelogram in $\mathbb{R}^2$ determined by x = (4, 0) and y = (1, 3).

6.6. Find the area of the parallelogram in $\mathbb{R}^2$ determined by x = (-2, -3) and y = (-3, 2).

6.7. Let $A = \begin{bmatrix} 1 & -2 \\ 2 & -4 \end{bmatrix}$. Use the product rule for determinants to show that there is no 2 by 2 matrix B such that $AB = \begin{bmatrix} 1 & 0 \\ 0 & 1 \end{bmatrix}$.

6.8. Let A be an n by n matrix, and let A^t be its transpose. Show that det $(A^t A) = (\det A)^2$.

6.9. If $(x_1, x_2, x_3) \in R^3$, let:

$$T(x_1, x_2, x_3) = \det \begin{bmatrix} x_1 & x_2 & x_3 \\ 1 & 2 & 3 \\ 4 & 5 & 6 \end{bmatrix}. \tag{1.16}$$

   a. Expand the determinant to find a formula for T (x_1, x_2, x_3).
   b. Show that equation (1.16) defines a linear transformation $T : R^3 \to R$.
   c. Find the matrix of T with respect to the standard bases.
   d. The matrix you found in part (c) should be a 1 by 3 matrix. Thinking of it as a vector in R^3, show that it is orthogonal to both (1, 2, 3) and (4, 5, 6), the second and third rows of the matrix used to define T (x_1, x_2, x_3).

PART-II
VECTOR-VALUED FUNCTIONS
OF ONE VARIABLE

2 PATHS AND CURVES

Chapter

This chapter is concerned with curves in $\mathbb{R}^n$. While we may have an intuitive sense of what a curve is, at least in $\mathbb{R}^2$ or R^3, the formal description here is somewhat indirect in that, rather than requiring a curve to have a defining equation, we describe it by how it is swept out, like the trace of a skywriter. Thus, in addition to studying geometric features of the curve, such as its length, we also look at quantities related to the skywriter's motion, such as its velocity and acceleration. The goal of the chapter is a remarkable result about curves in R^3 that describes measurements that characterize the geometry of a space curve completely. Along the way, we shall gain valuable experience applying vector methods.

The functions in this chapter take their values in $\mathbb{R}^n$, but they are functions of one variable. As a result, we treat the material as a continuation of first-year calculus without going back to redefine or redevelop concepts that are introduced there. At times, this assumed familiarity may lead to a rather loose treatment of certain basic topics, such as continuity, derivatives, and integrals. When we study functions of more than one variable, we shall go back and define these concepts carefully, and what we say then applies retroactively to what we cover here. We hope that any reader who becomes anxious about a possible lack of rigor will be willing to wait.

2.1 PARAMETRIZATIONS

Let I be an interval of real numbers, typically, I = [a, b], (a, b), or R.

Definition. A continuous function $\alpha : I \to \mathbb{R}^n$ is called a **path**. As t varies over I, $\alpha(t)$ traces out a **curve** C. More precisely:

C = {x $\in \mathbb{R}^n$: x = $\alpha(t)$ for some t in I}.

This is also known as the **image** of α. We say that α is a **parametrization** of the curve. We often refer to the input variable *t* as **time** and think of $\alpha(t)$ as the **position** of a moving object at time t. See Figure 2.1.

A path is a vector-valued function of one variable. To follow our notational convention, we should write the value in boldface as $\alpha(t)$, but for the most part we continue to use plainface, usually reserving boldface for a particular kind of vector-valued function that we begin studying in Chapter 8. For each t in I, $\alpha(t)$ is a point of $\mathbb{R}^n$, so we may write $\alpha(t) = (x_1(t), x_2(t), \ldots, xn(t))$, where each of the n coordinates is a real number that depends on t, i.e., a real-valued function of one variable.

Here are some of the standard examples of parametrized curves that we shall refer to frequently.

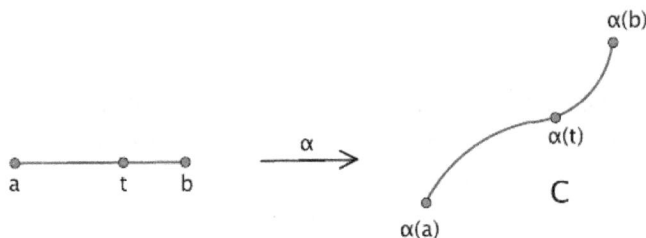

Figure 2.1: A parametrization $\alpha : [a, b] \to \mathbb{R}^n$ of a curve C

Example 2.1. Circles in $\mathbb{R}^2$: $x^2 + y^2 = a^2$, where a is the radius of the circle.

The equation of the circle can be rewritten as $\left(\frac{x}{a}\right)^2 + \left(\frac{y}{a}\right)^2 = 1$. To parametrize the circle, we use the identity $\cos^2 t + \sin^2 t = 1$ and let $\frac{x}{a} = \cos t$ and $\frac{y}{a} = \sin t$, or x=a cos t and y=a sin t. These expressions show that t has a geometric interpretation as the angle that (x, y) makes with the positive x-axis, as shown in Figure 2.2. The substitutions x = a cos t and y = a sin t satisfy the defining equation $x^2 + y^2 = a^2$ of the circle, so $\alpha : R \to \mathbb{R}^2$, parametrizes the circle, traversing it in the counterclockwise direction.

$$\alpha(t) = (a \cos t, a \sin t), \text{ a constant, } a > 0,$$

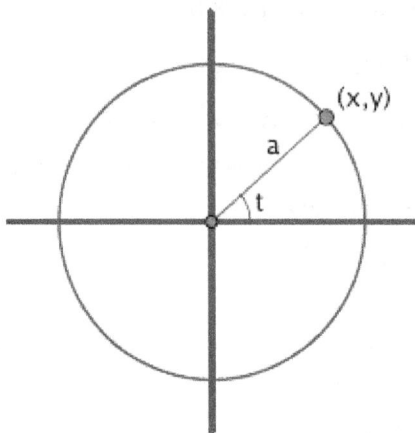

Figure 2.2: A circle of radius a

Example 2.2. Graphs $y = f(x)$ in $\mathbb{R}^2$.

As a concrete example, consider the curve described by $y = \sin x$ in $\mathbb{R}^2$, where $0 \le x \le \pi$. See Figure 2.3. It consists of points of the form $(x, \sin x)$, and it is traced out as x goes from $x = 0$ to $x = \pi$. Hence x serves as a natural parameter, and $\alpha : [0, \pi] \to \mathbb{R}^2$, is a parametrization of the curve.

$$\alpha(x) = (x, \sin x),$$

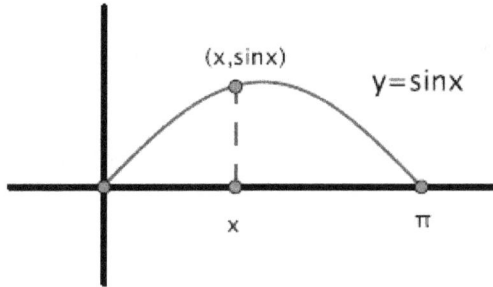

Figure 2.3: The graph $y = \sin x$, $0 \le x \le \pi$

Example 2.3. Helices in $\mathbb{R}^3$.

A helix winds around a circular cylinder, say of radius a. If the axis of the cylinder is the z-axis, then the x and y-coordinates along the helix satisfy the equation of the circle as in Example 2.1, while the z-coordinate changes at a constant rate. Thus we set $x = a \cos t$, $y = a \sin t$, and $z = bt$ for some constant b. That is, the helix is parametrized by $\alpha : \mathbb{R} \to \mathbb{R}^3$, where:

$$\alpha(t) = (a \cos t, a \sin t, bt), \text{ a, b constant, a > 0.}$$

If $b > 0$, the helix is said to be "right-handed" and if $b < 0$ "left-handed." Each type is shown in Figure 2.4.

Figure 2.4: Helices: right-handed (at left) and left-handed (at right)

Example 2.4. Lines in $\mathbb{R}^n$.

Given a point a and a nonzero vector v in $\mathbb{R}^n$, the line through a and parallel to v is parametrized by $\alpha : \mathbb{R} \to \mathbb{R}^n$, where:

$\alpha(t) = a + tv.$

See Figure 2.5. If a = $(a_1, a_2, \ldots, a_n)$ and v = $(v_1, v_2, \ldots, v_n)$, then $\alpha(t)$ = $(a_1, a_2, \ldots, a_n)$ + t$(v_1, v_2, \ldots, v_n)$ = $(a_1 + tv_1, a_2 + tv_2, \ldots, a_n + tv_n)$, so in terms of coordinates:

$$\begin{cases} x_1 &=& a_1 + tv_1 \\ x_2 &=& a_2 + tv_2 \\ &\vdots& \\ x_n &=& a_n + tv_n. \end{cases}$$

If the domain of α is restricted to an interval of finite length, then α parametrizes a finite segment of the line.

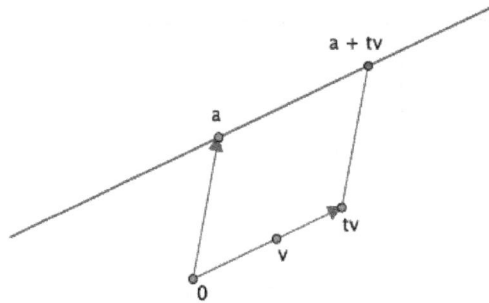

Figure 2.5: As t varies, α(t) = a + tv traces out a line.

Example 2.5. Find a parametrization of the line in R³ that passes through the points a = (1, 2, 3) and b = (4, 5, 6).

The line passes through the point a = (1, 2, 3) and is parallel to v = $\overrightarrow{ab}$ = b − a = (4, 5, 6) − (1, 2, 3) = (3, 3, 3). The setup is indicated in Figure 2.6. Therefore one parametrization is:

$$\alpha(t) = (1, 2, 3) + t(3, 3, 3) = (1 + 3t, 2 + 3t, 3 + 3t).$$

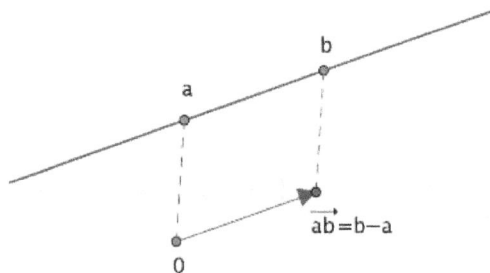

Figure 2.6: Parametrizing the line through two given points a and b

Any nonzero scalar multiple of v = (3, 3, 3) is also parallel to the line and could be used as part of a parametrization as well. For instance, w = (1, 1, 1) is such a multiple, and β(t) = (1, 2, 3) + t(1, 1, 1) = (1 + t, 2 + t, 3 + t) is another parametrization of the same line.

2.2. VELOCITY, ACCELERATION, SPEED, ARCLENGTH

We now introduce some basic aspects of calculus for paths and curves. We discuss derivatives of paths in this section and integrals of real-valued functions over paths in the next.

Definition. Given a path $\alpha : I \to \mathbb{R}^n$, the derivative of α is defined by:

$$
\begin{aligned}
\alpha'(t) &= \lim_{h \to 0} \frac{1}{h}\left(\alpha(t+h) - \alpha(t)\right) \\
&= \lim_{h \to 0} \frac{1}{h}\left((x_1(t+h), x_2(t+h), \dots, x_n(t+h)) - (x_1(t), x_2(t), \dots, x_n(t)) \right) \\
&= \lim_{h \to 0} \left(\frac{x_1(t+h) - x_1(t)}{h}, \frac{x_2(t+h) - x_2(t)}{h}, \dots, \frac{x_n(t+h) - x_n(t)}{h} \right) \\
&= (x'_1(t), x'_2(t), \dots, x'_n(t)),
\end{aligned}
$$

provided the limit exists. The derivative is also called the **velocity** v(t) of α.

Likewise, $\alpha''(t) = v'(t)$ is called the **acceleration**, denoted by a(t).

Example 2.6. For the helix parametrized by $\alpha(t) = (\cos t, \sin t, t)$:

- ⊙ velocity: v(t) = $\alpha'(t)$ = (- sin t, cos t, 1),
- ⊙ acceleration: a(t) = v'(t) = (- cos t, - sin t, 0).

Now, choose some time t_0, and, for any time t, let s(t) be the distance traced out along the path from $\alpha(t_0)$ to $\alpha(t)$. Then s(t) is called the arclength function, and we think of the rate of change of distance $\frac{ds}{dt}$ as the speed. Unfortunately, these terms should be defined more carefully, and, to get everything in the right logical order, it seems best to define the speed first.

To define what we expect $\frac{ds}{dt}$ to be, we make the intuitive approximation that the distance Δs along the curve between two nearby points $\alpha(t)$ and $\alpha(t + \Delta t)$ is approximately the straight line distance between them, as indicated in Figure 2.7. We assume that Δt is positive. If Δt is small:

"distance along curve" ≈ straight line distance, or

$\Delta s \approx \|\alpha(t + \Delta t) - \alpha(t)\|$.

Thus $\frac{\Delta s}{\Delta t} \approx \|\frac{\alpha(t+\Delta t) - \alpha(t)}{\Delta t}\|$. In the limit as Δt goes to 0, this approaches $\|\alpha'(t)\|$, the magnitude of the velocity. We take this as the definition of speed.

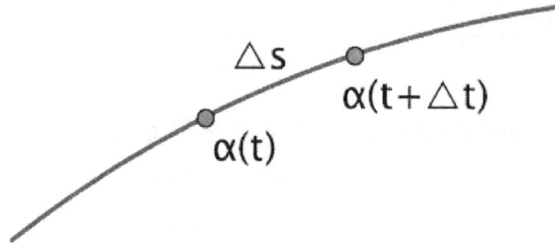

Figure 2.7: Δs ≈ ‖α(t + Δt) – α(t)‖

Definition. If $\alpha : I \to \mathbb{R}^n$ is a differentiable path, then its speed, denoted by v(t), is defined to be:

$$v(t) = \|\mathbf{v}(t)\|.$$

Note that the speed v(t) is a scalar quantity, whereas the velocity v(t) is a vector. Now, to define the length of the path from t = a to t = b, we integrate the speed.

Definition. The **arclength** from t = a to t = b is defined to be

$$\int_a^b v(t)\,dt = \int_a^b \|\mathbf{v}(t)\|\,dt.$$

The arclength function s(t) we considered above is then given by integrating the speed v(t), so, by the fundamental theorem of calculus, $\frac{ds}{dt} = v(t)$. Thus the definitions realize the intuitive relations with which we began.

Example 2.7. For the helix parametrized by α(t) = (cos t, sin t, t), find:

    a. the speed and
    b. the arclength from t = 0 to t = 4π, i.e., two full twists around the helix.

First, as just shown in Example 2.6, the velocity is v(t) = (– sin t, cos t, 1). Thus:

speed: $v(t) = \|\mathbf{v}(t)\| = \sqrt{\sin^2 t + \cos^2 t + 1} = \sqrt{2}.$

arclength: $\int_0^{4\pi} v(t)\,dt = \int_0^{4\pi}\sqrt{2}\,dt = \sqrt{2}\,t\big|_0^{4\pi} = 4\pi\sqrt{2}.$

Summary of definitions

Velocity v(t)=a'(t)

acceleration a(t)=a''(t)

speed v(t)= $\|\mathbf{v}(t)\| = \frac{ds}{dt}$

arclength= $\int_a^b v(t)\,dt$

2.3. INTEGRALS WITH RESPECT TO ARCLENGTH

Let C be a curve in $\mathbb{R}^n$, and let $f : C \to R$ be a real-valued function defined on C. To formulate the definition of the integral of f over C, we adapt the usual approach of first-year calculus for integrating a function over an interval, namely:

- ◉ Chop up the domain C into a sequence of short pieces of lengths $\Delta s_1, \Delta s_2, \ldots, \Delta s_k$.

- ◉ Choose a point $x_1, x_2, \ldots, x_k$ in each piece.

- ◉ Consider sums of the form $\sum_j f(x_j) \Delta s_j$.

- ◉ Take the limit of the sums as Δs_j goes to 0.

See Figure 2.8.

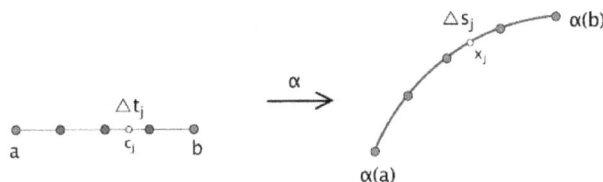

Figure 2.8: Defining the integral with respect to arclength

We can rewrite the sums in a way that leads to a normal first-year calculus integral over an interval by using a parametrization $\alpha : [a, b] \to \mathbb{R}^n$ of C. We assume that [a, b] can be subdivided into a sequence of consecutive subintervals of lengths $\Delta t_1, \Delta t_2, \ldots, \Delta t_k$ such that the subinterval of length Δt_j gets mapped to the curve segment of length Δs_j for each $j = 1, 2, \ldots, k$. Then the sum that models the integral can be written as:

$$\sum_j f(x_j) \Delta s_j = \sum_j f(\alpha(c_j)) \frac{\Delta s_j}{\Delta t_j} \Delta t_j.$$

where c_j is a value of the parameter for which $\alpha(c_j) = x_j$. In the limit as Δt_j goes to 0, these sums approach the integral $\int_a^b f(\alpha(t)) \frac{ds}{dt} dt$. In this last expression, $\frac{ds}{dt}$ is the speed, also denoted by $v(t)$. This intuition is formalized in the following definition.

Definition. Let $\alpha : [a, b] \to \mathbb{R}^n$ be a differentiable parametrization of a curve C, and let $v(t)$ denote its speed. If $f : C \to R$ is a continuous real-valued function, the **integral of f with respect to arclength**, denoted by $\int_\alpha f \, ds$, is defined to be:

$$\int_\alpha f \, ds = \int_a^b f(\alpha(t)) \, v(t) \, dt.$$

We also denote this integral by $\int_c f\ ds$, though this raises some issues that are addressed later when we have more practice with parametrizations and their effect on calculations. (See page 208.)

Example 2.8. If $f = 1$ (constant function), then the definition of the integral says $\int_c 1\ ds = \int_a^b v(t)\ dt$. On the other hand, this is also precisely the definition of arclength. Hence:

$$\int_C 1\ ds = \text{arclength of } C.$$

Example 2.9. Consider again the portion of the helix parametrized by $\alpha(t) = (\cos t, \sin t, t)$, $0 \le t \le 4\pi$. Find $\int_c (x + y + z)\ ds$.

Here, $f(x, y, z) = x + y + z$, so $f(\alpha(t)) = f(\cos t, \sin t, t) = \cos t + \sin t + t$. In other words, we read off the components of the parametrization to substitute $x = \cos t$, $y = \sin t$, and $z = t$ into the formula for f. From Example 2.7, $v(t) = \|v(t)\| = \sqrt{2}$. Thus:

$$\begin{aligned}
\int_C (x + y + z)\ ds &= \int_0^{4\pi} (\cos t + \sin t + t)\ \sqrt{2}\ dt \\
&= \sqrt{2}\left(\sin t - \cos t + \frac{1}{2}t^2\right)\Big|_0^{4\pi} \\
&= \sqrt{2}\left((0 - 1 + 8\pi^2) - (0 - 1 + 0)\right) \\
&= 8\sqrt{2}\ \pi^2.
\end{aligned}$$

2.4. THE GEOMETRY OF CURVES: TANGENT AND NORMAL VECTORS

With the basic notions involving paths and curves in $\mathbb{R}^n$ now in place, we direct most of the remainder of the chapter towards the special case of curves in R^3. It turns out that we are already in a position to derive a fairly substantial result about the geometry of such curves: we shall describe scalar measurements that determine whether two curves in R^3 are congruent in the same spirit that one learns in high school to test whether triangles in the plane are congruent. In a sense, this gives a classification of curves in three-dimensional space.

The main idea is to attach a three-dimensional coordinate system that travels along the curve. By measuring how fast the coordinate directions are changing, one recovers the geometry of the curve. In order for this to work, the coordinate directions need to have some intrinsic connection with the geometry of the curve.

We begin with the most natural direction related to the motion along the curve, namely, the tangent direction. Let $\alpha : I \to R^3$ be a differentiable parametrization of a curve, and consider the difference $\alpha(t + h) - \alpha(t)$ between two nearby points on the curve. As h goes to 0, this vector approaches the tangent direction at $\alpha(t)$, and hence so does the scalar multiple $\frac{1}{h}(\alpha(t + h) - \alpha(t))$. See Figure 2.9. In other words, the derivative $\alpha'(t) = \lim_{h \to 0} \frac{1}{h}(\alpha(t + h) - \alpha(t))$ points in the direction tangent to the curve.

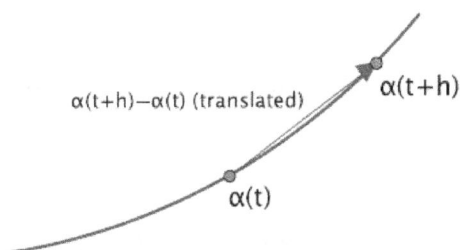

Figure 2.9: As h → 0, α(t + h) - α(t) approaches the tangent direction.

Definition. If α′(t)≠0, define:

$$T(t) = \frac{1}{\|\alpha'(t)\|}\alpha'(t).$$

This is a unit vector in the tangent direction at α(t), called the **unit tangent vector** (Figure 2.10).

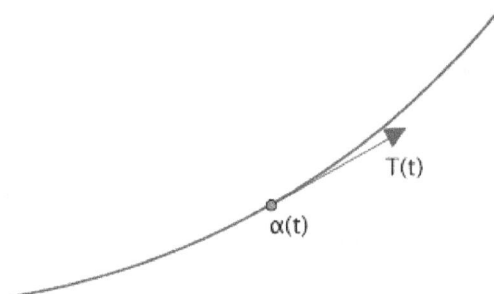

Figure 2.10: The unit tangent vector T(t), translated to start at α(t)[1]

The remaining two curve-related coordinate directions are orthogonal to this first one. In R³, there is a whole plane of orthogonal possibilities, but one of the possibilities turns out to be naturally distinguished. Identifying it requires some preliminary work.

Proposition 2.10 (Product rule for the dot product). If α, β : I → ℝⁿ are differentiable paths, then:

$$(\alpha \cdot \beta)' = \alpha' \cdot \beta + \alpha \cdot \beta'.$$

Proof. Note that α · β is real-valued, so by definition of the derivative in first-year calculus:

$$(\alpha \cdot \beta)'(t) = \lim_{h \to 0} \frac{1}{h}\big(\alpha(t+h) \cdot \beta(t+h) - \alpha(t) \cdot \beta(t)\big).$$

The expression within the limit can be put into more manageable form by using the trick of "adding zero," suitably disguised, in the middle:

$$\frac{1}{h}\big(\alpha(t+h) \cdot \beta(t+h) - \alpha(t) \cdot \beta(t)\big) = \frac{1}{h}\big(\alpha(t+h) \cdot \beta(t+h) - \alpha(t) \cdot \beta(t+h)$$

$$+ \alpha(t) \cdot \beta(t+h) - \alpha(t) \cdot \beta(t)\big)$$

$$= \frac{\alpha(t+h) - \alpha(t)}{h} \cdot \beta(t+h) + \alpha(t) \cdot \frac{\beta(t+h) - \beta(t)}{h}.$$

Now, taking the limit as h goes to 0 gives the desired result: $(\alpha \cdot \beta)'(t) = \alpha'(t) \cdot \beta(t) + \alpha(t) \cdot \beta'(t)$.

Here are two immediate consequences.

Corollary 2.11. $(\|\alpha\|^2)' = 2\alpha \cdot \alpha'$.

Proof. By Proposition 1.10 of Chapter 1, $\|\alpha\|^2 = \alpha \cdot \alpha$, so by the product rule, $(\|\alpha\|^2)' = (\alpha \cdot \alpha)' = \alpha' \cdot \alpha + \alpha \cdot \alpha' = 2\alpha \cdot \alpha'$.

Corollary 2.12. $\|\alpha\|$ is constant if and only if $\alpha \cdot \alpha' = 0$.

In words, a vector-valued function of one variable has constant magnitude if and only if it is orthogonal to its derivative at all times.

Proof. $\|\alpha\|$ is constant if and only if $\|\alpha\|2$ is constant. This in turns is true if and only if $(\|\alpha\|^2)' = 0$. By the preceding corollary, this is equivalent to saying that $\alpha \cdot \alpha' = 0$.

Returnsing to the geometry of curves, the unit tangent vector T(t) satisfies $\|T(t)\| = 1$, which is constant. Therefore T and T′ are orthogonal. The unit vector in the direction of T′ will be orthogonal to T, too. This gives us our second coordinate direction.

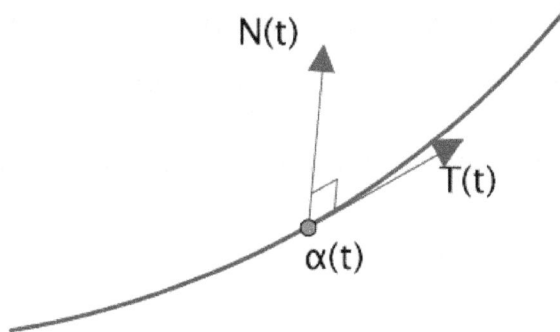

Figure 2.11: The unit tangent and principal normal vectors, T(t) and N(t)

Definition. If $T'(t) \neq 0$, then:

$$N(t) = \frac{1}{\|T'(t)\|} T'(t)$$

is called the **principal normal vector**. See Figure 2.11.

Example 2.13. For the helix parametrized by $\alpha(t) = (\cos t, \sin t, t)$, find the unit tangent $T(t)$ and the principal normal $N(t)$.

As calculated in Examples 2.6 and 2.7, $\alpha'(t) = (-\sin t, \cos t, 1)$ and $\|\alpha'(t)\| = v(t) = \sqrt{2}$. Thus:

$$T(t) = \frac{1}{\|\alpha'(t)\|} \alpha'(t) = \frac{1}{\sqrt{2}}(-\sin t, \cos t, 1).$$

Continuing with this,

$$T'(t) = \tfrac{1}{\sqrt{2}}(-\cos t, -\sin t, 0), \text{ and } \|T'(t)\| = \tfrac{1}{\sqrt{2}}\sqrt{\cos^2 t + \sin^2 t + 0^2} = \tfrac{1}{\sqrt{2}}.$$

Hence:

$$N(t) = \frac{1}{\|T'(t)\|} T'(t)$$
$$= \frac{1}{\frac{1}{\sqrt{2}}} \cdot \frac{1}{\sqrt{2}}(-\cos t, -\sin t, 0)$$
$$= (-\cos t, -\sin t, 0).$$

As a check, we verify that the unit tangent and principal normal are orthogonal, as predicted by the theory:

$$T(t) \cdot N(t) = \frac{1}{\sqrt{2}}((-\sin t)(-\cos t) + (\cos t)(-\sin t) + 1 \cdot 0)$$
$$= \frac{1}{\sqrt{2}}(\sin t \cos t - \cos t \sin t + 0) = 0.$$

The remaining step is to find a third and final coordinate direction, that is, a unit vector in R^3 orthogonal to both $T(t)$ and $N(t)$. There are two choices for such a vector, and, in the next section, we describe a systematic way to single out one of them, denoted by $B(t)$. Once this is done, we will have a basis of orthogonal unit vectors $(T(t), N(t), B(t))$ at each point of the curve that is naturally adapted to the motion along the curve.

2.5. THE CROSS PRODUCT

We remain focused on R^3. Recall that the standard basis vectors are called $i = (1, 0, 0)$, $j = (0, 1, 0)$, and $k = (0, 0, 1)$ and that every vector $x = (x_1, x_2, x_3)$ in R^3 can be written in terms of them:

$$x = (x_1, x_2, x_3) = (x_1, 0, 0) + (0, x_2, 0) + (0, 0, x_3) = x_1 i + x_2 j + x_3 k.$$

Our goal for the moment is this: given vectors v and w in R^3, find a vector, written v × w, that is orthogonal to both v and w. We concentrate initially not so much on what v × w actually is but rather on requiring it to have a certain critical property. Namely, whatever v × w is, we insist that it satisfy the following.

Key Property. For all x in R^3, $\mathbf{x} \cdot (\mathbf{v} \times \mathbf{w}) = \det \begin{bmatrix} x_1 & x_2 & x_3 \\ v_1 & v_2 & v_3 \\ w_1 & w_2 & w_3 \end{bmatrix}$, that is:

$$\boxed{\mathbf{x} \cdot (\mathbf{v} \times \mathbf{w}) = \det \begin{bmatrix} \mathbf{x} & \mathbf{v} & \mathbf{w} \end{bmatrix}} \tag{P}$$

where [x v w] denotes the 3 by 3 matrix whose rows are x, v, w.

For instance, i × j must satisfy:

$$\mathbf{x} \cdot (\mathbf{i} \times \mathbf{j}) = \det \begin{bmatrix} \mathbf{x} & \mathbf{i} & \mathbf{j} \end{bmatrix} = \det \begin{bmatrix} x_1 & x_2 & x_3 \\ 1 & 0 & 0 \\ 0 & 1 & 0 \end{bmatrix}$$
$$= x_1 \cdot (0 - 0) - x_2 \cdot (0 - 0) + x_3 \cdot (1 - 0)$$
$$= x_3 \quad \text{for all } \mathbf{x} = (x_1, x_2, x_3) \text{ in } \mathbb{R}^3.$$

But x · k = $(x_1, x_2, x_3) \cdot (0, 0, 1) = x_3$ for all x as well, so k satisfies the key property (P) required of i × j. We would expect that i × j = k.

Assuming for the moment that v × w can be defined in general to satisfy (P), our main goal follows immediately.

Proposition 2.14. v × w is orthogonal to v and w (Figure 2.12).

Proof. We take dot products and use the key property:

v · (v × w) = det [v v w] = 0 (two equal rows ⇒ det = 0).

Thus v × w is orthogonal to v. For the same reason, w · (v × w) = det [w v w] = 0.

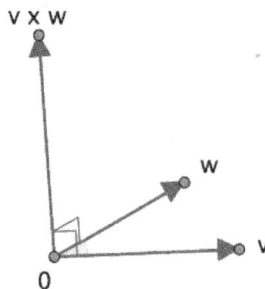

Figure 2.12: The cross product v × w is orthogonal to v and w.

It remains to define v × w in a way that satisfies (P). It turns out that there is no choice about how to do this. Set v × w = (c_1, c_2, c_3). The components c_1, c_2, c_3 are determined by (P). For instance:

$$c_1 = (1,0,0) \cdot (c_1, c_2, c_3) = \mathbf{i} \cdot (\mathbf{v} \times \mathbf{w}) = \det \begin{bmatrix} \mathbf{i} & \mathbf{v} & \mathbf{w} \end{bmatrix}$$

$$= \det \begin{bmatrix} 1 & 0 & 0 \\ v_1 & v_2 & v_3 \\ w_1 & w_2 & w_3 \end{bmatrix} = \det \begin{bmatrix} v_2 & v_3 \\ w_2 & w_3 \end{bmatrix}.$$

Similarly, c_2 and c_3 are determined by looking at $\mathbf{j} \cdot (\mathbf{v} \times \mathbf{w})$ and $\mathbf{k} \cdot (\mathbf{v} \times \mathbf{w})$, respectively.

Definition. Given vectors v and w in $\mathbb{R}^3$, their **cross product** is defined by:

$$\mathbf{v} \times \mathbf{w} = \left(\det \begin{bmatrix} v_2 & v_3 \\ w_2 & w_3 \end{bmatrix}, -\det \begin{bmatrix} v_1 & v_3 \\ w_1 & w_3 \end{bmatrix}, \det \begin{bmatrix} v_1 & v_2 \\ w_1 & w_2 \end{bmatrix} \right).$$

One can check that this is the same as:

$$\mathbf{v} \times \mathbf{w} = \det \begin{bmatrix} \mathbf{i} & \mathbf{j} & \mathbf{k} \\ v_1 & v_2 & v_3 \\ w_1 & w_2 & w_3 \end{bmatrix}. \tag{$\times$}$$

Working backwards, it is not hard to verify that the formula given by ($\times$) does indeed satisfy (P) (Exercise 5.8) so that we have actually accomplished something.

The odd-looking determinant in ($\times$) is somewhat illegitimate in that the entries in the top row are vectors, not scalars. It is meant to be calculated in the usual way by expansion along the top row. To understand how it works, perhaps it is best to do some examples.

Example 2.15. First, we reproduce our result for $\mathbf{i} \times \mathbf{j}$, putting i = (1, 0, 0) and j = (0, 1, 0) in the second and third rows of ($\times$) and expanding along the first row:

$$\mathbf{i} \times \mathbf{j} = \det \begin{bmatrix} \mathbf{i} & \mathbf{j} & \mathbf{k} \\ 1 & 0 & 0 \\ 0 & 1 & 0 \end{bmatrix} = \mathbf{i} \cdot (0 - 0) - \mathbf{j} \cdot (0 - 0) + \mathbf{k} \cdot (1 - 0) = \mathbf{k},$$

as expected.

Similarly, $\mathbf{j} \times \mathbf{k} = \mathbf{i}$ and $\mathbf{k} \times \mathbf{i} = \mathbf{j}$.

Example 2.16. If v = (1, 2, 3) and w = (4, 5, 6), then:

$$\mathbf{v} \times \mathbf{w} = \det \begin{bmatrix} \mathbf{i} & \mathbf{j} & \mathbf{k} \\ 1 & 2 & 3 \\ 4 & 5 & 6 \end{bmatrix}$$

$$= \mathbf{i} \cdot (12 - 15) - \mathbf{j} \cdot (6 - 12) + \mathbf{k} \cdot (5 - 8)$$

$$= -3\mathbf{i} + 6\mathbf{j} - 3\mathbf{k}$$

$$= (-3, 6, -3).$$

As a check against possible computational error, one can go back and confirm that the result is orthogonal to both v and w.

Example 2.17. With the same v and w as in the previous example, let u = (1, -2, 3). Then:

$$u \cdot (v \times w) = (1, -2, 3) \cdot (-3, 6, -3) = -3 - 12 - 9 = -24. \tag{2.1}$$

Suppose that we use the same three factors but in a permuted order, say w · (v × u). Taking advantage of the result of (2.1), what can you say about the value of this product without calculating the individual dot and cross products involved?

We use (P) and properties of the determinant:

$$
\begin{aligned}
w \cdot (v \times u) &= \det \begin{bmatrix} w & v & u \end{bmatrix} \quad \text{(property (P))} \\
&= -\det \begin{bmatrix} u & v & w \end{bmatrix} \quad \text{(row switch)} \\
&= -u \cdot (v \times w) \quad \text{(property (P))} \\
&= -(-24) \quad \text{(by (2.1))} \\
&= 24.
\end{aligned}
$$

The basic properties of the cross product are summarized below.

    1. w × v = -v × w

(Justification. Interchanging rows in (×) changes the sign of the determinant.)

    2. v × v = 0

(Justification. Equal rows in (×) implies the determinant zero.)

    3. (The length of the cross product.)

        $\|v \times w\| = \|v\|\,\|w\| \sin\theta$, where θ is the angle between v and w   (2.2)

(Justification. This appears as Exercise 5.9. It uses $v \cdot w = \|v\|\,\|w\| \cos\theta$.)

The length formula (2.2) for v × w has some useful consequences:

    a. v × w = 0 if and only if v = 0, w = 0, or θ = 0 or π. This follows immediately from (2.2). In other words, v × w = 0 if and only if v or w is a scalar multiple of the other.

    b. Consider the parallelogram determined by v and w. We can find its area by thinking of v as base and dropping a perpendicular from w to get the height, as in the case of Figure 1.10 in Chapter 1. See Figure 2.13.

$$\text{Area of parallelogram} = (\text{base})(\text{height})$$

$$= \|v\|(\|w\| \sin\theta)$$

$$= \|v \times w\|.$$

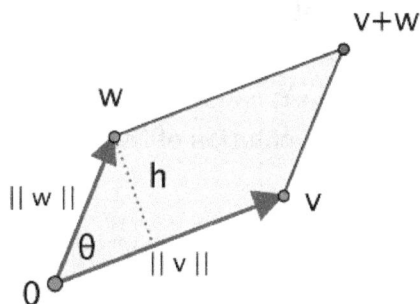

Figure 2.13: The parallelogram determined by v and w

In other words:

$\|v \times w\|$ is the area of the parallelogram determined by v and w.

    c. We can now describe the geometric interpretation of 3 by 3 determinants as volume that was mentioned in Chapter 1. Let u, v, and w be points in R^3 such that 0, u, v, and w are not coplanar. This determines a "paralellepiped," which is the three-dimensional analogue of a parallelogram in the plane, as shown in Figure 2.14. It consists of all points of the form $au + bv + cw$, where a, b, c are scalars such that $0 \le a \le 1$, $0 \le b \le 1$, and $0 \le c \le 1$. Then:

$|\det [u\ v\ w]|$ Is the volume of the parallelepiped determined by u, v and w.

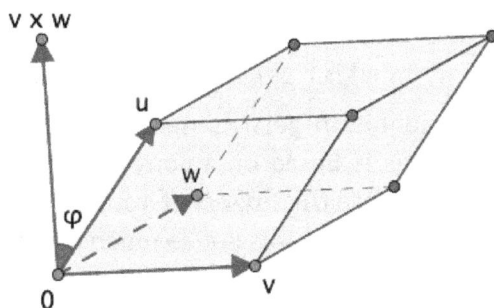

Figure 2.14: The parallelepiped determined by u, v, w in R^3

This follows from property (P), which we can use to reverse course and take what we've learned about dot and cross products to tell us about determinants. We leave the details for the exercises (namely, Exercise 5.13), though the relevant ingredients are labeled in Figure 2.14. In the figure, φ denotes the angle between u and $v \times w$.

    4. (The direction of the cross product.)

We have seen that the cross product $v \times w$ is orthogonal to v and w, but this does not pin down its direction completely. There are two possible orthogonal directions, each opposite the other. We present a criterion that distinguishes the possibilities in terms of v and w.

Definition. Let $(v_1, v_2, ..., v_n)$ be a basis of $\mathbb{R}^n$, where the vectors have been listed in a particular order. We say that $(v_1, v_2, ..., v_n)$ is **right -handed** if det $[v_1, v_2, ...$ $...,v_n] > 0$ and **left-handed** if det $[v_1, v_2, ..., v_n] > 0$ (As usual, $[v_1, v_2, ..., v_n]$ is the n by n matrix whose rows are $v_1, v_2, ..., v_n$.)

Example 2.18. Determine the orientation of the standard basis $(e_1, e_2, \ldots, e_n)$.

$$\det \begin{bmatrix} e_1 & e_2 & \ldots & e_n \end{bmatrix} = \det \begin{bmatrix} 1 & 0 & \ldots & 0 \\ 0 & 1 & \ldots & 0 \\ \vdots & \vdots & \ddots & \vdots \\ 0 & 0 & \ldots & 1 \end{bmatrix} = 1.$$

This is positive, so $(e_1, e_2, \ldots, e_n)$ is right-handed. By comparison, $(e_2, e_1, \ldots, e_n)$ is left-handed (switching two rows flips the sign of the determinant). The orientation of a basis depends on the order in which the basis vectors are written.

In $\mathbb{R}^3$, given v and w, what is the orientation of $(v, w, v \times w)$? Note that:

det $[v\ w\ v \times w] = +$ det $[v \times w\ v\ w]$ (two row switches)

$= (v \times w) \cdot (v \times w)$ (property (P) with $x = v \times w$)

$= \|v \times w\|^2.$ (2.3)

As shown earlier, $v \times w \neq 0$ provided that v and w are not scalar multiples of one another. If this is the case, then equation (2.3) implies that $[v, w, v \times w]$ is positive. Thus:

Proposition 2.19. If v and w are vectors in $\mathbb{R}^3$, neither a scalar multiple of the other, then the triple $(v, w, v \times w)$ is right-handed.

In $\mathbb{R}^3$, the right-handed orientation gets its name from a rule of thumb known as the "right-hand rule." This is based on a convention regarding the standard basis (i, j, k), namely, if you rotate the fingers of your right hand from i to j, your thumb points in the direction of k. This is not so much a fact as it is an agreement: whenever we draw R3, we agree to orient the positive x, y, and z-axes so that the right-hand rule for (i, j, k) is satisfied, as in Figure 2.15. As we continue with further material, figures in $\mathbb{R}^3$ become increasingly prominent, so it is probably good to state this convention explicitly.

More generally, given a basis (v, w, z) of $\mathbb{R}^3$, the plane P spanned by v and w separates $\mathbb{R}^3$ into two half-spaces, or "sides." Once we adopt the right-hand rule for (i, j, k), it turns out that (v, w, z) is right-handed in the sense of the definition if and only if, when you rotate the fingers of your right hand from v to w, your thumb lies on the same side of P as z. In particular, by Proposition 2.19, your thumb lies on the same side as $v \times w$. In this way, your right thumb gives the direction of $v \times w$.

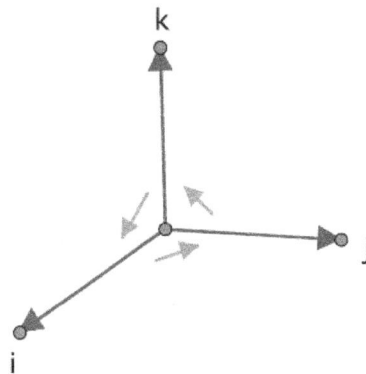

Figure 2.15: The circle of life: i × j = k, j × k = i, k × i = j

Proving the connection between the sign of a determinant and the right-hand rule would be too great a digression. One approach uses the continuity of the determinant and the geometry of rotations to show that, if the rule works for (i, j, k), then it works in general. In any case, the right-hand rule is sometimes a convenient way to find cross products geometrically. For instance, in addition to i × j = k, it allows us to see that j × k = i, not –i, and that k × i = j, as pictured in Figure 2.15.

2.6. THE GEOMETRY OF SPACE CURVES: FRENET VECTORS

We take up the following question: How can one tell when two curves in R^3 are congruent, that is, either curve can be translated, rotated, and/or reflected so that it fits exactly on top of the other? As mentioned earlier, the main idea is to construct a basis of orthogonal unit vectors that moves along the curve and then to study how the basis vectors change during the motion.

So let C be a curve in R^3, parametrized by a differentiable path $\alpha : I \to R^3$, where I is an interval in R. In Section 2.4, we defined two of the vectors in our moving basis:

⊙ the unit tangent vector $T(t) = \frac{1}{\|\alpha'(t)\|}\alpha'(t) = \frac{1}{v(t)}v(t)$ and

⊙ the principal normal vector $N(t) = \frac{1}{\|T'(t)\|}T'(t)$.

These definitions make sense only if $\alpha'(t) \neq 0$ and $T'(t) \neq 0$, so we assume that this is the case from now on. Both T(t) and N(t) are constructed to be unit vectors, and we showed before that T(t) · N(t) = 0. At this point, the third basis vector is easily defined.

Definition. The cross product B(t) = T(t) × N(t) is called the **binormal vector**.

As a cross product, B(t) is orthogonal to T(t) and N(t). Moreover, its length is $\|B(t)\| = \|T(t)\| \, \|N(t)\| \sin \theta = 1 \cdot 1 \cdot \sin \frac{\pi}{2} = 1$. Thus (T(t), N(t), B(t)) is a collection of orthogonal unit vectors. The vectors are called the Frenet vectors of α and are illustrated in Figure 2.16.

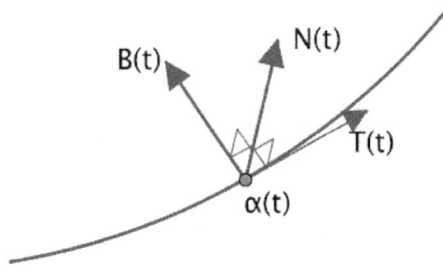

Figure 2.16: The Frenet vectors T(t), N(t), B(t)

Example 2.20 (The Frenet vectors of a helix). We shall use the helix with parametrization as a running example to illustrate new concepts as they arise. Here, a and b are constant, and a > 0. Recall that the helix is called right-handed if b > 0 and left-handed if b < 0. If b = 0, the curve collapses to a circle of radius a in the xy-plane.

$$\alpha(t) = (a \cos t, a \sin t, bt)$$

To find the Frenet vectors T(t), N(t), and B(t) of the helix, we simply and carefully follow the definitions.

Unittangent: $\alpha'(t) = (-a \sin t, a \cos t, b)$, so $\|\alpha'(t)\| = \sqrt{a^2 \sin^2 t + a^2 \cos^2 t + b^2} = \sqrt{a^2 + b^2}$. Thus:

$$\mathbf{T}(t) = \frac{1}{\|\alpha'(t)\|}\alpha'(t) = \frac{1}{\sqrt{a^2 + b^2}}(-a \sin t, a \cos t, b).$$

Principal normal: From the last calculation, $\mathbf{T}'(t) = \frac{1}{\sqrt{a^2+b^2}}(-a \cos t, -a \sin t, 0)$.

$\|\mathbf{T}'(t)\| = \frac{1}{\sqrt{a^2+b^2}}\sqrt{a^2 \cos^2 t + a^2 \sin^2 t + 0^2} = \frac{a}{\sqrt{a^2+b^2}}$. **Hence:**

$$\mathbf{N}(t) = \frac{1}{\|\mathbf{T}'(t)\|}\mathbf{T}'(t) = \frac{1}{\frac{a}{\sqrt{a^2+b^2}}} \cdot \frac{1}{\sqrt{a^2 + b^2}}(-a \cos t, -a \sin t, 0)$$

$$= \frac{1}{a}(-a \cos t, -a \sin t, 0)$$

$$= (-\cos t, -\sin t, 0).$$

Binormal: Lastly:

$$\mathbf{B}(t) = \mathbf{T}(t) \times \mathbf{N}(t) = \det \begin{bmatrix} \mathbf{i} & \mathbf{j} & \mathbf{k} \\ -\frac{a\sin t}{\sqrt{a^2+b^2}} & \frac{a\cos t}{\sqrt{a^2+b^2}} & \frac{b}{\sqrt{a^2+b^2}} \\ -\cos t & -\sin t & 0 \end{bmatrix}$$

$$= \mathbf{i} \cdot \left(\frac{b\sin t}{\sqrt{a^2+b^2}} \right) - \mathbf{j} \cdot \left(\frac{b\cos t}{\sqrt{a^2+b^2}} \right) + \mathbf{k} \cdot \left(\frac{a\sin^2 t}{\sqrt{a^2+b^2}} + \frac{a\cos^2 t}{\sqrt{a^2+b^2}} \right)$$

$$= \frac{1}{\sqrt{a^2+b^2}} (b\sin t, -b\cos t, a).$$

2.7 CURVATURE AND TORSION

We use the Frenet vectors (T, N, B) to make two geometric measurements:

- ⊙ curvature, the rate of turnsing, and
- ⊙ torsion, the rate of wobbling.

First, curvature measures how fast the tangent direction is changing, which is represented by $\|T'(t)\|$, the magnitude of the rate of change of the unit tangent vector. A given curve can be traced out in many different ways, however, so, if we want curvature to reflect purely on the geometry of the curve, we must take into account the effect of the parametrization. This is an important point, and it is an issue that we address more carefully later when we discuss the effect of parametrization on vector integrals over curves. (See Exercise 1.13 in Chapter 9.) For the time being, let's agree informally that we can normalize for the effect of the parametrization by dividing by the speed. For instance, if you traverse the same curve a second time, twice as fast before, the unit tangent will change twice as fast, too. Dividing $\|T'(t)\|$ by the speed maintains a constant ratio.

Definition. The curvature is defined by:

$$\kappa(t) = \frac{\|\mathbf{T}'(t)\|}{v(t)},$$

where v(t) is the speed.[2] Note that, by definition, curvature is a nonnegative quantity.

Example 2.21 (The curvature of a helix). For the helix α(t) = (a cos t, a sin t, bt), the ingredients that go into the definition of curvature were obtained as part of the work of Example 2.20, namely, $\|\mathbf{T}'(t)\| = \frac{a}{\sqrt{a^2+b^2}}$ and $v(t) = \|\alpha'(t)\| = \sqrt{a^2+b^2}$. Thus:

$$\text{Curvature of a helix: } \kappa(t) = \frac{\frac{a}{\sqrt{a^2+b^2}}}{\sqrt{a^2+b^2}} = \frac{a}{a^2+b^2}.$$

In particular, for a circle of radius a, in which case b = 0, the curvature is

$$\kappa = \frac{a}{a^2+0^2} = \frac{1}{a}.$$

Next, to measure wobbling, we look at the binormal B, which is orthogonal to the plane spanned by T and N. This plane is the "plane of motion" in the sense that it contains the velocity and acceleration vectors (see Exercise 9.4), so B′ can be thought of as representing the rate of wobble of that plane.

We prove in Lemma 2.24 below that B′(t) is always a scalar multiple of N(t), that is:

$$\mathbf{B}'(t) = c(t)\mathbf{N}(t),$$

where c(t) is a scalar that depends on t. For the moment, let's accept this as true. Then c(t) represents the rate of wobble. We again normalize by dividing by the speed. Moreover, the convention is to throw in a minus sign.

Definition. Given that B′(t) = c(t)N(t) as above, then the torsion is defined by:

$$\tau(t) = -\frac{c(t)}{v(t)},$$

where v(t) is the speed.

Example 2.22 (The torsion of a helix). To find the torsion of the helix α(t) = (a cos t, a sin t, bt), we again piggyback on the calculations of Example 2.20. For instance, using the result that $\mathbf{B}(t) = \frac{1}{\sqrt{a^2+b^2}}(b\sin t, -b\cos t, a)$, we obtain $\mathbf{B}'(t) = \frac{1}{\sqrt{a^2+b^2}}(b\cos t, b\sin t, 0)$. To find c(t), we want to manipulate this to be a scalar multiple of N(t) = (- cos t, - sin t, 0):

$$\mathbf{B}'(t) = \frac{1}{\sqrt{a^2+b^2}}(b\cos t, b\sin t, 0) = -\frac{b}{\sqrt{a^2+b^2}}(-\cos t, -\sin t, 0) = -\frac{b}{\sqrt{a^2+b^2}}\mathbf{N}(t).$$

Therefore $c(t) = -\frac{b}{\sqrt{a^2+b^2}}$, , so:

$$\text{Torsion of a helix: } \tau(t) = -\frac{c(t)}{v(t)} = -\frac{-\frac{b}{\sqrt{a^2+b^2}}}{\sqrt{a^2+b^2}} = \frac{b}{a^2+b^2}.$$

Note that the torsion is positive if the helix is right-handed (b > 0) and negative if left-handed (b < 0). In the case of a circle (b = 0), the torsion is zero. This reflects the fact that planar curves don't wobble in space at all.

To complete the discussion of torsion, it remains to justify the claim that B′ is a scalar multiple of N. The argument uses a product rule for the cross product.

Proposition 2.23 (Product rules). Let I be an interval of real numbers, and let α, β: I → $\mathbb{R}^n$ be differentiable paths and $f: I \to R$ a differentiable real-valued function. Then:

1. (Dot product) (α · β)′ = α′ · β + α · β′.

2. (Cross product) For paths α, β in R3, $(\alpha \times \beta)' = \alpha' \times \beta + \alpha \times \beta'$.

3. (Scalar multiplication) $(f\alpha)' = f'\alpha + f\alpha'$.

Proof. The first of these was proven earlier (Proposition 2.10), and the others are the same, swap- ping in the appropriate type of product. We leave them as exercises (Exercise 7.3).

Lemma 2.24. The derivative B′ of the binormal is always a scalar multiple of the principal normal N.

Proof. We show that B′ is orthogonal (a) to B and (b) to T. The only vectors orthogonal to both are precisely the scalar multiples of N, so the lemma follows.

   a. B is a unit vector, so ‖B‖ is constant. Hence B and its derivative B′ are always orthogonal (Corollary 2.12).

   b. By definition, B = T × N, so using the product rule:

$$B' = T' \times N + T \times N'$$
$$= (\|T'\| N) \times N + T \times N' \quad \text{(definition of N)}$$
$$= \|T'\| 0 + T \times N' \quad (v \times v = 0 \text{ always})$$
$$= T \times N'.$$

As a cross product, B′ is orthogonal to T (and to N′ for that matter).

2.8. THE FRENET-SERRET FORMULAS

We saw in Examples 2.21 and 2.22 that, for the helix $\alpha(t) = (a \cos t, a \sin t, bt)$, $a > 0$, the curvature and torsion are given by:

$$\kappa = \frac{a}{a^2 + b^2} \quad \text{and} \quad \tau = \frac{b}{a^2 + b^2}.$$

respectively. For example, if $\alpha(t) = (2 \cos t, 2 \sin t, t)$, i.e., $a = 2$ and $b = 1$, then $\kappa = \frac{2}{5}$ and $\tau = \frac{1}{5}$.

In particular, helices have constant curvature and torsion.

We are about to see that the geometry of a curve in R^3 is determined by its curvature and torsion. For instance, it turns out that any curve with constant curvature and torsion is necessarily congruent to a helix. The proof is based on the following formulas, which describe precisely how the Frenet vectors change within the coordinate systems that they determine.

Proposition 2.25 (Frenet-Serret formulas). Let $\alpha : I \to R^3$ be a differentiable path in R^3 with speed v, Frenet vectors (T, N, B), curvature κ, and torsion τ. Assume that v $/= 0$ and T′ $\neq 0$ so that the Frenet vectors are defined for all t. Then:

   1. T′ = κvN,

2. $N' = -\kappa vT + \tau vB,$

3. $B' = -\tau vN.$

Proof. The first and third of these are basically the definitions of curvature and torsion, so we prove them first. To prove (1), we know that, by definition, $N = \frac{1}{\|T'\|}T'$ and $\kappa = \frac{\|T'\|}{v}$. Thus:

$$T' = \|T'\| N = \kappa vN.$$

For (3), we know that $B' = cN$ and $\tau = -\frac{c}{v}$ for some scalar-valued function c. Hence c = -τv, and $B' = -\tau vN.$

Lastly, for (2), the right-hand rule gives N = B × T (see Figure 2.16). Therefore, according to the product rule and the two Frenet-Serret formulas that were just proven:

$$N' = B' \times T + B \times T'$$
$$= -\tau vN \times T + B \times (\kappa vN).$$

But, again by the right-hand rule, N × T = -B and B × N = -T. Substituting into the last expression gives $N' = -\kappa vT + \tau vB$, completing the proof.

2.9. THE CLASSIFICATION OF SPACE CURVES

Our discussion of curves in R3 culminates with a result on the connection between curvature and torsion—that is, turnsing and wobbling—and the intrinsic geometry of a curve. On the one hand, moving a curve rigidly in space does not change geometric features of the curve such as its length. It is plausible that this extends to not changing the curvature or torsion as well, and we begin by elaborating on this point.

For example, suppose that a path α(t) is translated by a constant amount c to obtain a new path β(t) = α(t) + c. Then β'(t) = α'(t), so α and β have the same velocity and therefore the same speed. It follows from the definition that they also have the same unit tangent vector T(t). Following along with further definitions, they then have the same principal normal N(t), hence the same binormal B(t), and finally the same curvature κ(t) and torsion τ (t). In other words, speed, curvature, and torsion are preserved under translations.

The same conclusion applies to rotations, though we don't really have the tools to give a rigorous proof at this point. So we try an intuitive explanation. Suppose that α is rotated in R³ to obtain a new path β. The velocities α'(t) and β'(t) are related by the same rotation, and then so are the respective Frenet vectors (T(t), N(t), B(t)). So the Frenet vectors of the two paths are not the same, but, since they are rotated versions of one another, the corresponding vectors change at the same rates. In particular, the speed, curvature, and torsion are the same.

Our main theorem is a converse to these considerations. That is, we show that, if two paths α and β have the same speed, curvature, and torsion, then they differ by a translation and/or rotation in the sense that there is a composition of translations and rotations $F : R^3 \to R^3$ that transforms α into β, i.e., $F(\alpha(t)) = \beta(t)$ for all t. Hence measuring the three scalar quantities v, κ, τ suffices to determine the geometry of a space curve. It is striking and satisfying that such a complete answer is possible. In the spirit of the theorems in plane geometry about congruent triangles (SAS, ASA, SSS, etc.), we call the theorem the "v$\kappa\tau$ theorem."

Theorem 2.26 (v$\kappa\tau$ theorem). Let I be an interval of real numbers, and let α, β : I $\to R^3$ be differentiable paths in R^3 such that v and T' are nonzero so that the Frenet vectors are defined for all t. Then either path can be translated and rotated so that it fits exactly on top of the other if and only if they have:

- ◉ the same speed $v(t)$,
- ◉ the same curvature $\kappa(t)$, and
- ◉ the same torsion $\tau(t)$.

In particular, two paths with the same speed, curvature, and torsion are congruent to each other.

Proof. That translations and rotations preserve speed, curvature, and torsion was discussed above. For the converse, assume that α and β have the same speed, curvature, and torsion. We proceed in three stages to move α on top of β.

Step 1. A translation.

Choose now and for the rest of the proof a point a in the interval I. We first translate α so that $\alpha(a)$ moves to $\beta(a)$. In other words, we shift α by the constant vector d = $\beta(a)$ - $\alpha(a)$ to get a new path $\gamma(t) = \alpha(t) + d = \alpha(t) + \beta(a) - \alpha(a)$. Then:

$$\gamma(a) = \alpha(a) + \beta(a) - \alpha(a) = \beta(a).$$

Moreover, as a translate, γ has the same speed, curvature, and torsion as α and hence as β as well.

Step 2. Two rotations.

Let x_0 denote the common point $\gamma(a) = \beta(a)$. The unit tangents to γ and β at x_0 may not be equal, but we can rotate γ about x_0 until they are. Then, rotate γ again using this tangent direction as axis of rotation until the principal normals at x0 coincide, too. The binormals are then automatically the same, since they are the cross products of T and N. Moreover, neither rotation changes the speed, curvature, or torsion.

Call this final path $\widetilde{\alpha}$. In this way, we have translated and rotated α into a path α satisfying the following three conditions:

$$\widetilde{\alpha}(a) = \beta(a) = \mathbf{x}_0,$$

at the point $\mathbf{x}_0$, $\widetilde{\alpha}$ and β have the same Frenet vectors, and

$\widetilde{\alpha}$ and β have the same speed, curvature, and torsion for all t.

Step 3. A calculation.

We show that $\widetilde{\alpha} = \beta$. Since $\widetilde{\alpha}$ has been obtained from a by translation and rotation, the theorem follows.

The argument uses the Frenet-Serret formulas, which we repeat for convenience:

$$\mathbf{T}' = \kappa v \mathbf{N}, \qquad \mathbf{N}' = -\kappa v \mathbf{T} + \tau v \mathbf{B}, \qquad \mathbf{B}' = -\tau v \mathbf{N}.$$

We denote the Frenet vectors of $\widetilde{\alpha}$ by $(\widetilde{\mathbf{T}}, \widetilde{\mathbf{N}}, \widetilde{\mathbf{B}})$ and those of β by $(\mathbf{T}, \mathbf{N}, \mathbf{B})$. Using similar notation for the speed, curvature, and torsion, we have $\widetilde{v} = v$, $\widetilde{\kappa} = \kappa$, and $\widetilde{\tau} = \tau$ by construction.

Let θ be the angle between $\widetilde{\mathbf{T}}(t)$ and $\mathbf{T}(t)$. Then $\widetilde{\mathbf{T}}(t) \cdot \mathbf{T}(t) = \|\widetilde{\mathbf{T}}(t)\| \, \|\mathbf{T}(t)\| \cos \theta = 1 \cdot 1 \cdot \cos \theta = \cos \theta$. Therefore $\widetilde{\mathbf{T}}(t) \cdot \mathbf{T}(t)$ is less than or equal to 1 always and is equal to 1 if and only if $\theta = 0$, that is, if and only if $\widetilde{\mathbf{T}}(t) = \mathbf{T}(t)$. The same comments apply to N (t) · N(t) and B (t) · B(t).

Thus, if we define $f: I \to R$ by

$$f(t) = \widetilde{\mathbf{T}}(t) \cdot \mathbf{T}(t) + \widetilde{\mathbf{N}}(t) \cdot \mathbf{N}(t) + \widetilde{\mathbf{B}}(t) \cdot \mathbf{B}(t),$$

then the previous remarks imply:

$$f(t) \begin{cases} \leq 3 & \text{for all } t, \\ = 3 & \text{if and only if } \widetilde{\mathbf{T}}(t) = \mathbf{T}(t), \ \widetilde{\mathbf{N}}(t) = \mathbf{N}(t), \text{ and } \widetilde{\mathbf{B}}(t) = \mathbf{B}(t). \end{cases}$$

For instance, $f(a) = 3$ by Step 2.

$\kappa = \kappa$, and $\tau = \tau$:

Now, by the product rule, $f' = (\widetilde{\mathbf{T}}' \cdot \mathbf{T} + \widetilde{\mathbf{T}} \cdot \mathbf{T}') + (\widetilde{\mathbf{N}}' \cdot \mathbf{N} + \widetilde{\mathbf{N}} \cdot \mathbf{N}') + (\widetilde{\mathbf{B}}' \cdot \mathbf{B} + \widetilde{\mathbf{B}} \cdot \mathbf{B}')$. We use the Frenet-Serret formulas to substitute for every derivative in sight, keeping in mind that $\widetilde{v} = v$, $\widetilde{\kappa} = \kappa$, and $\widetilde{\tau} = \tau$:

$$f' = \left(\kappa v \widetilde{\mathbf{N}} \cdot \mathbf{T} + \widetilde{\mathbf{T}} \cdot \kappa v \mathbf{N} \right) + \left((-\kappa v \widetilde{\mathbf{T}} + \tau v \widetilde{\mathbf{B}}) \cdot \mathbf{N} + \widetilde{\mathbf{N}} \cdot (-\kappa v \mathbf{T} + \tau v \mathbf{B}) \right)$$

$$+ \left((-\tau v \widetilde{\mathbf{N}}) \cdot \mathbf{B} + \widetilde{\mathbf{B}} \cdot (-\tau v \mathbf{N}) \right)$$

$$= (\kappa v \widetilde{\mathbf{N}} \cdot \mathbf{T} + \kappa v \widetilde{\mathbf{T}} \cdot \mathbf{N}) + (-\kappa v \widetilde{\mathbf{T}} \cdot \mathbf{N} + \tau v \widetilde{\mathbf{B}} \cdot \mathbf{N} - \kappa v \widetilde{\mathbf{N}} \cdot \mathbf{T} + \tau v \widetilde{\mathbf{N}} \cdot \mathbf{B})$$

$$- (\tau v \widetilde{\mathbf{N}} \cdot \mathbf{B} + \tau v \widetilde{\mathbf{B}} \cdot \mathbf{N}).$$

Miraculously, the terms cancel in pairs so that $f' = 0$ for all t.

Hence f is a constant function. Since $f(a) = 3$, it follows that $f(t) = 3$ for all t. As noted above, this in turns implies that $\widetilde{\mathbf{T}}(t) = \mathbf{T}(t)$ for all t (and $\widetilde{\mathbf{N}}(t) = \mathbf{N}(t)$, $\widetilde{\mathbf{B}}(t) = \mathbf{B}(t)$ as well, but we don't need them).

By definition of the unit tangent, this gives $\frac{1}{v(t)}\tilde{\alpha}'(t) = \frac{1}{v(t)}\beta'(t)$. so $\tilde{\alpha}'(t) = \beta'(t)$ for all t.

Consequently $\tilde{\alpha}$ and β differ by a constant vector: $\tilde{\alpha}(t) = \beta(t) + \mathbf{c}$. Since $\tilde{\alpha}(a) = \mathbf{x}_0 = \beta(a)$, the constant is $\mathbf{c} = 0$. Thus $\alpha(t) = \beta(t)$ for all t. We have succeeded in translating and rotating α to reach a path α that lies on top of β.

Note that the theorem does not say what happens when a path is reflected in a plane, that is, how the speed, curvature, and torsion of a path are related to those of its mirror image. For a clue as to what happens in this case, see Exercise 9.14.

2.10. EXERCISES FOR CHAPTER 2

Section 1 Parametrizations

In Exercises 1.1–1.8, sketch the curve parametrized by the given path. Indicate on the curve the direction in which it is being traversed.

1.1. $\alpha : [0, 2] \to \mathbb{R}^2$, $\alpha(t) = (t, t^2)$

1.2. $\alpha : [0, 6\pi] \to \mathbb{R}^2$, $\alpha(t) = (t \cos t, t \sin t)$

1.3. $\alpha : R \to \mathbb{R}^2$, $\alpha(t) = (e^t, e^{-t})$ (Hint: How are the x and y-coordinates related?)

1.4. $\alpha : R \to R^3$, $\alpha(t) = (\cos t, \sin t, t)$

1.5. $\alpha : R \to R^3$, $\alpha(t) = (\cos t, \sin t, -t)$

1.6. $\alpha : R \to R^3$, $\alpha(t) = (\sin t, \cos t, t)$

1.7. $\alpha : R \to R^3$, $\alpha(t) = (\sin t, \cos t, -t)$

1.8. $\alpha : [0, 2\pi] \to R^3$, $\alpha(t) = (\cos t, \sin t, \cos 2t)$

In Exercises 1.9–1.12, find a parametrization of the given curve.

1.9. The portion of the parabola $y = x^2$ in $\mathbb{R}^2$ with $-1 \le x \le 1$

1.10. The portion of the curve $xy = 1$ in $\mathbb{R}^2$ with $1 \le x \le 2$

1.11. The circle $x^2 + y^2 = a^2$ in $\mathbb{R}^2$ traversed once in the clockwise direction

1.12. A right-handed helix in R^3 that:

  ⊙ lies on the circular cylinder of radius 2 whose axis is the z-axis,

  ⊙ completes one complete twist as it rises from $z = 0$ to $z = 1$.

1.13. Find a parametrization of the line in R^3 that passes through the point $a = (1, 2, 3)$ and is parallel to $v = (4, 5, 6)$.

1.14. Let ℓ be the line in R^3 parametrized by $\alpha(t) = (1 + 2t, -3t, 4 + 5t)$. Find a point that lies on ℓ and a vector that is parallel to ℓ.

1.15. Find a parametrization of the line in R^3 that passes through the points $a = (1, 0, 0)$ and $b = (0, 0, 1)$.

1.16. Find a parametrization of the line in $\mathbb{R}^3$ that passes through the points a = (1, -2, 3) and b = (-4, 5, -6).

1.17. Let ℓ be the line in $\mathbb{R}^n$ parametrized by $\alpha(t) = a + tv$, and let p be a point in $\mathbb{R}^n$. Suppose that q is the foot of the perpendicular dropped from p to ℓ (Figure 2.17).

  a. Since q lies on ℓ, there is a value of t such that q = a + tv. Find a formula for t in terms of a, v, and p. (Hint: ℓ lies in a direction perpendicular to p - q.)
  b. Find a formula for the point q.

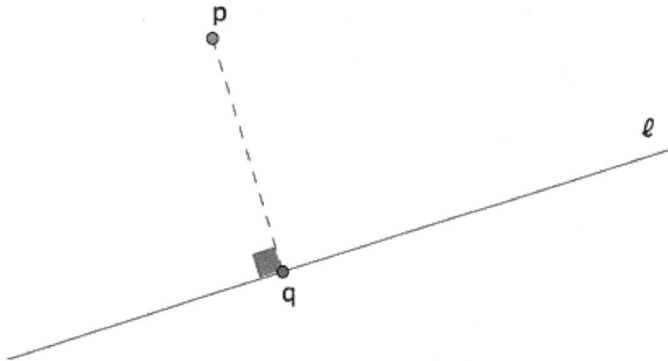

Figure 2.17: The foot of the perpendicular from p to ℓ

1.18. Let ℓ be the line in $\mathbb{R}^3$ parametrized by $\alpha(t) = (1 + t, 1 + 2t, 1 + 3t)$, and let p = (0, 0, 4). Find the foot of the perpendicular dropped from p to ℓ.

1.19. Let ℓ and m be lines in $\mathbb{R}^3$ parametrized by:

$$\alpha(t) = \mathbf{a} + t\mathbf{v} \quad \text{and} \quad \beta(t) = \mathbf{b} + t\mathbf{w},$$

respectively, where a = (a_1, a_2, a_3), v = (v_1, v_2, v_3), b = (b_1, b_2, b_3), and w = (w_1, w_2, w_3).

  a. Explain why the problem of determining whether ℓ and m intersect in $\mathbb{R}^3$ amounts to solving a system of three equations in two unknowns t_1 and t_2.
  b. Determine whether the lines parametrized by $\alpha(t) = (-1, 0, 1) + t(1, -3, 2)$ and $\beta(t) = (1, 5, 4) + t(-1, 1, 6)$ intersect by writing down the corresponding system of equations, as described in part (a), and solving the system.
  c. Repeat for the lines parametrized by $\alpha(t) = (2, 1, 0) + t(0, 0, 1)$ and $\beta(t) = (0, 1, 3) + t(1, 0, 0)$.

Section 2 Velocity, acceleration, speed, arclength

2.1. Let $\alpha(t) = (t^3, 3t^2, 6t)$.

  a. Find the velocity, speed, and acceleration.
  b. Find the arclength from t = 0 to t = 4.

2.2. Let $\alpha(t) = (\frac{1}{2}t^2, 2t, \frac{4}{3}t^{3/2})$.

    a. Find the velocity, speed, and acceleration.

    b. Find the arclength from t = 2 to t = 4.

2.3. Let $\alpha(t)$ = (cos 2t, sin 2t, $4t^{3/2}$).

    a. Find the velocity, speed, and acceleration.

    b. Find the arclength from t = 1 to t = 2.

2.4. Let a and b be real numbers, a > 0, and let $\alpha(t)$ = (a cos t, a sin t, bt).

    a. Find the velocity, speed, and acceleration.

    b. Find the arclength from t = 0 to t = 2π.

2.5. Find a parametrization of the line through the points a = (1, 1, 1) and b = (2, 3, 4) that traverses the line with constant speed 1.

Section 3 Integrals with respect to arclength

In Exercises 3.1–3.2, evaluate the integral with respect to arclength for the curve C with the given parametrization α.

3.1. $\int_C xyz$ ds, where $\alpha(t)$ = $(t^3, 3t^2, 6t)$, $0 \le t \le 1$

3.2. $\int_C z$ ds, where $\alpha(t)$ = (cos t + t sin t, sin t - t cos t, t), $0 \le t \le 4$

3.3. Let C be the line segment in $\mathbb{R}^2$ from (1, 0) to (0, 1). With as little calculation as possible, find the values of the following integrals.

    a. $\int_C 5\,ds$

    b. $\int_C (x+y)\,ds$

3.4. Let C be the line segment in $\mathbb{R}^2$ parametrized by $\alpha(t)$ = (t, t), $0 \le t \le 1$.

    1. Evaluate $\int_C (x + y)$ ds.

    2. Find a triangle in R3 whose base is the segment C (in the xy-plane) and whose area is represented by the integral in part (a). (Hint: Area under a curve.)

Section 4 The geometry of curves: tangent and normal vectors

Note: Further exercises on this material appear in Section 9.

In Exercises 4.1–4.2, find the unit tangent T(t) and the principal normal N(t) for the given path α. Then, verify that T(t) and N(t) are orthogonal.

4.1. $\alpha(t)$ = (sin 4t, cos 4t, 3t)

4.2. $\alpha(t)$ = $(t, t^2, 0)$

4.3. Let α be a differentiable path in $\mathbb{R}^n$.

    a. If the speed is constant, show that the velocity and acceleration vectors are always orthogonal. (Hint: Consider $\|v(t)\|^2$.)

b. Conversely, if the velocity and acceleration are always orthogonal, show that the speed is constant.

4.4. Let C be a curve in $\mathbb{R}^n$ traced out by a differentiable path $\alpha : (a, b) \to \mathbb{R}^n$ defined on an open interval (a, b). Assume that there is a point $\alpha(t0)$ on C that is closer to the origin than any other point of C. Prove that the tangent vector $\alpha'(t0)$ is orthogonal to $\alpha(t0)$. Draw a sketch that illustrates the conclusion.

Section 5 The cross product

In Exercises 5.1–5.4, (a) find the cross product v × w and (b) verify that it is orthogonal to v and w.

5.1. v = (1, 1, 1), w = (1, 3, 5)

5.2. v = (1, 2, 3), w = (2, 4, 6)

5.3. v = (1, -2, 4) , w = (-2, 1, 3)

5.4. v = (2, 0, -1), w = (0, -2, 1)

5.5. Find a nonzero vector in R^3 that points in a direction perpendicular to the plane that contains the origin and the points p = (1, 1, 1) and q = (2, 1, -3).

5.6. Find a nonzero vector in R^3 that points in a direction perpendicular to the plane that contains the points p = (1, 0, 0), q = (0, 1, 0), and r = (0, 0, 1).

5.7. Let v = (-1, 1, -2), w = (4, -1, -1), and u = (-3, 2, 1).

    a. Find v × w.

    b. What is the area of the parallelogram determined by v and w?

    c. Find u · (v × w). Then, use your answer to find v · (u × w) and (u × v) · w.

    d. What is the volume of the parallelepiped determined by u, v, and w?

5.8. Verify that the cross product v × w as defined in equation (×) satisfies the key property that x . (v × w) = det [w v w] for all x in R^3.

5.9. Let v = (v_1, v_2, v_3) and w = (w_1, w_2, w_3) be vectors in R^3.

    a. Use the definitions of the dot and cross products in terms of coordinates to prove that:

$$\|v \times w\|^2 = \|v\|^2\|w\|^2 - (v \cdot w)^2.$$

    b. Use part (a) to give a proof that $\|v \times w\| = \|v\| \, \|w\| \sin \theta$, where θ is the angle between v and w.

5.10. Do there exist nonzero vectors v and w in R^3 such that v · w = 0 and v × w = 0? Explain.

5.11. In R^3, let ℓ be the line parametrized by $\alpha(t) = a + tv$, and let p be a point.

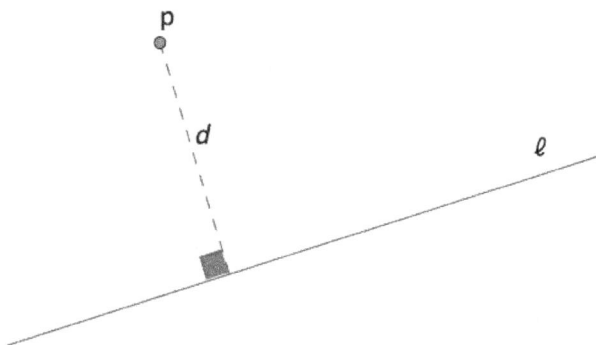

Figure 2.18: The distance from a point to a line

a. Show that the distance from p to ℓ is given by the formula:

$$d = \frac{\|(\mathbf{p} - \mathbf{a}) \times \mathbf{v}\|}{\|\mathbf{v}\|}.$$

See Figure 2.18.

  b. Find the distance from the point p = (1, 1, -1) to the line parametrized by $\alpha(t)$ = (2, 1, 2) + t(1, -2, 2).

5.12. True or false: u × (v × w) = (u × v) × w for all u, v, w in R^3. Either give a proof or find a counterexample.

5.13. Let u, v, and w be points in R^3 such that 0, u, v, and w are not coplanar. Prove that the volume of the parallelepiped determined by u, v, and w is equal to |det [u v w]| . (Hint: See Figure 2.14, where φ is the angle between u and v × w. Keep in mind that u and v × w may lie on opposite sides of the plane spanned by v and w.)

5.14. Let (v, w, z) be a basis of R^3, and let θ be the angle between v × w and z, where $0 \le \theta \le \pi$, as usual. Use the definition of handedness to show that $\theta < \frac{\pi}{2}$ < if (v, w, z) is right-handed and $\theta > \frac{\pi}{2}$ if left-handed. Draw pictures to illustrate both situations. (Hint: Consider z · (v × w).)

5.15. Strictly speaking, the cross product is defined only in R^3, but there is an analogous operation in R^4 if three factors are allowed. That is, given vectors u, v, w in R4, the product u × v × w is a vector in R4 that satisfies the property:

$$\mathbf{x} \cdot (\mathbf{u} \times \mathbf{v} \times \mathbf{w}) = \det \begin{bmatrix} \mathbf{x} & \mathbf{u} & \mathbf{v} & \mathbf{w} \end{bmatrix}$$

for all x in R^4, where the determinant of the 4 by 4 matrix whose rows are x, u, v, w.

  a. Assuming for the moment that such a product u×v×w exists, show that it is orthogonal to each of u, v, and w.

b. If $u \times v \times w \neq 0$, is the quadruple $(u, v, w, u \times v \times w)$ right-handed or left-handed?

c. Describe how you might find a formula for $u \times v \times w$, and use it to compute the product if $u = (1, -1, 1, -1)$, $v = (1, 1, -1, -1)$, and $w = (-1, 1, 1, -1)$.

5.16. In general, the "cross product" in $\mathbb{R}^n$ is a function of $n - 1$ factors. In $\mathbb{R}^2$, this means that there is only one factor.

a. Given a vector $v = (a, b)$ in $\mathbb{R}^2$, find a reasonable formula for "$v\times$", and explain how you came up with it.

b. Draw a sketch that illustrates v and your candidate for $v\times$.

Section 6 The geometry of space curves: Frenet vectors

Note: Further exercises on this material appear in Section 9.

In Exercises 6.1–6.2, find the binormal $B(t)$ for the given path α. These problems are continuations of Exercises 4.1–4.2.

6.1. $\alpha(t) = (\sin 4t, \cos 4t, 3t)$

6.2. $\alpha(t) = (t, t^2, 0)$

6.3. Consider the line in $\mathbb{R}^3$ parametrized by $\alpha(t) = (1 + t, 1 + 2t, 1 + 3t)$. What happens when you try to find its Frenet vectors $(T(t), N(t), B(t))$?

Section 7 Curvature and torsion

Note: Further exercises on this material appear in Section 9.

7.1. Find a parametrization of a curve that has constant curvature $\frac{1}{2}$ and constant torsion 0.

7.2. Find a parametrization of a curve that has constant curvature $\frac{1}{2}$ and constant torsion $\frac{1}{4}$.

7.3. Prove the product rule for cross products: If α and β are differentiable paths in $\mathbb{R}^3$, then:

$$(\alpha \times \beta)' = \alpha' \times \beta + \alpha \times \beta'.$$

Section 9 The classification of space curves

9.1. Consider the curve parametrized by:

$$\alpha(t) = (\cos t, -\sin t, t).$$

a. Find the speed $v(t)$.

b. Find the unit tangent $T(t)$.

c. Find the principal normal $N(t)$.

d. Find the binormal $B(t)$.

e. Find the curvature $\kappa(t)$.

    f. Find the torsion τ (t).

    g. The curve traced out by α is congruent to a curve with a parametrization of the form β(t) = (a cos t, a sin t, bt), where a and b are constant and a > 0. Find the values of a and b. Then, sketch the curves traced out by α and β, and describe a motion of R³ that takes the path α and puts it right on top of β.

9.2. Let α be the path defined by:

$$\alpha(t) = (6\cos t, 8\cos t, 10\sin t).$$

    a. Find the speed v(t).

    b. Find the unit tangent T(t).

    c. Find the principal normal N(t).

    d. Find the binormal B(t).

    e. Find the curvature κ(t).

    f. Find the torsion τ (t).

    g. Based on your answers to parts (e) and (f), what can you say about the type of curve traced out by α?

9.3. Let α be the path defined by:

$$\alpha(t) = (t + \sqrt{3}\sin t, \ \sqrt{3}\,t - \sin t, \ 2\cos t).$$

The following calculations are kind of messy, but the conclusion may be unexpected.

    a. Find the speed v(t).

    b. Find the unit tangent T(t).

    c. Find the principal normal N(t).

    d. Find the binormal B(t).

    e. Find the curvature κ(t).

    f. Find the torsion τ (t).

    g. Based on your answers to parts (e) and (f), what can you say about the type of curve traced out by α?

In Exercises 9.4–9.8, α is a path in R³ with velocity v(t), speed v(t), acceleration a(t), and Frenet vectors T(t), N(t), and B(t). You may assume that v(t) ≠ 0 and T'(t) ≠ 0 for all t so that the Frenet vectors are defined.

9.4. Prove the following statements.

    a. v = vT

    b. a = v'T + κv²N (Hence v' is the tangential component of acceleration and κv² the normal component.)

9.5. Prove that v · a = vv'.

9.6. Show that the curvature is given by $\kappa = \dfrac{\|\mathbf{v}\times\mathbf{a}\|}{v^3}$.

9.7. Let C be the curve in R³ parametrized by α(t) = (t, t², t³). Use the formula from the previous exercise to find the curvature at the point (1, 1, 1).

According to Exercise 9.4, the velocity and acceleration vectors lie in the plane spanned by T and N. Thus, like the binormal vector B, the cross product v × a is orthogonal to that plane. It follows that the unit vector in the direction of v × a must be ±B. Show that in fact it is B. In other words, show that $B = \frac{v \times a}{\|v \times a\|}$.

9.9. Let α(t) = (2 cos t, 2 sin t, 2 cos t).

    a. Find the curvature κ(t). (You may use the result of Exercise 9.6 if you wish.)

    b. For this curve, what does v × a tell you about the torsion τ (t)?

9.10. Let α be a path in R3 such that v ≠ 0 and T'≠ 0, and let:

$$w = \tau v T + \kappa v B.$$

Prove the Darboux **formulas**:

$$w \times T = T',$$
$$w \times N = N'.$$
$$w \times B = B'.$$

9.11. Suppose that the Darboux vector w = τvT + κvB from the previous exercise is a constant vector.

    a. Prove that τv and κv are both constant. (Hint: Calculate w', and use the Frenet-Serret formulas. Note that T and B need not be constant even if w is.)

    b. If α has constant speed, prove that α is congruent to a helix.

9.12. Let α be a path in R^3 that lies on the sphere of radius a centered at the origin, that is:

$$\|\alpha(t)\| = a \quad \text{for all } t.$$

Let T(t) be the unit tangent, N(t) the principal normal, v(t) the speed, and κ(t) the curvature. Assume that v(t)≠ 0 and T'(t)≠ 0 for all t so that the Frenet vectors are defined.

    a. Show that α(t) · T(t) = 0 for all t.

    b. Show that κ(t)α(t) · N(t) = -1 for all t. (Hint: Differentiate part (a).)

    c. Show that $\kappa(t) \geq \frac{1}{a}$ for all t.

9.13. Let α be a path in R^3 that has:

    ⊙ constant speed v = 1,

    ⊙ positive torsion τ (t), and

    ⊙ binormal vector $B(t) = (-\frac{\sqrt{2}}{2}, \frac{\sqrt{2}}{2}\sin 2t, \frac{\sqrt{2}}{2}\cos 2t)$.

    a. Find, in whatever order you wish, the unit tangent T(t), the principal normal N(t), the curvature κ(t), and the torsion τ (t).

    b. In addition, if α(0) = (0, 0, 0), find a formula for α(t) as a function of t. (Hint: Use your formula for T(t).)

9.14. In this problem, we determine the effect of a reflection, specifically the reflection in the xy- plane, on the Frenet vectors, the speed, the curvature, and the torsion of a path. To fix some notation, if p = (x, y, z) is a point in R^3, let p∗ = (x, y, -z) denote its reflection in the xy-plane.

Now, let α(t) = (x(t), y(t), z(t)) be a path in R^3 whose Frenet vectors are defined for all t, and let β(t) be its reflection:

$$\beta(t) = \alpha(t)^* = (x(t), y(t), -z(t)).$$

    a. Show that α and β have the same speed v.

    b. Show the unit tangent vectors of α and β are related by $T_\beta = T^*_\alpha$, i.e., they are also reflections of one another.

    c. Similarly, show that the principal normals are reflections: $N_\beta = N^*_\alpha$. (Hint: To organize the calculation, start by writing $T_\alpha = (t_1, t_2, t_3)$. Then, what does T_β look like?)

    d. Show that α and β have the same curvature: $\kappa_\alpha = \kappa_\beta$.

    e. Show that the binormals are related in the following way: if $B_\alpha = (b_1, b_2, b_3)$, then $B_\beta = (-b_1, -b_2, b_3) = -B^*_\alpha$.

    f. Describe how the torsions τ_α and τ_β of the paths are related. (Hint: Use parts (c) and (e).)

PART-III
REAL-VALUED FUNCTIONS

3 Chapter

REAL-VALUED FUNCTIONS: PRELIMINARIES

Thus far, we have studied functions α for which the input is a real number t and the output is a vector $\alpha(t) = (x_1(t), x_2(t), \ldots, x_n(t))$. The techniques from calculus that we used were basically familiar from first-year calculus. Now, we reverse the roles and consider functions where the input is a vector $x = (x_1, x_2, \ldots, x_n)$ and the output is a real number $y = f(x) = f(x_1, x_2, \ldots, x_n)$. Thinking of the coordinates $x_1, x_2, \ldots, x_n$ as variables, these are real-valued functions of n variables. More formally, they are functions $f: A \to R$, where the domain A is a subset of $\mathbb{R}^n$. When more than one variable is involved, we shall need methods that go beyond first-year calculus.

Example 3.1. We have seen that a helix can be parametrized by $\alpha(t) = (a \cos t, a \sin t, bt)$. The values of a and b give the radius of the cylinder around which the helix winds and the rate at which it rises, respectively. If a and b vary, the size and shape of the helix change.

The curvature of the helix is given by $\kappa = \frac{a}{a^2+b^2}$. We can think of this as a real-valued function that describes how the curvature depends on the geometric parameters a and b. In a way, it's like a function on the set of helices, but, strictly speaking, it is a function $\kappa: A \to R$ whose domain as far as helices go is $A = \{(a, b) \in \mathbb{R}^2 : a > 0\}$, a subset of $\mathbb{R}^2$. Likewise, the torsion of a helix, $\tau = \frac{b}{a^2+b^2}$, is a real-valued function of the same two variables with the same domain.

3.1. GRAPHS AND LEVEL SETS

In order to analyze a function, it seems like a good idea to try to visualize its behavior somehow. It is easy enough to come up with ways of doing this in principle, but it takes a lot of practice to do it effectively when there are two or more variables.

For instance, to graph a real-valued function of n variables, we replace $y = f(x)$ in the one-variable case with $x_{n+1} = f(x_1, x_2, \ldots, x_n)$.

Definition. Let A be a subset of $\mathbb{R}^n$, and let $f: A \to R$ be a function. The **graph** of f is the set:

$$\{(x_1, x_2, \ldots, x_{n+1}) \in \mathbb{R}^{n+1} : (x_1, x_2, \ldots, x_n) \in A \text{ and } x_{n+1} = f(x_1, x_2, \ldots, x_n)\}.$$

Note that the graph is a subset of $\mathbb{R}^{n+1}$, so we can hope to draw it only if n equals 1 or 2, i.e., when there are one or two variables. In the case of two variables, which is our focus, the domain A is a subset of $\mathbb{R}^2$, and the graph is the surface in R3 that satisfies the equation z = f (x, y) and whose projection on the xy-plane is A.

Example 3.2. Let $f : \mathbb{R}^2 \to$ R be given by $f(x, y) = \sqrt{x^2 + y^2}$. To sketch the graph $z = \sqrt{x^2 + y^2}$, we slice with various well-chosen planes and then try to reconstruct an image of the surface from the cross-sections.

For instance, consider the cross-sections with the three coordinate planes. The yz-plane is where x = 0, so the equation of the intersection of the graph with this plane is $z = \sqrt{0^2 + y^2} = |y|$ and x = 0, a V -shaped curve. Similarly, the cross-section with the xz-plane is z = |x|, y = 0, another V -shaped curve at right angles to the first. These cross-sections are shown in Figure 3.1. Lastly, the cross-section with the xy-plane, where $0 = \sqrt{x^2 + y^2}, z = 0$. This consists of the single point (0, 0, 0).

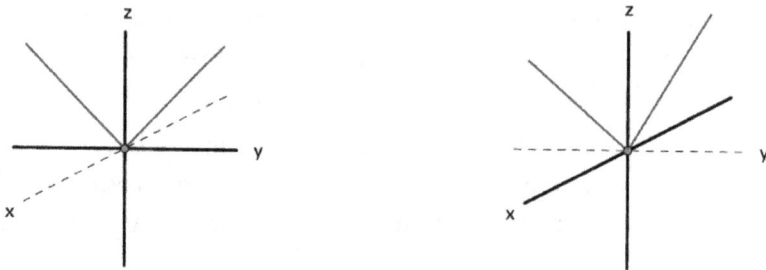

Figure 3.1: Two cross-sections with coordinate planes: z = |y|, x = 0 (left) and z = |x|, y = 0 (right)

In general, cross-sections with horizontal planes z = c are also useful. Here, they are described by $c = \sqrt{x^2 + y^2}$, or $x^2 + y^2 = c^2$, where c ≥ 0. This is a circle of radius c in the plane z = c. As c increases, the circles get bigger. Some of them are shown in Figure 3.2.

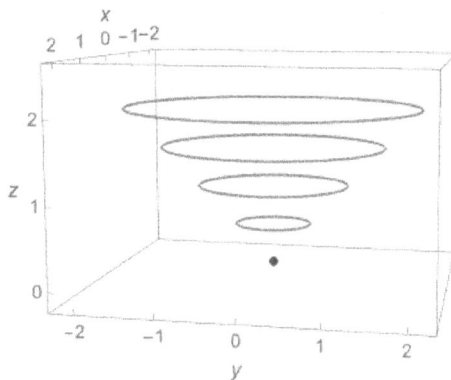

Figure 3.2: Some horizontal cross-sections

By putting all this information together, we can assemble a pretty good picture of the whole graph $z = \sqrt{x^2 + y^2}$. It is the circular cone with vertex at the origin pictured in Figure 3.3.

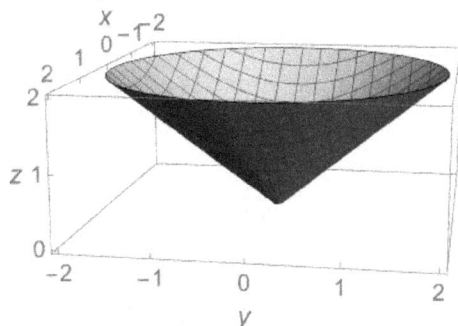

Figure 3.3: A circular cone: the graph $z = \sqrt{x^2 + y^2}$

The information about horizontal cross-sections in Figure 3.2 can also be presented by projecting the cross-sections onto the xy-plane and presenting them like a topographical map. As we saw above, the cross-section with z = c is described by the points (x, y) such that $x^2 + y^2 = c^2$, a circle of radius c. When projected onto the xy-plane, this is called a **level curve**. Sketching a representative sample of level curves in one picture and labeling them with the corresponding value of c is called a **contour map**. In this case, it consists of concentric circles about the origin. See Figure 3.4. The function f is constant on each circle and increases uniformly as you move out.

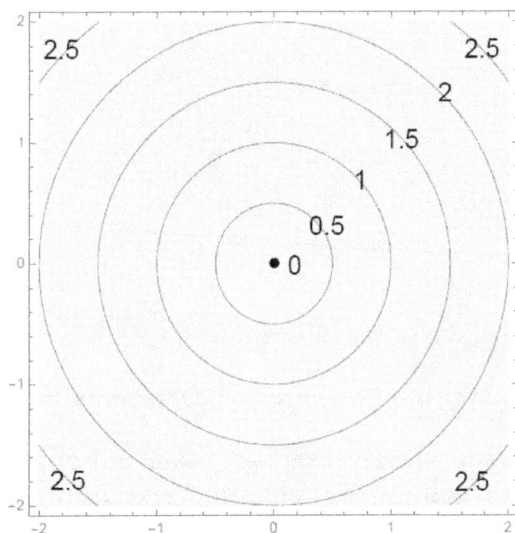

Figure 3.4: Level curves of $f(x,y) = \sqrt{x^2 + y^2}$ **corresponding to c = 0, 0.5, 1, 1.5, 2, 2.5**

Definition. Let A be a subset of $\mathbb{R}^n$, and let $f: A \to R$ be a real-valued function. Given a real number c, the level set corresponding to c is $\{x \in A : f(x) = c\}$. It is the set of all points in the domain at which f has a value of c. For functions of two variables, level sets are also called level curves and, for functions of three variables, **level surfaces**.

A level set is a subset of the domain A, which is contained in $\mathbb{R}^n$. So we can hope to draw level sets only for functions of one, two, or three variables.

Example 3.3. Let $f: \mathbb{R}^2 \to R$ be given by $f(x, y) = x^2 - y^2$. Sketch some level sets and the graph of f.

Level sets: We set $f(x, y) = x2 - y2 = c$ and choose a few values of c. For example:

Level set

c = 1	$x^2 - y^2 = 1$, a hyperbola
c = 0	$x^2 - y_2 = 0$, or $y = \pm x$, two lines
c = -1	$x^2 - y^2 = -1$, or $y^2 - x^2 = 1$, a hyperbola

These curves and a couple of others are sketched in Figure 3.5.

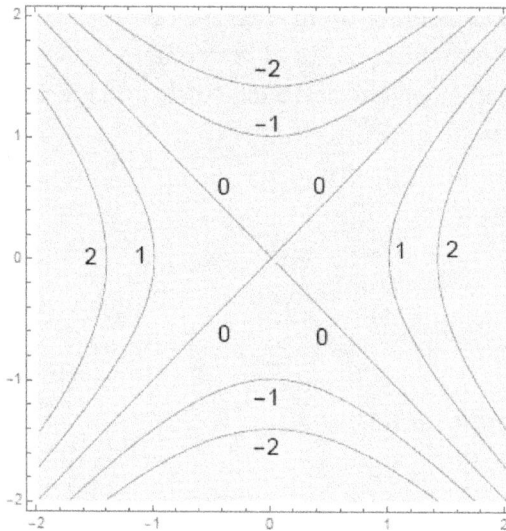

Figure 3.5: Level curves of f (x, y) = $x^2 - y^2$ corresponding to c = -2, -1, 0, 1, 2

The graph: For the graph, the level sets are taken out of the xy-plane and raised to height z = c. Furthermore, identifying the cross-sections with the coordinate planes provides some additional framework.

Coordinate plane	Equation of cross-section in that plane
yz-plane: x = 0	$z = y^2$, a downward parabola
xz-plane: y = 0	$z = x^2$, an upward parabola

These are shown in Figure 3.6.

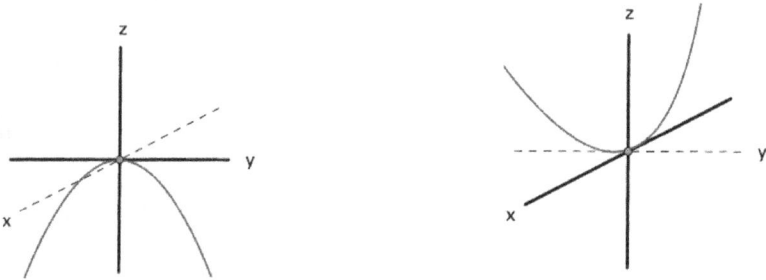

Figure 3.6: Two cross-sections with coordinate planes: $z = -y^2$, x = 0 (left) and z = x^2, y = 0 (right)

Reconstructing the graph from these pieces gives a saddle-like surface, also known as a hyperbolic paraboloid. See Figure 3.7. It has the distinctive feature that, in the cross-section with the yz-plane, the origin looks like a local maximum, while, in the xz-plane, it looks like a local minimum. (This last information can be deduced from the contour map, too. See Figure 3.5.)

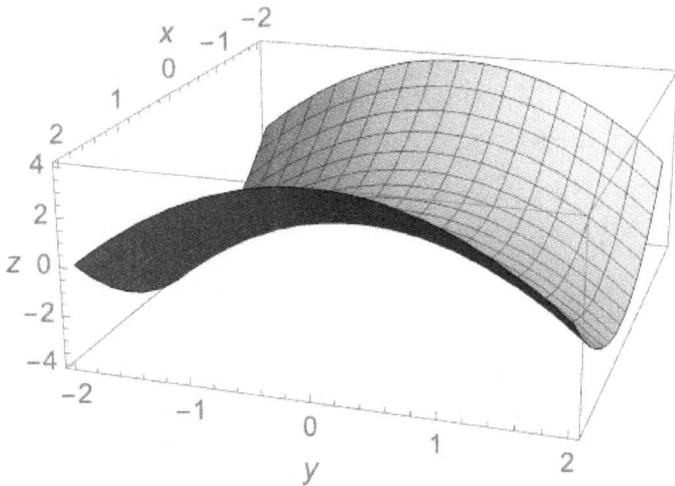

Figure 3.7: A saddle surface: the graph $z = x^2 - y^2$

3.2. MORE SURFACES IN R³

Not all surfaces in R³ are graphs $z = f(x, y)$. We look at a few examples of some other surfaces and the equations that define them.

Example 3.4. A sphere of radius a centered at the origin.

A point (x, y, z) lies on the sphere if and only if ‖(x, y, z)‖ = a, in other words, if and only if $\sqrt{x^2 + y^2 + z^2} = a$. Hence:

$$\text{Equation of a sphere: } x^2 + y^2 + z^2 = a^2$$

The sphere of radius 1 is shown in Figure 3.8, left.

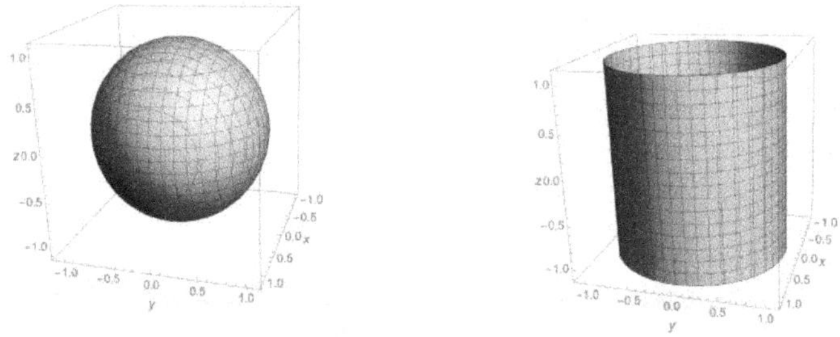

Figure 3.8: The unit sphere x² + y² + z² = 1 (left) and the cylinder x² + y² = 1 (right)

Example 3.5. A circular cylinder of radius a whose axis is the z-axis.

Here, as long as (x, y) satisfies the condition for being on the circle of radius a, z can be anything.

Thus:

$$\text{Equation of a circular cylinder: } x^2 + y^2 = a^2$$

An example is shown on the right of Figure 3.8.

Example 3.6. Sketch the surface x² + y² - z² = 1.

This is not the graph z = f (x, y) of a function: given an input (x, y), there can be two choices of z that satisfy the equation. In fact, the surface is the union of two graphs, namely, $z = \sqrt{x^2 + y^2 - 1}$ and $z = -\sqrt{x^2 + y^2 - 1}$. Nevertheless, as was the case with graphs, we can try to construct a sketch by looking at cross-sections with various well-chosen planes. For example:

Cross-sectional plane	Equation of cross-section in that plane
yz-plane: x = 0	y2 - z2 = 1, a hyperbola
xz-plane: y = 0	x2 - z2 = 1, a hyperbola
horizontal plane: z = c	x2 + y2 = 1 + c2, a circle of radius $\sqrt{1 + c^2}$

See Figure 3.9.

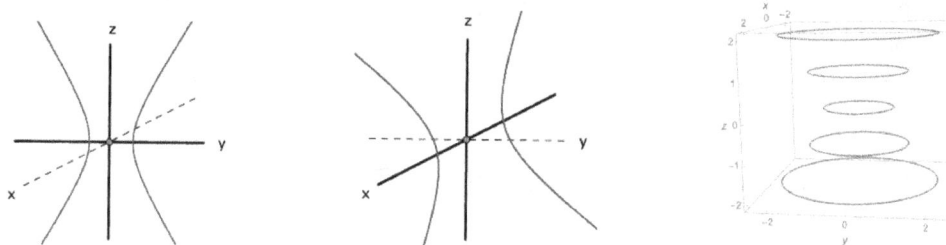

Figure 3.9: Cross-sections with the yz-plane (left), xz-plane (middle), and horizontal planes (right)

These cross-sections fit together to form a surface that looks like a nuclear cooling tower, as shown in Figure 3.10.

In fact, each of these last three surfaces is a level set of a function of three variables. The sphere is the level set of $f(x, y, z) = x^2 + y^2 + z^2$ corresponding to $c = a^2$; the cylinder the level set of $f(x, y, z) = x^2 + y^2$ corresponding to $c = a^2$; and the cooling tower the level set of $f(x, y, z) = x^2 + y^2 - z^2$ corresponding to $c = 1$.

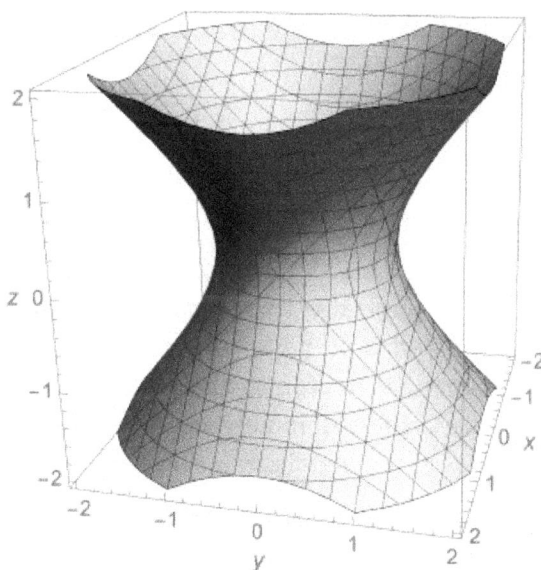

Figure 3.10: The cooling tower $x^2 + y^2 - z^2 = 1$

This suggests a way to visualize the behavior of functions of three variables, which would require four dimensions to graph. Namely, determine the level sets given by $f(x, y, z) = c$. Then the way these surfaces morph into one another as c varies reflects which points get mapped to which values. For example, the function $f(x, y, z) = x^2 + y^2 + z^2$ is constant on its level sets, which are spheres, and the larger the sphere, the greater the value of f.

Example 3.7. A graph of a function $f(x, y)$ of two variables can be regarded as a level set, too, but of a function of three variables. For let $F(x, y, z) = f(x, y) - z$. Then the graph $z = f(x, y)$ is the level set of F corresponding to c = 0. Both are defined by the condition $F(x, y, z) = f(x, y) - z = 0$.

3.3. THE EQUATION OF A PLANE IN R³

Suppose that a real-valued function of three variables $T : R^3 \to R$ happens to be a linear transformation. As with any linear transformation, T is represented by a matrix, in this case, a 1 by 3 matrix:

$$T(x, y, z) = \begin{bmatrix} A & B & C \end{bmatrix} \begin{bmatrix} x \\ y \\ z \end{bmatrix} = Ax + By + Cz.$$

The level set of T corresponding to a value D is the set of all (x, y, z) in R³ such that:

$$Ax + By + Cz = D. \tag{3.1}$$

To understand what this level set is, we rewrite (3.1) as a dot product, (A, B, C) · (x, y, z) = D, or, writing the typical point (x, y, z) of R³ as x:

n · x = D,

where n = (A, B, C). Suppose that we happen to know one particular point p on the level set. Then n · p = D, so all points on the level set satisfy:

$$\mathbf{n} \cdot \mathbf{x} = D = \mathbf{n} \cdot \mathbf{p}. \tag{3.2}$$

In other words, n · (x - p) = 0. This says that x lies on the level set if and only if x - p is orthogonal to n. Geometrically, this is true if and only if x lies in the plane that both passes through p and is perpendicular to n. This is illustrated in Figure 3.11. We say that n is a **normal vector** to the plane. Thus the level sets of the linear transformation T are planes, all having as normal vector the vector n whose components are the entries of the matrix that represents T .

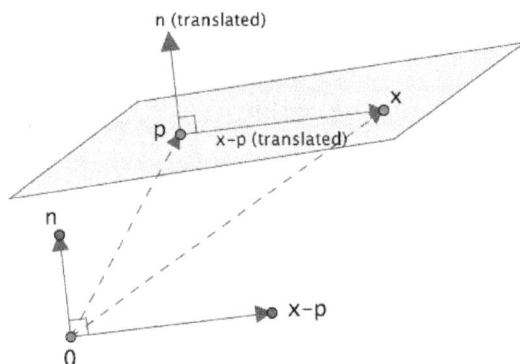

Figure 3.11: The plane through p with normal vector n

For future reference, we record the results of (3.1) and (3.2).

Proposition 3.8.

In $\mathbb{R}^3$, the equation of the plane through p with normal vector n is:

$$\mathbf{n} \cdot \mathbf{x} = \mathbf{n} \cdot \mathbf{p}, \quad where \ \mathbf{x} = (x, y, z).$$

Alternatively, the equation Ax + By + Cz = D is the equation of a plane with normal vector n = (A, B, C).

Example 3.9. Find an equation of the plane containing the point p = (1, 2, 3) and the line parametrized by $\alpha(t)$ = (4, 5, 6) + t(7, 8, 9). See Figure 3.12.

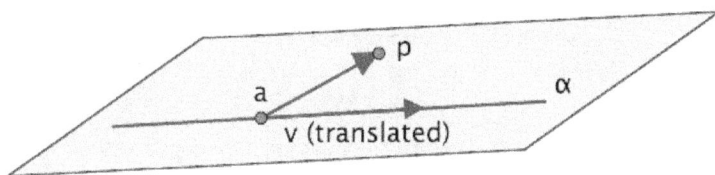

Figure 3.12: The plane determined by a given point and line

We use the form $\mathbf{n} \cdot \mathbf{x} = \mathbf{n} \cdot \mathbf{p}$. As a point in the desired plane, we can take p = (1, 2, 3). From its parametrization, the line α passes through the point a = (4, 5, 6) and is parallel to v = (7, 8, 9).

Hence, as normal vector to the plane, we can use the cross product:

$$\overrightarrow{\mathbf{ap}} \times \mathbf{v} = (\mathbf{p} - \mathbf{a}) \times \mathbf{v} = \big((1, 2, 3) - (4, 5, 6)\big) \times (7, 8, 9)$$

$$= (-3, -3, -3) \times (7, 8, 9)$$

$$= \det \begin{bmatrix} \mathbf{i} & \mathbf{j} & \mathbf{k} \\ -3 & -3 & -3 \\ 7 & 8 & 9 \end{bmatrix}$$

$$= (-3, 6, -3)$$

$$= -3(1, -2, 1).$$

The scalar multiple n = (1, -2, 1) is also a normal vector, and, since it's a little simpler, it's the one we use. Substituting into $\mathbf{n} \cdot \mathbf{x} = \mathbf{n} \cdot \mathbf{p}$ gives:

$$(1, -2, 1) \cdot (x, y, z) = (1, -2, 1) \cdot (1, 2, 3),$$

$$x - 2y + z = 1 - 4 + 3,$$

$$\text{or} \quad x - 2y + z = 0.$$

Example 3.10. Are the planes x + y - z = 5 and x - 2y + 3z = 6 perpendicular?

The angle between two planes is the same as the angle between their normal vectors (Figure 3.13), so it suffices to check if the normals are orthogonal. Reading off the coefficients from the equations that define the planes, the normals are $n_1 = (1, 1, -1)$ and $n_2 = (1, -2, 3)$. Then $n_1 \cdot n_2 = 1 - 2 - 3 = -4$, which is nonzero. The planes are not perpendicular.

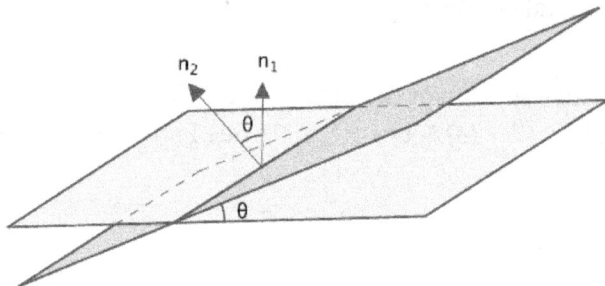

Figure 3.13: The angle between two planes

3.4. OPEN SETS

We prepare for the important concept of continuity. In first-year calculus, a real-valued function f of one variable is said to be continuous if its graph has no holes or breaks. Intuitively, whenever x approaches a point a, the value $f(x)$ approaches the value $f(a)$. This can be written succinctly as $\lim_{x \to a} f(x) = f(a)$.

Unfortunately, this idea is difficult to formulate in a rigorous way. The notions of continuity and limit are closely related. In many respects, limits are more fundamental, but we shall work more directly with continuity. Hence we base our discussion on the definition of a continuous function and modify it later to incorporate limits.

We begin by introducing the most natural type of subset of $\mathbb{R}^n$ for taking limits. These are sets in which there is a cushion around every point so that one can approach the point from all possible directions.

Definition. Let a be a point of $\mathbb{R}^n$, and assume that r > 0. The open ball with center a and radius r, denoted by B(a, r), is defined to be:

$$B(\mathbf{a}, r) = \{\mathbf{x} \in \mathbb{R}^n : \|\mathbf{x} - \mathbf{a}\| < r\}.$$

In other words, it is the set of all points within r units of a.

Example 3.11. In R, an open ball is an open interval (a - r, a + r). In $\mathbb{R}^2$, it is a disk, the region in the interior of a circle. These are shown in Figure 3.14. In R3, an open ball is a solid ball, the region in the interior of a sphere.

Figure 3.14: Open balls in R and $\mathbb{R}^2$

Definition. A subset U of $\mathbb{R}^n$ is called **open** if, given any point a of U , there exists a positive real number r such that $B(a, r) \subset U$.

So to show that a set U is open, one starts with an arbitrary point a of U and finds a value of r so that the open ball about a of radius r stays entirely within U . The value of r depends typically on the conditions that define the set U as well as on the point a. For the time being, we shall be content to find a concrete expression for r and accept geometric intuition as justification that it works. If, for some point a of U , there is no value of r that works, then U is not open.

Example 3.12. In $\mathbb{R}^2$, let U be the open ball $B((0, 0), 1)$ of radius 1 centered at the origin. This is the set of all x in $\mathbb{R}^2$ such that $\|x\| < 1$. We verify that this open ball is in fact an open set. Let a be a point of U , and let $r = 1 - \|a\|$. Note that $r > 0$ since $\|a\| < 1$. Then $B(a, r) \subset U$, so, given any a, we've found an r that works. See Figure 3.15. Hence U is an open set.

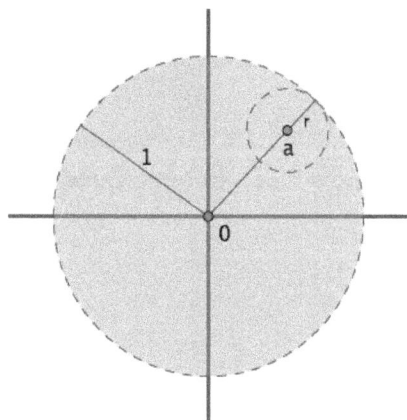

Figure 3.15: Open balls are open sets.

To those who would like a more detailed justification that $B(a, r) \subset U$, please see Exercise 7.2. Note that the expression $r = 1 - \|a\|$ gets smaller as a approaches the boundary of the disk. That this is necessary makes sense from the geometry of the situation. By modifying the argument slightly, one can show that every open ball in $\mathbb{R}^n$ is an open set in $\mathbb{R}^n$ (Exercise 7.3).

Example 3.13. Prove that the set U = $\{(x, y) \in \mathbb{R}^2 : x > 0, y < 0\}$ is open in $\mathbb{R}^2$.

This is the fourth quadrant of the plane, not including the coordinate axes. Let a be a point of U . Write a = (c, d), so c > 0 and d < 0. The closest point on the boundary of U lies on one of the axes, either c or $|d|$ units away (Figure 3.16). We choose r to be whichever of these is smaller, that is, r = min{c, $|d|$}. Then B(a, r) $\subset$ U . Thus U is open.

Example 3.14. Is U = $\{(x, y) \in \mathbb{R}^2 : x > 0, y \leq 0\}$ open in $\mathbb{R}^2$?

This is the same set as in the previous example with the positive x-axis added in. The change is enough to keep the set from being open. For instance, the point a = (1, 0) is in U , but every open ball about a includes points outside of U , namely, points in the upper half-plane. This is shown in Figure 3.17. Thus no value of r works for this choice of a.

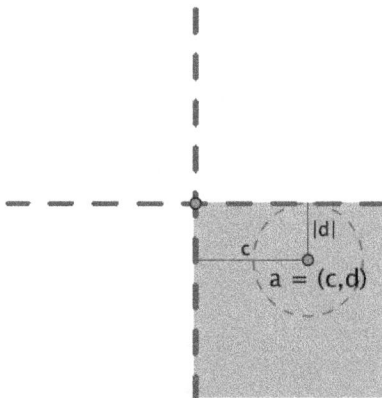

Figure 3.16: An open quadrant

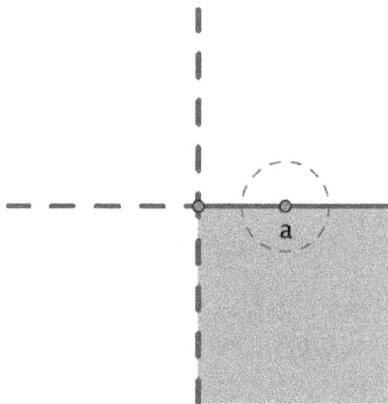

Figure 3.17: A nonopen quadrant

Example 3.15. If U and V are open sets in $\mathbb{R}^n$, prove that their intersection $U \cap V$ is open, too. Let a be a point of $U \cap V$. Then $a \in U$, and, since U is open, there exists a positive number r_1 such that $B(a, r_1) \subset U$. Likewise, $a \in V$, and there exists an $\mathbb{R}^2$ such that $B(a, \mathbb{R}^2) \subset V$. Let $r = \min\{r_1, \mathbb{R}^2\}$ (Figure 3.18). Then $B(a, r) \subset B(a, r_1) \subset U$ and $B(a, r) \subset B(a, \mathbb{R}^2) \subset V$, so $B(a, r) \subset U \cap V$. In other words, given any a in $U \cap V$, there's an r that works.

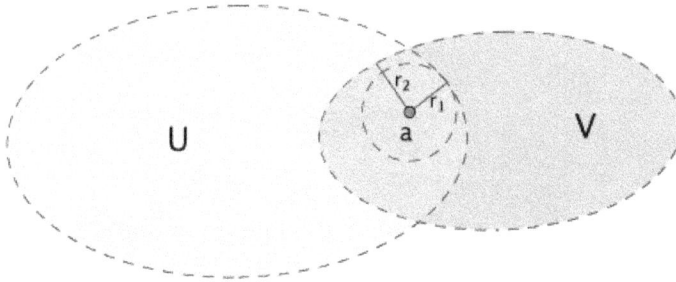

Figure 3.18: The intersection of two open sets

By similar reasoning, the intersection $U_1 \cap U_2 \cap \cdots \cap U_k$ of a finite number of open sets in $\mathbb{R}^n$ is open in $\mathbb{R}^n$. On the other hand, an infinite intersection $U_1 \cap U_2 \cap \cdots$ need not be open. For instance, in $\mathbb{R}^2$, consider the sequence of concentric open balls $U_1 = B((0,0), 1)$, $U_2 = B((0,0), \frac{1}{2})$, $U_3 = B((0,0), \frac{1}{3})$. etc. Each Un is open, but the only point common to all of them is the origin. Thus $U_1 \cap U_2 \cap = \{(0, 0)\}$. This is no longer an open set.

There is also a notion of closed set, though the definition may not be what one would guess.

Definition. A subset K of $\mathbb{R}^n$ is called closed if $\mathbb{R}^n - K = \{x \in \mathbb{R}^n : x \square/ K\}$ is open. The set $\mathbb{R}^n - K$ is called the complement of K in $\mathbb{R}^n$.

Example 3.16. Let $K = \{x \in \mathbb{R}^2 : \|x\| \leq 1\}$. This is the open ball centered at the origin of radius 1 together with its boundary, the unit circle, as shown in Figure 3.19. Its complement $\mathbb{R}^2 - K$ is the set of all points x such that $\|x\| > 1$, which is an open set. (Briefly, given a in $\mathbb{R}^2 - K$, then $r = \|a\| - 1$ works in the definition of open set.) Hence K is closed.

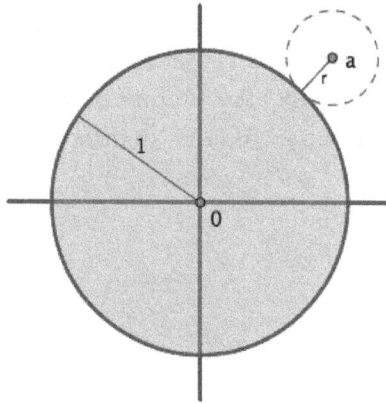

Figure 3.19: Closed balls are closed sets.

More generally, if $a \in \mathbb{R}^n$ and $r \geq 0$, then $K(a, r) = \{x \in \mathbb{R}^n : \|x - a\| \leq r\}$ is called a closed ball. These are closed subsets of $\mathbb{R}^n$ as well by a similar argument.

Example 3.17. Let $K = \{x = (x, y) \in \mathbb{R}^2 :$ either $\|x\| < 1$ or $x^2 + y^2 = 1$ where $y \geq 0\}$. This is the same as the closed ball of the preceding example except that the lower semicircle has been deleted. See Figure 3.20.

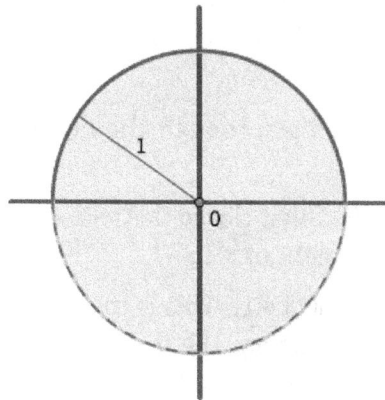

Figure 3.20: A set that is neither open nor closed

K is no longer closed: for instance, the point $a = (0, -1)$ belongs to $\mathbb{R}^2 - K$, but no open ball about a is contained entirely within $\mathbb{R}^2 - K$. At the same time, K is not open either: the point $a = (0, 1)$ belongs to K, but no open ball about a stays within K.

3.5 CONTINUITY

We are ready for continuity. Intuitively, the idea is that a function f is continuous at a point a if $\lim_{x \to a} f(x) = f(a)$. That is, as x gets close to a, $f(x)$ gets close to $f(a)$. We make this precise by expressing the requirement in terms of open balls.

Definition. Let U be an open set in $\mathbb{R}^n$, and let $f : U \to R$ be a real-valued function. We say that f is **continuous at a point** a of U if, given any open ball $B(f(a), \epsilon)$ about $f(a)$, there exists an open ball $B(a, \delta)$ about a such that:

$$f(B(a, \delta)) \subset B(f(a), \epsilon).$$

In other words, if $x \in B(a, \delta)$, then $f(x) \in B(f(a), \epsilon)$ (Figure 3.21). Equivalently, writing out the definition of open ball, f is continuous at a if, given any $\epsilon > 0$, there exists a $\delta > 0$ such that:

$$\text{if } \|x - a\| < \delta, \text{ then } |f(x) - f(a)| < \epsilon.$$

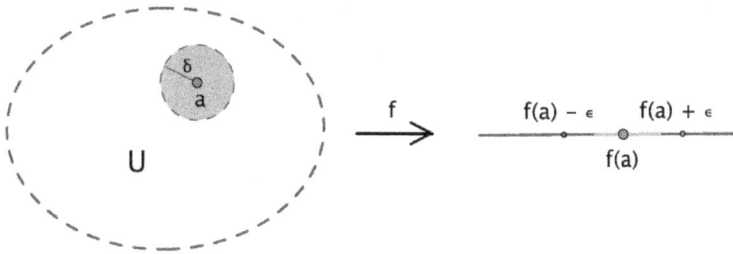

Figure 3.21: f maps the open ball B(a, δ) about a into the open ball (f (a) - ϵ, f (a) + ϵ) about f (a).

The definition says that f is continuous at a if you can guarantee that $f(x)$ will be as close to $f(a)$ as you want ("within ϵ") by making sure that x is close enough to a ("within δ"). The strategy for proving that a function is continuous is similar formally to proving that a set is open. There, one starts with a point a and tries to find a radius r. Here, one starts with an ϵ and tries to find a δ.

It may seem that using the open ball $B(f(a), \epsilon)$ in the definition is an unnecessarily confusing way to write the interval $(f(a) - \epsilon, f(a) + \epsilon)$, but we have in mind the generalization of continuity to vector-valued functions in Chapter 6. The definition phrased in terms of open balls extends naturally to the more general case, as we shall see.

Definition. A function $f : U \to R$ is simply called continuous if it is continuous at every point of its domain U .

Example 3.18. Let $f : \mathbb{R}^2 \to R$ be defined by $f(x, y) = x$. In other words, f is the projection of the xy-plane onto the x-axis. Is f a continuous function?

Let a be a point of $\mathbb{R}^2$. To get a sense of what answer to expect, we consider $\lim_{x \to a} f(x)$. (Of course, we haven't defined rigorously what a limit is yet, so this is meant to be completely informal.) Write x = (x, y) and a = (c, d). Then:

$$\lim_{\mathbf{x}\to\mathbf{a}} f(\mathbf{x}) = \lim_{(x,y)\to(c,d)} f(x,y) = \lim_{(x,y)\to(c,d)} x = c.$$

At the same time, $f(a) = f(c, d) = c$, which agrees. Thus we expect intuitively that f is continuous at a.

To prove this rigorously, let $\epsilon > 0$ be given. We want to find a radius δ so that the open ball B (c, d), δ) is projected inside the open interval B(f (c, d), ϵ) = B(c, ϵ) = (c - ϵ, c + ϵ). From the geometry of the projection, taking $\delta = \epsilon$ works. See Figure 3.22. That is, if (x, y) $\in$ B((c, d), ϵ), then its x-coordinate satisfies c - ϵ < x < c + ϵ. In other words, c - ϵ < f (x, y) < c + ϵ, or f (x, y) $\in$ (c - ϵ, c + ϵ). Therefore $\delta = \epsilon$ works in the definition of continuity.

Actually, δ could be anything less than ϵ, too, but the definition requires only that we find one δ that works, not all of them.

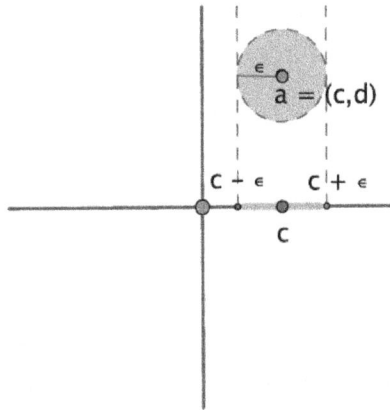

Figure 3.22: For the projection, choosing $\delta = \epsilon$ works.

Since f is continuous at every point a of $\mathbb{R}^2$, it is a continuous function.

Example 3.19. Let $f(x,y) = \frac{xy}{x^2+y^2}$ for all (x, y) $\neq$ (0, 0). Is it possible to assign a value to f(0,0) so that f is continuous at (0, 0)? See Figure 3.23.

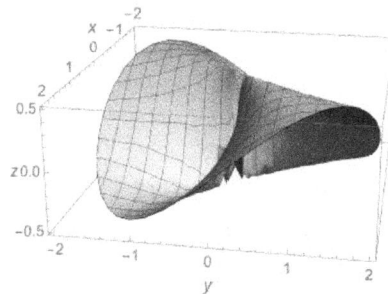

Figure 3.23: The graph $z = \frac{xy}{x^2+y^2}$ **: can it be made continuous at (0, 0)?**

Say we set $f(0, 0) = c$. First, the intuitive approach: we want to know if it's possible to choose c so that:

$$\lim_{(x,y)\to(0,0)} f(x,y) = c.$$

To get a feel for this, we try approaching the origin in a couple of different ways. For instance, suppose that we approach along the x-axis. Then $f(x,0) = \frac{x \cdot 0}{x^2+0^2} = 0$, so:

$$\lim_{(x,0)\to(0,0)} f(x,0) = \lim_{(x,0)\to(0,0)} 0 = 0.$$

On the other hand, if we approach along the line y = x, then $f(x,x) = \frac{x \cdot x}{x^2+x^2} = \frac{1}{2}$, so:

$$\lim_{(x,x)\to(0,0)} f(x,x) = \lim_{(x,x)\to(0,0)} \frac{1}{2} = \frac{1}{2}.$$

Thus $f(x, y)$ can approach different values depending on how you approach the origin. As a result, intuitively, the limit does not exist, so f cannot be continuous regardless of what c is.

To prove this rigorously, we need to articulate what it means for the criterion for continuity to fail. The definition says that a function is continuous if, for every ϵ, there exists a δ. The negation of this is that there is some ϵ for which there is no δ. We try to find such a bad ϵ.

Our intuitive calculations above showed that there are points arbitrarily close to the origin where $f = 0$ and other points arbitrarily close where $f = \frac{1}{2}$. We choose ϵ so that these values of f cannot both be within ϵ of $f(0, 0) = c$.

For this, let $\epsilon = \frac{1}{8}$. Then $B(f(0,0), \epsilon) = B(c, \frac{1}{8}) = (c - \frac{1}{8}, c + \frac{1}{8})$, an interval of length $\frac{1}{4}$. As just noted, in any open ball $B((0,0), \delta)$ about the origin, there will be points (x, 0) where $f(x, 0) = 0$ and points (x, x) where $f(x,x) = \frac{1}{2}$. It's impossible to fit both these values inside an interval of length $\frac{1}{4}$. So for $\epsilon = \frac{1}{8}$, there is no $\delta > 0$ such that $f(B((0,0), \delta)) \subset (c - \frac{1}{8}, c + \frac{1}{8})$. Thus f is not continuous at (0, 0) no matter what the value of $f(0, 0)$ is.

The $\epsilon\delta$-definition of continuity is awkward and difficult to work with. One of its virtues, however, is that, not only can it be used to verify that a function is continuous, but it also gives an unambiguous condition for proving that a function is not. That is the moral of the preced

Example 3.20. In the same spirit, let $f(x, y) = \frac{x^3+y^3}{x^2+y^2}$ for all (x, y) ≠ (0, 0). The graph of f is shown in Figure 3.24. Is it possible to assign a value to $f(0, 0)$ so that f is continuous at (0, 0)?

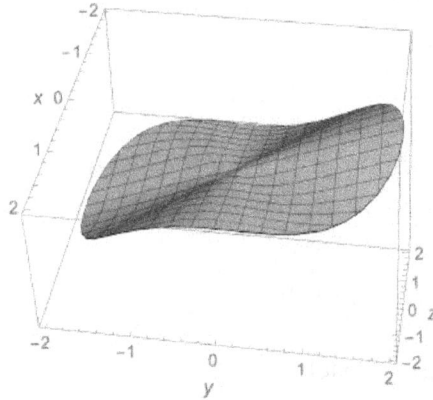

Figure 3.24: The graph $z = \frac{x^3+y^3}{x^2+y^2}$

As before, intuitively, in order to be continuous, f (0, 0) should equal $\lim_{(x,y)\to(0,0)} \frac{x^3+y^3}{x^2+y^2}$, if the limit exists. We try approaching the origin along various lines:

Along the x-axis $\qquad\qquad f(x,0) = \frac{x^3+0^3}{x^2+0^2} = x \to 0$ as $x \to 0$

Along the y-axis $\qquad\qquad f(0,y) = \frac{0^3+y^3}{0^2+y^2} = y \to 0$ as $y \to 0$

Along the line y = mx $\qquad f(x,mx) = \frac{x^3+m^3x^3}{x^2+m^2x^2} = \frac{1+m^3}{1+m^2}x \to 0$ as $x \to 0$

We get a consistent answer of 0, but this does not prove that the limit is 0. We have not exhausted all possible ways of approaching the origin, and being close to the origin is different from being close to it along any individual curve. Nevertheless, the evidence suggests that choosing f (0, 0) = 0 might make f continuous, and this gives us something concrete to shoot for.

So set $f(0, 0) = 0$, and let $\epsilon > 0$ be given. We want to find a $\delta > 0$ such that, if ‖(x, y)-(0, 0)‖ < δ, then |f(x, y) – 0| < ε. This is satisfied automatically when (x, y) = (0, 0) for any value of δ since f (0, 0) = 0, so we assume (x, y)≠ (0, 0) and look for a δ such that, if ‖(x, y)‖ < δ, then $\left|\frac{x^3+y^3}{x^2+y^2}\right| < \epsilon$.

To hunt for a connection between the quantities (x, y) and $\left|\frac{x^3+y^3}{x^2+y^2}\right|$, we introduce polar coordinates: x = r cos θ and y = r sin θ, where $r = \sqrt{x^2+y^2} = \|(x,y)\|$. See Figure 3.25.

Then, if (x, y)≠(0, 0):

$$\frac{x^3+y^3}{x^2+y^2} = \frac{r^3(\cos^3\theta + \sin^3\theta)}{r^2} = r(\cos^3\theta + \sin^3\theta).$$

Since | cos³ θ + sin³ θ| ≤ 2, this implies that:

$$\left|\frac{x^3+y^3}{x^2+y^2}\right| = |r(\cos^3\theta + \sin^3\theta)| \le 2r = 2\sqrt{x^2+y^2} = 2\|(x,y)\|.$$

Thus, if $\|(x, y)\| < a$, then $\left|\frac{x^3+y^3}{x^2+y^2}\right| \leq 2\|(x,y)\| < 2a$.

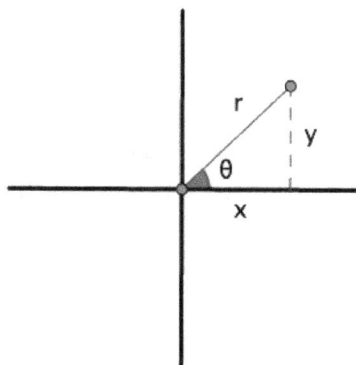

Figure 3.25: Polar coordinates r and θ

So let $\delta = \frac{\epsilon}{2}$. With this choice, if $\|(x, y)\| < \delta$, then $|f(x, y)| < 2\delta = \epsilon$, i.e., f maps $B((0, 0), \delta)$ inside $(-\epsilon, \epsilon)$. In other words, given any $\epsilon > 0$, we've found a $\delta > 0$ that satisfies the definition of continuity, namely, $\delta = \frac{\epsilon}{2}$. Therefore setting $f(0, 0) = 0$ makes f continuous at $(0, 0)$.

3.6. SOME PROPERTIES OF CONTINUOUS FUNCTIONS

The definition of continuity makes for unimpeachable arguments, but, as we lea$\mathbb{R}^n$ more about continuous functions, we might prefer to build on what we lea$\mathbb{R}^n$ and not have to start from scratch with the definition every time. We present a couple of general properties that are especially useful. These results are behind the working principle that most functions that look continuous really are continuous. One need not bring out the ε's and δ's.

Proposition 3.21. Let $f, g : U \to R$ be real-valued functions defined on an open set U in $\mathbb{R}^n$. If f and g are continuous at a point a of U, then so are:

$$\begin{cases} f+g, \\ cf \text{ for any scalar } c, \\ fg, \text{ the product of } f \text{ and } g, \\ \frac{f}{g}, \text{ the quotient, assuming that } g(\mathbf{a}) \neq 0. \end{cases}$$

Before proving this result, let's apply it to some examples.

Example 3.22. We showed earlier in Example 3.18 that the projection onto the x-axis $f : \mathbb{R}^2 \to R$, $f(x, y) = x$, is continuous. Likewise, the projection onto the y-axis g $: \mathbb{R}^2 \to R$, $g(x, y) = y$, is continuous. It then follows immediately from the proposition that functions like x + y, 3x, xy, $x^2 = x \cdot x$, $x^3 - 4x^2y^2 + 5y$, and, at points other than (0, 0), $\frac{xy}{x^2+y^2}$ are all continuous.

Proof. We prove the statement about the sum $f + g$ and leave the rest for the exercises. (See Exercises 7.4–7.7.) For this, we want to argue that, as x gets close to a, $f(x) + g(x)$ gets close to $f(a) + g(a)$. Since f(x) and g(x) get close to $f(a)$ and g(a) individually, this conclusion seems clear, and it's just a matter of feeding the intuition into the formal definition. The key ingredient in the argument below is the "triangle inequality."

Let $\epsilon > 0$ be given. We need to come up with a $\delta > 0$ such that $f(x) + g(x) \in B(f(a) + g(a), \epsilon)$ whenever $x \in B(a, \delta)$, in other words, a δ such that:

$$|(f(\mathbf{x}) + g(\mathbf{x})) - (f(\mathbf{a}) + g(\mathbf{a}))| < \epsilon \text{ whenever } \mathbf{x} \in B(\mathbf{a}, \delta).$$

Note that:

$$|(f(\mathbf{x}) + g(\mathbf{x})) - (f(\mathbf{a}) + g(\mathbf{a}))| = |(f(\mathbf{x}) - f(\mathbf{a})) + (g(\mathbf{x}) - g(\mathbf{a}))|. \qquad (3.3)$$

The triangle inequality says that, for any real numbers r and s, $|r + s| \leq |r| + |s|$. This seems reasonable, but it has a generalization to $\mathbb{R}^n$ that is important enough that we discuss it separately in the next section. Accepting that it is true for the time being, we find from equation (3.3) that:

$$|(f(\mathbf{x}) + g(\mathbf{x})) - (f(\mathbf{a}) + g(\mathbf{a}))| \leq |f(\mathbf{x}) - f(\mathbf{a})| + |g(\mathbf{x}) - g(\mathbf{a})|. \qquad (3.4)$$

This separates the f contribution and the g contribution into terms that can be manipulated independently.

Since f is continuous at a, taking the positive quantity $\frac{\epsilon}{2}$ as the given input in the definition of continuity, there is a $\delta_1 > 0$ such that $|f(x) - f(a)| < \epsilon$ whenever $x \in B(a, \delta1)$. Similarly, there is a $\delta_2 > 0$ such that $|g(\mathbf{x}) - g(\mathbf{a})| < \frac{\epsilon}{2}\epsilon$ whenever $x \in B(a, \delta_2)$. Let $\delta = \min\{\delta1, \delta2\}$. Then the last two inequalities remain true for all x in B(a, δ), and substituting them into (3.4) gives:

$$|(f(\mathbf{x}) + g(\mathbf{x})) - (f(\mathbf{a}) + g(\mathbf{a}))| < \frac{\epsilon}{2} + \frac{\epsilon}{2} = \epsilon \quad \text{whenever } \mathbf{x} \in B(\mathbf{a}, \delta).$$

This shows that $f + g$ is continuous at a.

Proposition 3.23. Compositions of continuous functions are continuous. That is, let U be an open set in $\mathbb{R}^n$ and V an open set in R, and let $f : U \to R$ and $g : V \to R$ be functions such that $f(x) \in V$ for all x in U. (This assumption guarantees that the composition $g \circ f : U \to R$ is defined.) Then, if f and g are continuous, so is $g \circ f$.

Again, we look at some examples first.

Example 3.24. We accept without comment that the familiar functions of one variable from first year calculus that were said to be continuous there really are continuous. This includes $|x|$, $\sqrt[n]{x}$, sin x, cos x, e^x, and In x. The proposition then implies that function like $\sqrt{x^2 + y^2}$, $\sin(x + y)$, and e^{xy} are continuous. For example, the first is the composition $(x, y) \mapsto x^2 + y^2 \mapsto \sqrt{x^2 + y^2}$, which is a composition of continuous steps.

Proof. Let a be a point of U , and let $\epsilon > 0$ be given. We want to find an open ball about a that is mapped by $g \circ f$ inside the interval $B(g(f(a)), \epsilon) = g(f(a)) - \epsilon, g(f(a)) + \epsilon$. We work our way backwards from $g(f(a))$ to a using the definition of continuity twice to find such a ball. The ingredients are depicted in Figure 3.26.

First, since g is continuous at the point $f(a)$, there exists a $\delta' > 0$ such that:

$$g(B(f(\mathbf{a}), \delta')) \subset B(g(f(\mathbf{a})), \epsilon).$$

But then since f is continuous at a, treating δ' as the given input, there exists a $\delta > 0$ such that:

$$f(B(\mathbf{a}, \delta)) \subset B(f(\mathbf{a}), \delta').$$

Linking these two steps:

$$(g \circ f)(B(\mathbf{a}, \delta)) = g(f(B(\mathbf{a}, \delta))) \subset g(B(f(\mathbf{a}), \delta')) \subset B(g(f(\mathbf{a})), \epsilon).$$

Thus, given an ϵ, we've found a δ that works.

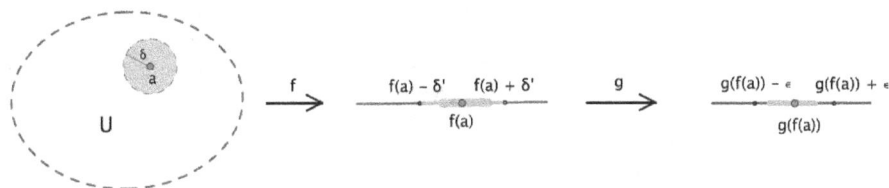

Figure 3.26: The composition of continuous functions: given an ϵ, a δ' exists (g is continuous), then a δ exists (f is continuous).

3.7. THE CAUCHY-SCHWARZ AND TRIANGLE INEQUALITIES

We now discuss two fundamental inequalities about vectors in $\mathbb{R}^n$, one involving the dot product and the other involving vector addition.

Theorem 3.25 (Cauchy-Schwarz inequality).

$$|\mathbf{v} \cdot \mathbf{w}| \leq \|\mathbf{v}\| \|\mathbf{w}\|$$

for all v, w in $\mathbb{R}^n$.

Proof. Recall that $v \cdot w = \|v\| \|w\| \cos \theta$, where θ is the angle between v and w (see Theorem 1.12). Thus, since $|\cos \theta| \leq 1$, we have $|v \cdot w| = \|v\| \|w\| |\cos \theta| \leq \|v\| \|w\|$.

From this follows a generalization of the property of R that was alluded to in the proof of Proposition 3.21.

Theorem 3.26 (Triangle inequality).

$$\|\mathbf{v} + \mathbf{w}\| \leq \|\mathbf{v}\| + \|\mathbf{w}\|$$

for all v, w in $\mathbb{R}^n$.

Proof. It's equivalent to square both sides and prove that $\|v + w\|^2 \leq (\|v\| + \|w\|)^2$. But:

$$\|v + w\|^2 = (v + w) \cdot (v + w)$$
$$= v \cdot v + v \cdot w + w \cdot v + w \cdot w$$
$$= \|v\|^2 + 2v \cdot w + \|w\|^2. \tag{3.5}$$

Also, $v \cdot w \leq |v \cdot w| \leq \|v\|\,\|w\|$, where the second inequality is Cauchy-Schwarz. Substituting this into (3.5) gives:

$$\|v + w\|^2 \leq \|v\|^2 + 2\|v\|\,\|w\| + \|w\|^2 = (\|v\| + \|w\|)^2.$$

The connection with triangles is illustrated in Figure 3.27.

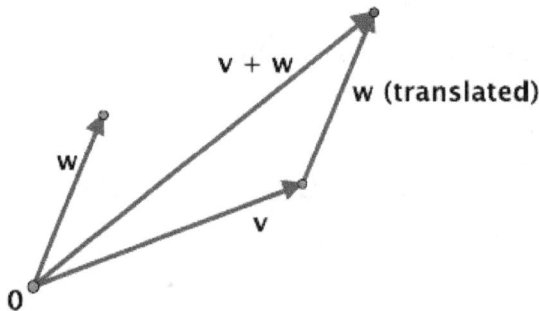

Figure 3.27: The geometry of the triangle inequality: the length $\|v + w\|$ of one side of a triangle is less than or equal to the sum $\|v\| + \|w\|$ of the lengths of the other two sides.

3.8. LIMITS

We used our intuition about limits as a way to think about continuity. Now that continuity has been defined rigorously, we turns the tables and use continuity to give a formal definition of limits. One technical, though important, point is that, in determining how f (x) behaves as x approaches a, what is happening right at a does not matter. The value of f (a), or even whether f is defined there, is irrelevant.

Definition. Let U be an open set in $\mathbb{R}^n$, and let a be a point of U . If f is a real-valued function that is defined on U , except possibly at the point a, then we say that $\lim_{x \to a} f(x)$ exists if there is a number L such that the function $\tilde{f} : U \to \mathbb{R}$ defined by

$$\tilde{f}(x) = \begin{cases} f(x) & \text{if } x \neq a, \\ L & \text{if } x = a \end{cases}$$

is continuous at a. When this happens, we write $\lim_{x \to a} f(x) = L$.

By carefully applying the definition of continuity to the function $\tilde{f}$, the definition of limit can be restated using ϵ's and δ's, which is its more customary form. Namely, $\lim_{x \to a} f(x) = L$ means:

Given any $\epsilon > 0$, there exists a $\delta > 0$ such that $|f(x) - L| < \epsilon$ whenever $\|x - a\| < \delta$, except possibly when $x = a$.

We shall never use this formulation, however.

If f is defined at a, then it's basically a tautology from the definition that $\lim_{x \to a} f(x) = f(a)$ if and only if f is continuous at a, thus completing the intuitive connection between the two concepts with which we began.

Example 3.27. In Example 3.20, we showed that the function given by $f(x,y) = \frac{x^3+y^3}{x^2+y^2}$, which is defined for $(x, y) \neq (0, 0)$, could be made continuous at $(0, 0)$ by setting $f(0, 0) = 0$. (Strictly speaking, this extended function is the one called $\tilde{f}$ in the definition.) Hence, by definition:

$$\lim_{(x,y) \to (0,0)} \frac{x^3 + y^3}{x^2 + y^2} = 0.$$

Similarly, in Example 3.19, for the function $f(x,y) = \frac{xy}{x^2+y^2}$, there is no value that can be assigned to $f(0, 0)$ that makes f continuous at $(0, 0)$. In this case:

$$\lim_{(x,y) \to (0,0)} \frac{xy}{x^2 + y^2} \quad \text{does not exist.}$$

Because of the close connection between the two concepts, properties of continuous functions have analogues for limits. Here is the version of Proposition 3.21 for limits.

Proposition 3.28. Assume that $\lim_{x \to a} f(x)$ and $\lim_{x \to a} g(x)$ both exist. Then:

1. $\lim_{x \to a} (f(x) + g(x)) = (\lim_{x \to a} f(x)) + (\lim x \to a\ g(x))$.
2. $\lim_{x \to a} (cf(x)) = c (\lim_{x \to a} f(x))$ for any scalar c.
3. $\lim_{x \to a} f(x)g(x) = (\lim_{x \to a} f(x)) (\lim_{x \to a} g(x))$.
4. $\lim_{x \to a} \frac{f(x)}{g(x)} = \frac{\lim_{x \to a} f(x)}{\lim_{x \to a} g(x)}$, provided that $\lim_{x \to a} g(x) \neq 0$.

We leave the proofs for the exercises (Exercise 8.1).

3.9. EXERCISES FOR CHAPTER 3

Section 1 Graphs and level sets

For the functions $f(x, y)$ in Exercises 1.1–1.7: (a) sketch the cross-sections of the graph $z = f(x, y)$ with the coordinate planes, (b) sketch several level curves of f, labeling each with the corresponding value of c, and (c) sketch the graph of f.

1.1. $f(x, y) = x^2 + y^2$

1.2. $f(x, y) = x^2 + y^2 + 1$

1.3. $f(x, y) = y - x^2$

1.4. $f(x, y) = x + y$

1.5. $f(x, y) = |x| + |y|$

1.6. $f(x, y) = xy$

1.7. $f(x, y) = \frac{y}{x^2 + 1}$

1.8. The graph of the function $f(x, y) = x^3 - 3xy^2$ is called a **monkey saddle**. You will need to bring one along whenever you invite a monkey to go riding with you. The level sets of f are not easy to sketch directly, but it is still possible to get a reasonable idea of what the graph looks like.

    a. Sketch the level set corresponding to c = 0.

        (*Hint:* $x^3 - 3xy^2 = x(x + \sqrt{3}y)(x - \sqrt{3}y)$.)

    b. Draw the region of the xy-plane in which $f(x, y) > 0$ and the region in which $f(x, y) < 0$. (Hint: Part (a) might help.)

    c. Use the information from parts (a) and (b) to make a rough sketch of the monkey saddle.

1.9. It is possible to construct an origami approximation of the saddle surface z = $x^2 - y^2$ of Example 3.3 by folding a piece of paper. The resulting model is called a **hypar**. (The saddle surface itself is known as a hyperbolic paraboloid.) See Figure 3.28. Construct your own hypar by starting with a square piece of paper and following the instructions appended to the end of the chapter.

Figure 3.28: A hypar

Section 2 More surfaces in R³

In Exercises 2.1–2.4, sketch the surface in R³ described by the given equation.

2.1. $x^2 + y^2 + z^2 = 4$

2.2. $x^2 + y^2 = 4$

2.3. $x^2 + z^2 = 4$

2.4. $x^2 + \frac{y^2}{9} + \frac{z^2}{4} = 1$

In Exercises 2.5–2.9, sketch the level sets corresponding to the indicated values of c for the given function $f(x, y, z)$ of three variables. Make a separate sketch for each individual level set.

2.5. $f(x, y, z) = x^2 + y^2 + z^2$, c = 0, 1, 2

2.6. $f(x, y, z) = x^2 + y^2$, c = 0, 1, 2

2.7. $f(x, y, z) = x^2 + y^2 - z$, c = -1, 0, 1

2.8. $f(x, y, z) = x^2 + y^2 - z^2$, c = -1, 0, 1

2.9. $f(x, y, z) = x^2 - y^2 - z^2$, c = -1, 0, 1

2.10. Can the saddle surface $z = x^2 - y^2$ in R^3 be described as a level set? If so, how? What about the parabola $y = x^2$ in $\mathbb{R}^2$?

Section 3 The equation of a plane

3.1. Find an equation of the plane through the point p = (1, 2, 3) with normal vector n = (4, 5, 6).

3.2. Find an equation of the plane through the point p = (2, -1, 3) with normal vector n = (1, -4, 5).

3.3. Find a point that lies on the plane x + 3y + 5z = 9 and a normal vector to the plane.

3.4. Find a point that lies on the plane x + z = 1 and a normal vector to the plane.

3.5. Are the planes x - y + z = 8 and 2x + y - z = -1 perpendicular? Are they parallel?

3.6. Are the planes x - y + z = 8 and 2x - 2y + 2z = -1 perpendicular? Are they parallel?

3.7. Find an equation of the plane that contains the points p = (1, 0, 0), q = (0, 2, 0), and r = (0, 0, 3).

3.8. Find an equation of the plane that contains the points p = (1, 1, 1), q = (2, -1, 2), and r = (-1, 2, 0).

3.9. Find an equation of the plane that contains the point p = (1, 1, 1) and the line parametrized by $\alpha(t) = (1, 0, 1) + t(2, 3, -4)$.

3.10. Find a parametrization of the line of intersection of the planes x + y + z = 3 and x-y + z = 1.

3.11. What point on the plane x - y + 2z = 3 is closest to the point q = (2, 1, -1)? (Hint: Find a parametrization of an appropriate line through q.)

3.12. Two planes in R^3 are perpendicular to each other. Their line of intersection is described by the parametric equations:

$$x = 2 - t, \quad y = -1 + t, \quad z = -4 + 2t.$$

If one of the planes has equation 2x + 4y - z = 4, find an equation for the other plane.

3.13. Consider the plane x + y + z = 1.

    a. Sketch the cross-sections with the three coordinate planes.

    b. Sketch and describe in words the portion of the plane that lies in the "first octant" of R^3, that is, the part where x ≥ 0, y ≥ 0, and z ≥ 0.

3.14. Sketch and describe some of the level sets of the function $f(x, y, z) = x + y + z$.

3.15. Let π_1 and π_2 be parallel planes in R3 given by the equations:

$$\pi_1 : Ax + By + Cz = D_1 \text{ and } \pi_2 : Ax + By + Cz = D_2.$$

    a. If p_1 and p_2 are points on π_1 and π_2, respectively, show that:

$$n \cdot (p_2 - p_1) = D_2 - D_1.$$

    b. Show that the perpendicular distance d between π_1 and π_2 is given by:

$$d = \frac{|D_2 - D_1|}{\sqrt{A^2 + B^2 + C^2}}.$$

Find the perpendicular distance between the planes x + y + z = 1 and x + y + z = 5.

3.16. Let α be a path in R^3 that has constant torsion $\tau = 0$. You may assume that $\alpha'(t) \neq 0$ and $T'(t) \neq 0$ for all t so that the Frenet vectors of α are always defined.

Prove that the binormal vector B is constant.

Fix a time t = t_0, and let $x_0 = \alpha(t0)$. Show that the function is a constant function. What is the value of the constant?

$$f(t) = B \cdot (\alpha(t) - x_0)$$

Prove that $\alpha(t)$ lies in a single plane for all t. In other words, curves that have constant torsion 0 must be planar. (Hint: Use part (b) to identify a point and a normal vector for the plane in which the path lies.)

Section 4 Open sets

In Exercises 4.1–4.7, let U be the set of points (x, y) in $\mathbb{R}^2$ that satisfy the given conditions. Sketch U , and determine whether it is an open set. Your arguments should be at a level of rigor comparable to those given in the text.

4.1. 0 < x < 1 and 0 < y < 1

4.2. |x| < 1 and |y| < 1

4.3. |x| ≤ 1 and |y| < 1

4.4. $x^2 + y^2 = 1$

4.5. $1 < x^2 + y^2 < 4$

4.6. $y > x$

4.7. $y \geq x$

4.8. Is $\mathbb{R}^2 - \{(0, 0)\}$, the plane with the origin removed, an open set in $\mathbb{R}^2$? Justify your answer.

4.9. Prove that the union of any collection of open sets in $\mathbb{R}^n$ is open. That is, if $\{U_\alpha\}$ is a collection of open sets in $\mathbb{R}^n$, then $U_\alpha U_\alpha$ is open.

Section 5 Continuity

5.1. Let $f : \mathbb{R}^2 \to \mathbb{R}$ be the function:

$$f(x, y) = \begin{cases} \frac{x^3 y}{x^2 + y^2} & \text{if } (x, y) \neq (0, 0), \\ 0 & \text{if } (x, y) = (0, 0). \end{cases}$$

a. Let a be a positive real number. Show that, if $\|(x, y)\| < a$, then $|f(x, y)| < a^2$.
b. Use the $\epsilon\delta$-definition of continuity to determine whether f is continuous at $(0, 0)$.

5.2. Consider the function $f(x, y)$ defined by:

$$f(x, y) = \frac{x^2 y^4}{(x^2 + y^4)^2} \quad \text{if } (x, y) \neq (0, 0).$$

Determine what happens to the value of $f(x, y)$ as (x, y) approaches the origin along:

a. the x-axis,
b. the y-axis,
c. the line $y = mx$,
d. the parabola $x = y^2$.
e. Is it possible to assign a value to $f(0, 0)$ so that f is continuous at $(0, 0)$? Justify your answer using the $\epsilon\delta$-definition of continuity.

5.3. Let $f(x, y) = \frac{x^2 - y^2}{x^2 + y^2}$ if $(x, y) \neq (0, 0)$. Is it possible to assign a value to $f(0, 0)$ so that f is continuous at $(0, 0)$? Justify your answer using the $\epsilon\delta$-definition.

5.4. $f(x, y) = \frac{x^3 + 2y^3}{x^2 + y^2}$ if $(x, y) \neq (0, 0)$. Is it possible to assign a value to $f(0, 0)$ so that f is continuous at $(0, 0)$? Justify your answer using the $\epsilon\delta$-definition.

Section 6 Some properties of continuous functions

6.1. Let $a = (1, 2)$. Show that the function $f : \mathbb{R}^2 \to \mathbb{R}$ given by $f(x) = a \cdot x$ is continuous.

In Exercises 5.1–5.2, you determined the continuity of the following two functions at $(0, 0)$. Now, consider points (x, y) other than $(0, 0)$. Determine all such points at which f is continuous.

6.2. $f(x,y) = \frac{x^3 y}{x^2+y^2}$ if $(x,y) \neq (0,0)$, $f(0,0) = 0$

6.3. $f(x,y) = \frac{x^2 y^4}{(x^2+y^4)^2}$ if $(x,y) \neq (0,0)$

6.4. Let U be an open set in $\mathbb{R}^n$, and let $f : U \to R$ be a function that is continuous at a point a of U .

If f (a) > 0, show that there exists an open ball B = B(a, r) centered at a such that $f(x) > \frac{f(a)}{2}$ for all x in B. (Hint: $\epsilon = \frac{f(a)}{2}$.)

Similarly, if f (a) < 0, show that there exists an open ball B = B(a, r) such that $f(x) < \frac{f(a)}{2}$ for all x in B.

In particular, if $f(a) \neq 0$, there is an open ball B centered at a throughout which $f(x)$ has the same sign as $f(a)$.

6.5. Let U be the set of all points (x, y) in $\mathbb{R}^2$ such that y > sin x, that is:

$$U = \{(x,y) \in \mathbb{R}^2 : y > \sin x\}.$$

Prove that U is an open set in $\mathbb{R}^2$. (Hint: Apply the previous exercise to an appropriate continuous function.)

Section 7 The Cauchy-Schwarz and triangle inequalities

7.1. Let v and w be vectors in $\mathbb{R}^n$.

    a. Show that $\|v\| - \|w\| \leq \|v - w\|$. (Hint: v = (v - w) + w.)

    b. Show that $\|v\| - \|w\| \leq \|v - w\|$.

7.2.   a. Let a be a point of $\mathbb{R}^n$. If r > 0, show that B(a, r) ⊂ B(0, ‖a‖ + r). Draw a picture that illustrates the result in $\mathbb{R}^2$. (Hint: To prove a set inclusion of this type, you must show that, if x ∈ B(a, r), then x ∈ B(0, ‖a‖+r), that is, if ‖x-a‖ < r, then ‖x-0‖ < ‖a‖+r.)

    B. In $\mathbb{R}^2$, if a ∈ B((0, 0), 1) and r = 1 - ‖a‖, show that B(a, r) ⊂ B((0, 0), 1). (Recall that this inclusion was used in Example 3.12 to prove that B((0, 0), 1) is an open set.)

7.3. Let a be a point of $\mathbb{R}^n$, and let r be a positive real number. Prove that the open ball B(a, r) is an open set in $\mathbb{R}^n$.

Exercises 7.4–7.7 provide the missing proofs of parts of Proposition 3.21 about properties of continuous functions. Here, U is an open set in $\mathbb{R}^n$, f, g : U → R are real-valued functions defined on U , and a is a point of U .

7.4. If f is continuous at a and c is a scalar, prove that cf is continuous at a as well. (Hint: One approach is to consider the two cases c = 0 and c≠ 0 separately. Another is to start by showing that |cf (x) - cf (a)| ≤ (|c| + 1) |f (x) - f (a)|.)

7.5. If f is continuous at a, prove that there exists a δ > 0 such that, if x ∈ B(a, δ), then |f (x)| < |f (a)| + 1. (Hint: Use Exercise 7.1.)

7.6. If f and g are continuous at a, prove that their product fg is continuous at a. (Hint: Use an "add zero" trick and the previous exercise to show that there is a $\delta_1 >$ 0 such that, if x $\in$ B(a, δ1), then

$$\left| f(\mathbf{x})g(\mathbf{x}) - f(\mathbf{a})g(\mathbf{a}) \right| \le (|f(\mathbf{a})|+1)|g(\mathbf{x})-g(\mathbf{a})| + (|g(\mathbf{a})|+1)|f(\mathbf{x})-f(\mathbf{a})|.)$$

7.7. Assume that f and g are continuous at a and that g(a)$\neq$0.

    a. Prove that $\frac{1}{g}$ is continuous at a. (Hint: Use Exercise 6.4 to show that there exists a δ1 > 0 such that, if x $\in$ B (a, δ_1), then $\left| \frac{1}{g(\mathbf{x})} - \frac{1}{g(\mathbf{a})} \right| \le \frac{2}{|g(\mathbf{a})|^2}|g(\mathbf{x}) - g(\mathbf{a})|$.)

    b. Prove that the quotient $\frac{f}{g}$ is continuous at a. (*Hint:* $\frac{f}{g} = f \cdot \frac{1}{g}$.)

7.8. Let $f: U \to \mathbb{R}$ be a real-valued function defined on an open set U in $\mathbb{R}^n$, and let g : U $\to$ R be given by g(x) = |f (x)|.

    a. If f is continuous at a point a of U, is g necessarily continuous at a as well? Either give a proof or find a counterexample.

    b. Conversely, if g is continuous at a, is f as well? Again, give a proof or find a counterexample.

Section 8 Limits

8.1. Prove parts 1 and 3 of Proposition 3.28 about limits of sums and products. (Hint: By using the corresponding properties of continuous functions in Proposition 3.21, you should be able to avoid ϵ's and δ's entirely.)

8.2. Let U be an open set in $\mathbb{R}^n$, x_0 a point of U , and f a real-valued function defined on U , except possibly at x_0, such that $\lim_{x \to x0} f(x) = L$. Let I be an open interval in R, and let $\alpha : I \to U$ be a continuous path in U that passes through x0, i.e., $\alpha(t_0) = x_0$ for some t_0 in I. Consider the function g : I $\to$ R given by:

$$g(t) = \begin{cases} f(\alpha(t)) & \text{if } \alpha(t) \ne \mathbf{x}_0, \\ L & \text{if } \alpha(t) = \mathbf{x}_0. \end{cases}$$

Show that $\lim_{t \to t0}$ g(t) = L.

8.3. (Sandwich principles.) Let U be an open set in $\mathbb{R}^n$, and let a be a point of U .

    a. Let $f, g, h : U \to R$ be real-valued functions such that $f(x) \le g(x) \le h(x)$ for all x in U . If f (a) = h(a)—let's call the common value c—and if f and h are continuous at a, prove that g(a) = c and that g is continuous at a as well.

    b. Let f, g, h be real-valued functions defined on U , except possibly at the point a, such that $f(x) \le g(x) \le h(x)$ for all x in U , except possibly when x = a. If $\lim_{x \to a} f(x) = \lim_{x \to a} h(x) = L$, prove that $\lim_{x \to a} g(x) = L$, too.

Addendum: Folding a hypar (Exercise 1.9)

1.   Take your square piece of paper, and fold and unfold along each of the diagonals.

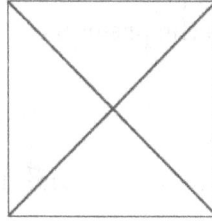

Step 1

2. Turns the paper over.
3. Fold the bottom edge of the paper to the center point, but only crease the part between the diagonals. Unfold back to the square.

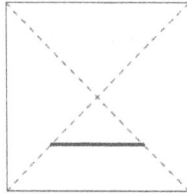

Step 3

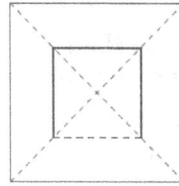

Step 4

4. Repeat for the other three sides.
5. Now, fold the bottom edge up to the upper crease line that you just made, 3/4 of the way up, again only creasing the part between the diagonals. This is a fairly short crease. Unfold back to the square.

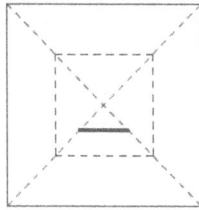

Step 5

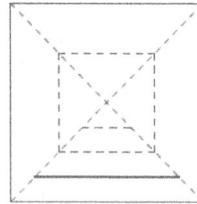

Step 6

6. Fold the bottom edge to the lower crease line (the one you made in step 3), once again only creasing between the diagonals. Unfold back to the square.
7. Repeat for all four sides. At this point, the square should be divided into four concentric square rings. See the figure on the next page.
8. Turns the paper over.
9. The goal now is to create additional creases halfway between the ones constructed so far, for a total of eight concentric square rings. You can do this by folding the bottom edge up to previously constructed folds, but use only every other fold. That is, fold the bottom edge up to the existing creases 7/8, 5/8, 3/8,

and 1/8 of the way to the top, again only creasing between the diagonals. Do this for all four sides.

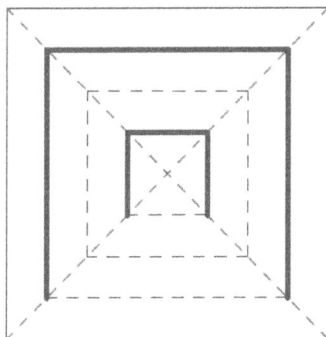

Step 7

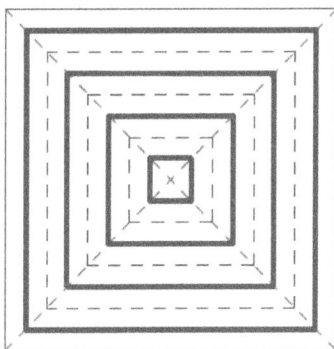

Step 9

10. Turns the paper over. The creases should alternate as you move from one concentric square to the next: mountain, valley, mountain, valley, ...

11. Now, try to fold all the creases. It's easiest to start from the outer ring and work your way in, pinching in at the corners and trying to compress the faces of adjacent rings together as you move towards the center. Half of the short diagonal creases between the rings need to be reversed so that the rings nest inside each other, though the paper can be coaxed to do this without too much trouble. In the end, you should get something shaped like an X, or a four-armed starfish.

12. Open up the arms of the starfish. The paper will naturally assume the hypar shape.

4

Chapter

REAL-VALUED FUNCTIONS: DIFFERENTIATION

We now extend the notion of derivative from real-valued functions of one variable, as in first-year calculus, to real-valued functions of n variables $f : U \to R$, where U is an open set in $\mathbb{R}^n$. The most natural choice would be to copy the one-variable definition verbatim and let the derivative at a point a be $\lim_{x \to a} \frac{f(x) - f(a)}{x - a}$. Unfortunately, this involves dividing by a vector, which does not make sense.

One potential remedy would be to consider $\lim_{x \to a} \frac{f(x) - f(a)}{\|x - a\|}$ instead. Here, it turns out that the limit fails to exist even for very simple functions. For example, for the function $f(x) = x$ of one variable, this would be $\lim_{x \to a} \frac{x - a}{|x - a|}$ which approaches +1 from the right and –1 from the left. So we must look harder to find a formulation of the one-variable derivative that can be generalized.

4.1 THE FIRST-ORDER APPROXIMATION

In first-year calculus, we think of a function f as being differentiable at a point a if it has a good tangent line approximation:

$$f(x) \approx \ell(x) \text{ when x is near a,}$$

where $y = \ell(x) = mx + b$ is the tangent line to the graph of f at the point $(a, f(a))$. See Figure 4.1.

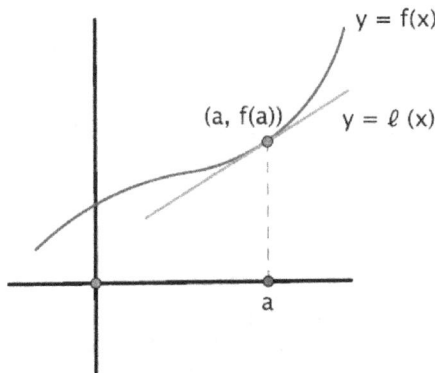

Figure 4.1: Approximating a differentiable function f (x) by one of its tangent lines ℓ(x)

In fact, the tangent line is given by

$$\ell(x) = f(a) + f'(a) \cdot (x - a). \tag{4.1}$$

so m = f'(a) and b = f (a) - f'(a) · a.

By definition, $f'(a) = \lim_{x \to a} \frac{f(x)-f(a)}{x-a}$, .assuming the limit exists. This can be rewritten as

$$0 = \lim_{x \to a} \left(\frac{f(x)-f(a)}{x-a} - f'(a) \right) = \lim_{x \to a} \frac{f(x)-f(a)-f'(a)\cdot(x-a)}{x-a},$$

or, using equation (4.1), as

$$\lim_{x \to a} \frac{f(x) - \ell(x)}{x - a} = 0. \tag{4.2}$$

In particular, when x is near a, f (x) - ℓ(x) must be much much smaller than x - a in order for $\left| \frac{f(x)-\ell(x)}{x-a} \right|$ to be near 0. Note that f (x) - ℓ(x) is the error in using the tangent line to approximate f , so, in order for f to be differentiable at a, not only must this error go to 0 as x approaches a, it must go to 0 much faster than x - a. This is the principle we generalize in moving to functions of more than one variable.

Definition. A function $\ell : \mathbb{R}^n \to \mathbb{R}m$ is called an **affine function** if it has the form $\ell(x) = T (x)+b$, where:

⊙ $T : \mathbb{R}^n \to \mathbb{R}m$ is a linear transformation and

⊙ b is a vector in Rm.

Let U be an open set in $\mathbb{R}^n$ and $f : U \to \mathbb{R}$ a real-valued function. Provisionally, we'll say that f is differentiable at a point a of U if there exists a real-valued affine function $\ell : \mathbb{R}^n \to \mathbb{R}$, $\ell(x) = T (x) + b$, such that:

$$\lim_{x \to a} \frac{f(x) - \ell(x)}{\|x - a\|} = 0. \tag{4.3}$$

This is the generalization of equation (4.2).

We can pin down the best choice of ℓ much more precisely. For instance, as ℓ is meant to approximate f near a, it should be true that f (a) = ℓ(a), that is, f (a) = T (a) + b. Thus b = f (a) - T (a), and ℓ(x) = T (x) + f (a) - T (a). Using the fact that T is linear, this becomes:

$$\ell(x) = f(a) + T(x - a).$$

This happens to have the same form as the tangent line (4.1), which seems like a good sign. Substituting into equation (4.3), we say, still provisionally, that f is differentiable at a if there exists a linear transformation $T : \mathbb{R}^n \to \mathbb{R}$ such that:

$$\lim_{x \to a} \frac{f(x) - f(a) - T(x - a)}{\|x - a\|} = 0. \tag{4.4}$$

There is really only one choice for T as well. For, as a linear transformation, T is represented by a matrix, in this case, a 1 by n matrix $A = [m_1 \; m_2 \ldots m_n]$, where mj $= T(e_j)$ for j = 1, 2, ..., n. (See Proposition 1.7.) We shall determine the values of the entries m_j.

Suppose that x approaches a in the x1-direction, i.e., let $x = a + he_1$, and let h go to 0. Then $x - a = he_1$, and $\|x - a\| = |h|$. Assume for a moment that h > 0. In order for equation (4.4) to hold, it must be true that:

$$\lim_{h \to 0} \frac{f(a + he_1) - f(a) - T(he_1)}{h} = 0;$$

or, using the linearity of T :

$$\lim_{h \to 0} \frac{f(a + he_1) - f(a) - hT(e_1)}{h} = 0. \tag{4.5}$$

If h < 0, then the denominator in equation (4.4) is $|h| = -h$. The minus sign can be factored out and canceled, so equation (4.5) still holds. This can be solved for $T(e_1)$:

$$m_1 = T(e_1) = \lim_{h \to 0} \frac{f(a + he_1) - f(a)}{h}$$
$$= \lim_{h \to 0} \frac{f(a_1 + h, a_2, \ldots, a_n) - f(a_1, a_2, \ldots, a_n)}{h}.$$

This limit of a difference quotient is the definition of the ordinary one-variable derivative where $x_2, \ldots, x_n$ are held fixed at $x_2 = a_2, \ldots, x_n = a_n$ and only x_1 varies. It is called the partial derivative $\frac{\partial f}{\partial x_1}(a)$.

Similarly, by approaching a in the x2, . . . , xn directions, we find that $m_2 = \frac{\partial f}{\partial x_2}(a), \ldots, m_n = \frac{\partial f}{\partial x_n}(a)$ and hence that $A = \left[\frac{\partial f}{\partial x_1}(a) \quad \frac{\partial f}{\partial x_2}(a) \quad \ldots \quad \frac{\partial f}{\partial x_n}(a) \right]$.

Definition. For a real-valued function $f : U \to R$ defined on an open set U in $\mathbb{R}^n$ and a point a of U :

If j = 1, 2, . . . , n, the partial derivative of f at a with respect to x_j is defined by:

$$\frac{\partial f}{\partial x_j}(a) = \lim_{h \to 0} \frac{f(a + he_j) - f(a)}{h}.$$

Note that $a + he_j = (a_1, \ldots, a_j + h, \ldots, a_n)$, so $a + he_j$ and a differ only in the jth coordinate. Thus the partial derivative is defined by the one-variable difference quotient for the derivative with variable x_j.

Other common notations for the partial derivative are f_{x_i} (a) and $(D_j f)$(a).

Df (a) is defined to be the 1 by n matrix $Df(a) = \left[\frac{\partial f}{\partial x_1}(a) \quad \frac{\partial f}{\partial x_2}(a) \quad \cdots \quad \frac{\partial f}{\partial x_n}(a)\right]$. The corresponding vector $\nabla f(a) = (\frac{\partial f}{\partial x_1}(a), \frac{\partial f}{\partial x_2}(a), \ldots, \frac{\partial f}{\partial x_n}(a))$ in $\mathbb{R}^n$ is called the gradient of f. (The intent of the notation might be clearer if these objects were denoted by (Df)(a) and (∇f)(a), respectively, but the extra parentheses are usually omitted.)

Example 4.1. Let $f(x, y) = x^3 + 2x^2y - 3y^2$, and let a = (2, 1). To find $\frac{\partial f}{\partial x}$(a), we fix y = 1 and differentiate with respect to x. Since $f(x, 1) = x^3 + 2x^2 - 3$, this gives $\frac{\partial f}{\partial x}(x, 1) = 3x^2 + 4x$ and then $\frac{\partial f}{\partial x}(2, 1) = 12 + 8 = 20.$

Similarly, for the partial derivative at a with respect to

$$y, \ f(2, y) = 8 + 8y - 3y^2, \text{ so } \frac{\partial f}{\partial y}(2, y) = 8 - 6y \text{ and } \frac{\partial f}{\partial y}(2, 1) = 8 - 6 = 2.$$

In practice, this is not how partial derivatives are calculated usually. Instead, one works directly with the general formula for f . For instance, to find the partial derivative with respect to x, differentiate with respect to x, thinking of all other variables—in this case, only y—as constant:

$$\frac{\partial f}{\partial x} = 3x^2 + 4xy - 0 = 3x^2 + 4xy, \text{ so } \frac{\partial f}{\partial x}(2, 1) = 12 + 8 = 20$$

as before. Likewise, $\frac{\partial f}{\partial y} = 2x^2 - 6y$, and $\frac{\partial f}{\partial y}(2, 1) = 8 - 6 = 2.$

In any case, Df (a) = 20 2 , and ∇f (a) = (20, 2).

We now have all the ingredients needed to state the formal definition of differentiability, as motivated by equation (4.4).

Definition. Let U be an open set in $\mathbb{R}^n$, $f: U \to R$ a real-valued function, and a a point of U . Then f is said to be **differentiable at a** if:

$$\lim_{x \to a} \frac{f(x) - f(a) - Df(a) \cdot (x - a)}{\|x - a\|} = 0.$$

When this happens, the matrix Df (a) is called the derivative of f at a. It is also known as the **Jacobian matrix**. The affine function $\ell(x) = f$ (a)+Df (a)·(x-a) is called the **first-order affine approximation** of f at a. (Note the resemblance to the one-variable tangent line approximation (4.1).)

The product in the numerator of the definition is the matrix product:

$$Df(a) \cdot (x - a) = \left[\frac{\partial f}{\partial x_1}(a) \quad \frac{\partial f}{\partial x_2}(a) \quad \cdots \quad \frac{\partial f}{\partial x_n}(a)\right] \begin{bmatrix} x_1 - a_1 \\ x_2 - a_2 \\ \vdots \\ x_n - a_n \end{bmatrix}.$$

It could also be written as a dot product: ∇f (a) · (x - a).

The definition generalizes equation (4.2) for one variable in that it says that f is differentiable at a if the error in using the first-order approximation ℓ goes to 0 faster than ||x - a|| as x approaches a. In particular, the first-order approximation is a good approximation. This is worth emphasizing: the existence of a good first-order approximation is often the most important way to think of what it means for a function to be differentiable. Note that the derivative $Df(a)$ itself is no longer just a number the way it is in first-year calculus, but rather a matrix or, even better, the linear part of a good affine approximation of f near a.

Example 4.2. Let $f: \mathbb{R}^2 \to \mathbb{R}$ be $f(x, y) = x^2 + y^2$, and let a = (1, 2). Is f differentiable at a?

We $\frac{\partial f}{\partial x} = 2y$, he various elements that go into the definition. First, $\frac{\partial f}{\partial x} = 2x$ and $\frac{\partial f}{\partial y} = 2y$, so $Df(x, y) = 2x\ 2y]$ and $Df(a) = Df(1, 2) = 2\ 4]$. Also, $f(a) = 5$, and $x - a = \begin{bmatrix} x - 1 \\ y - 2 \end{bmatrix}$. Thus we need to check if:

$$\lim_{(x,y)\to(1,2)} \frac{x^2 + y^2 - 5 - \begin{bmatrix} 2 & 4 \end{bmatrix} \begin{bmatrix} x - 1 \\ y - 2 \end{bmatrix}}{\sqrt{(x - 1)^2 + (y - 2)^2}} = 0.$$

Let's simplify the numerator, completing the square at an appropriate stage:

$$x^2 + y^2 - 5 - \begin{bmatrix} 2 & 4 \end{bmatrix} \begin{bmatrix} x - 1 \\ y - 2 \end{bmatrix} = x^2 + y^2 - 5 - (2(x - 1) + 4(y - 2))$$

$$= x^2 + y^2 - 5 - (2x - 2 + 4y - 8)$$

$$= x^2 - 2x + y^2 - 4y + 5$$

$$= (x^2 - 2x + 1) - 1 + (y^2 - 4y + 4) - 4 + 5$$

$$= (x - 1)^2 + (y - 2)^2.$$

Therefore:

$$\lim_{(x,y)\to(1,2)} \frac{x^2 + y^2 - 5 - \begin{bmatrix} 2 & 4 \end{bmatrix} \begin{bmatrix} x - 1 \\ y - 2 \end{bmatrix}}{\sqrt{(x - 1)^2 + (y - 2)^2}} = \lim_{(x,y)\to(1,2)} \frac{(x - 1)^2 + (y - 2)^2}{\sqrt{(x - 1)^2 + (y - 2)^2}}$$

$$= \lim_{(x,y)\to(1,2)} \sqrt{(x - 1)^2 + (y - 2)^2}.$$

The function in this last expression is continuous, as it is a composition of sums and products of continuous pieces. Thus the value of the limit is simply the value of the function at the point (1, 2). In other words:

$$\lim_{(x,y)\to(1,2)} \frac{x^2 + y^2 - 5 - \begin{bmatrix} 2 & 4 \end{bmatrix} \begin{bmatrix} x - 1 \\ y - 2 \end{bmatrix}}{\sqrt{(x - 1)^2 + (y - 2)^2}} = \sqrt{(1 - 1)^2 + (2 - 2)^2} = 0.$$

This shows that the answer is: yes, f is differentiable at a = (1, 2).

In addition, the preceding calculations show that the first-order approximation of $f(x, y) = x^2 + y^2$ at a = (1, 2) is:

$$\ell(x, y) = f(\mathbf{a}) + Df(\mathbf{a}) \cdot (\mathbf{x} - \mathbf{a})$$

$$= 5 + \begin{bmatrix} 2 & 4 \end{bmatrix} \begin{bmatrix} x - 1 \\ y - 2 \end{bmatrix}$$

$$= 2x + 4y - 5. \tag{4.6}$$

For instance, (x, y) = (1.05, 1.95) is near a, and f (1.05, 1.95) = 1.05² + 1.95² = 4.905 while ℓ(1.05, 1.95) = 2(1.05) + 4(1.95) – 5 = 4.9. The values are pretty close.

The same sort of reasoning can be used to show that, more generally, the function $f(x, y) = x^2 + y^2$ is differentiable at every point of $\mathbb{R}^2$. This is Exercise 1.16.

Example 4.3. Let $f(x,y) = \begin{cases} \frac{xy}{x^2+y^2} & \text{if } (x, y) \neq (0,0), \\ 0 & \text{if } (x, y) = (0, 0) \end{cases}$ Its graph is shown in Figure 4.2. Is f differentiable at a = (0, 0)?

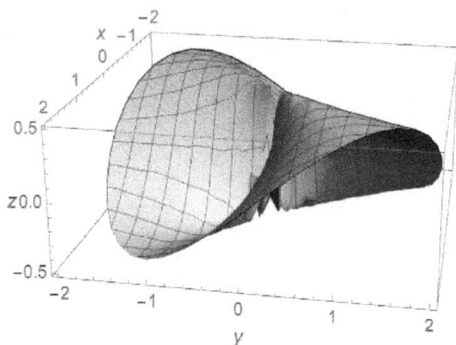

Figure 4.2: The graph $z = \dfrac{xy}{x^2+y^2}$

Using the definition requires knowing the partial derivatives at (0, 0). To calculate them, we could apply the quotient rule to the formula that defines f, but the results would be valid only when (x, y)/= (0, 0), which is precisely what we don't need. So instead we go back to the basic idea behind partial derivatives, namely, that they are one-variable derivatives where all variables but one are held constant. For example, to find $\frac{\partial f}{\partial x}(0,0)$, fix y = 0 in $f(x, y)$ and look at the resulting function of x. By the formula for f, $f(x, 0) = \frac{x \cdot 0}{x^2 + 0^2} = 0$ if x ≠ 0. This expression also holds when x = 0: by definition, $f(0, 0) = 0$. Thus $f(x, 0) = 0$ for all x. Differentiating this with respect to x gives $\frac{\partial f}{\partial x}(x,0) = \frac{d}{dx}(0) = 0$ whence $\frac{\partial f}{\partial x}(0,0) = 0$. A symmetric calculation result in $\frac{\partial f}{\partial y}(0,0) = 0$. Hence $Df(0, 0) = [0\ 0]$.

Moreover, $\mathbf{x} - \mathbf{a} = \begin{bmatrix} x - 0 \\ y - 0 \end{bmatrix} = \begin{bmatrix} x \\ y \end{bmatrix}$. and $\|\mathbf{x} - \mathbf{a}\| = \sqrt{x^2 + y^2}$. Thus, according to the definition of differentiability, we need to check if:

$$\lim_{(x,y)\to(0,0)} \frac{\frac{xy}{x^2+y^2} - 0 - \begin{bmatrix} 0 & 0 \end{bmatrix} \begin{bmatrix} x \\ y \end{bmatrix}}{\sqrt{x^2 + y^2}} = 0,$$

i.e., if $\lim_{(x,y)\to(0,0)} \frac{xy}{(x^2+y^2)^{3/2}} = 0$.

For this, suppose we approach the origin along the line y = x. Then we are looking at the behavior of $\frac{x \cdot x}{(x^2+x^2)^{3/2}} = \frac{x^2}{2^{3/2}|x|^3} = \frac{1}{2^{3/2}|x|}$. This blows up as x goes to 0. It certainly doesn't approach 0. Hence there's no way $\lim_{(x,y)\to(0,0)} \frac{xy}{(x^2+y^2)^{3/2}}$, can equal 0 (in fact, the limit does not exist), so f is not differentiable at (0, 0).

This example shows that a function of more than one variable can fail to be differentiable at a point even though all of its partial derivatives exist there. This differs from the one-variable case.

4.2 CONDITIONS FOR DIFFERENTIABILITY

We now look for ways to tell if a function is differentiable without having to do all the work of going through the definition. As we saw in the last example, calculating the partial derivatives is not enough. That's bad news, because calculating partial derivatives is pretty easy. The good news is that we are about to obtain results that show that the examples in the last section could have been solved much more quickly using means other than the definition.

For the remainder of this section:

U denotes an open set in $\mathbb{R}^n$, $f: U \to R$ is a real-valued function defined on U , and a is a point of U .

We won't keep repeating this.

For instance, in Example 4.3 above, we asked if the function were differentiable at (0, 0). In fact, we had shown before that f is not continuous at (0, 0) (see Example 3.19 in Chapter 3). In first-year calculus, discontinuous functions cannot be differentiable. If the same were true for functions of more than one variable, we could have saved ourselves a lot of work. We state the result in contrapositive form.

$$f(x, y) = \begin{cases} \frac{xy}{x^2+y^2} & \text{if } (x, y) \neq (0,0), \\ 0 & \text{if } (x, y) = (0,0) \end{cases} \tag{4.7}$$

Proposition 4.4. If f is differentiable at a, then f is continuous at a.

Proof. We just outline the idea and leave the details for the exercises (Exercise 2.3). If f is differentiable at a, then $\lim_{x\to a} (f(x) - \ell(x)) = 0$, where ℓ is the first-order approximation at a. In other words:

$$\lim_{x \to a} (f(\mathbf{x}) - f(\mathbf{a}) - Df(\mathbf{a}) \cdot (\mathbf{x} - \mathbf{a})) = 0.$$

The matrix product term $Df(\mathbf{a}) \cdot (\mathbf{x} - \mathbf{a})$ goes to 0 as x approaches a (it's a continuous function of x), so the last line becomes $\lim_{x \to a}(f(\mathbf{x}) - f(\mathbf{a}) - 0) = 0$. Hence $\lim_{x \to a} f(\mathbf{x}) = f(\mathbf{a})$. As a result, f is continuous at a.

Example 4.5. Returnsing to function (4.7) above, we could have simply said from the start that f is not continuous at (0, 0) and therefore is not differentiable there either, avoiding the definition of differentiability entirely.

Conversely, we know from first-year calculus that a function can be continuous without being differentiable. The absolute value function $f(x) = |x|$ is the standard example. It is continuous, but, because its graph has a coRⁿer, it is not differentiable at the origin. Sometimes, we can use the failure of a one-variable derivative to tell us about the differentiability of a function of more than one variable.

Proposition 4.6. If some partial derivative $\frac{\partial f}{\partial x_j}$ does not exist at a, then f is not differentiable at a.

Proof. This follows simply because if $\frac{\partial f}{\partial x_j}(\mathbf{a})$ doesn't exist, then neither does $Df(\mathbf{a})$, so the condition for differentiability cannot be satisfied.

Example 4 .7. One analogue of the absolute value function is the function $f: \mathbb{R}^2 \to \mathbb{R}, \; f(x, y) = \|(x, y)\| = \sqrt{x^2 + y^2}$. We have seen that its graph is a cone with vertex at the origin (Example 3.2)

To find $\frac{\partial f}{\partial x}$ at the origin, we fix y = 0 to get $f(x, 0) = \sqrt{x^2} = |x|$. The derivative with respect to x does not exist at x = 0, so, by definition, $\frac{\partial f}{\partial x}(0, 0)$ does not exist either. Thus f is not differentiable at (0, 0) (though it is continuous there).

We have saved the most useful criterion for differentiability for last. To understand why it is true, however, we take a slight detour.

4.3 THE MEAN VALUE THEOREM

We devote this section to reviewing an important, and possibly underappreciated, result from first-year calculus.

Theorem 4.8 (Mean Value Theorem). Let $f : [a, b] \to R$ be a continuous function that is differentiable on the open interval (a, b). Then there exists a point c in (a, b) such that:

$$f(b) - f(a) = f'(c) \cdot (b - a).$$

The theorem relates the difference in function values $f(b) - f(a)$ to the difference in domain values b - a. The relation can be useful even if one cannot completely pin down the location of the point c (in fact, one rarely tries). For instance, if $f(x) =$

sin x, then $|f'(c)| = |\cos c| \le 1$ regardless of what c is. Hence the mean value theorem gives $|\sin b - \sin a| \le |b - a|$ for all a, b. By the same reasoning, $|\cos b - \cos a| \le |b - a|$ as well.

There is an n-variable version of the mean value theorem that follows from results we prove later (see Exercise 5.5), but for now we show how the one-variable theorem can be used to study a difference in function values for a function of two variables by allowing only one variable to change at a time and using partial derivatives.

Example 4.9. Let $f(x, y)$ be a continuous real-valued function of two variables that is defined on an open ball B in $\mathbb{R}^2$ and whose partial derivatives $\frac{\partial f}{\partial x}$ and $\frac{\partial f}{\partial y}$ exist at every point of B. Let $a = (a_1, a_2)$ and $b = (b_1, b_2)$ be points of B. We show that there exist points p and q in B such that $\|p - a\| \le \|b - a\|$, $\|q - a\| \le \|b - a\|$, and:

$$f(b) - f(a) = \frac{\partial f}{\partial x}(p) \cdot (b_1 - a_1) + \frac{\partial f}{\partial y}(q) \cdot (b_2 - a_2). \tag{4.8}$$

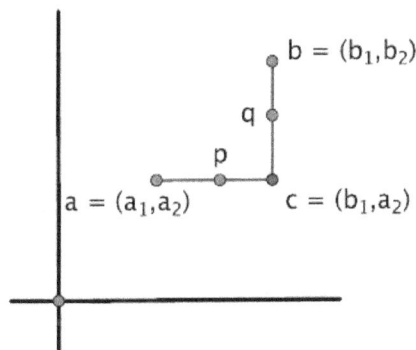

Figure 4.3: Going from a to b by way of c

See Figure 4.3.

To prove this, we introduce the intermediate point $c = (b_1, a_2)$ and think of going from a to b by first going from a to c and then from c to b, as in Figure 4.3. Only one variable varies along each of the segments, so by writing the mean value theorem applies to each of the differences on the right.

$$f(b) - f(a) = (f(b) - f(c)) + (f(c) - f(a)), \tag{4.9}$$

That is, in the second difference, using x as variable and $y = a_2$ fixed, the mean value theorem implies that there exists a point $p = (p_1, a_2)$ on the line segment between a and c such that:

$$f(c) - f(a) = \frac{\partial f}{\partial x}(p) \cdot (b_1 - a_1). \tag{4.10}$$

Similarly, for the first difference, using y as variable and x = b_1 fixed, there exists a point q = (b1, q2) on the line segment between c and b such that:

$$f(\mathbf{b}) - f(\mathbf{c}) = \frac{\partial f}{\partial y}(\mathbf{q}) \cdot (b_2 - a_2).$$
(4.11)

Substituting (4.10) and (4.11) into equation (4.9) results in $f(\mathbf{b}) - f(\mathbf{a}) = \frac{\partial f}{\partial y}(\mathbf{q}) \cdot (b_2 - a_2) + \frac{\partial f}{\partial x}(\mathbf{p}) \cdot (b_1 - a_1)$, which is the desired conclusion (4.8).

4.4. THE C¹ TEST

We present the result that often gives the simplest way to show that a function is differentiable. It uses partial derivatives.

Definition. Let U be an open set in $\mathbb{R}^n$. A real-valued function $f: U \to R$ all of whose partial derivatives $\frac{\partial f}{\partial x_1}, \frac{\partial f}{\partial x_2} \cdots \cdots \frac{\partial f}{\partial x_n}$ are continuous on U is said to be of class C^1, or continuously differentiable, on U .

Theorem 4.10 (The C^1 test). If f is of class C^1 on U, then f is differentiable at every point of U.

Proof. We give the proof in the case of two variables, that is, $U \subset \mathbb{R}^2$. This contains the main new idea, and we leave for the reader the generalization to n variables, which is a relatively straightforward extension.

We use the definition of differentiability. Let a be a point of U . We first analyze the error $f(x) - f(a) - Df(a) \cdot (x - a)$ in the first-order approximation of f at a. Since U is an open set and we are interested only in what happens when x is near a, we may assume that x lies in an open ball about a that is contained in U . Say a = (c, d) and x = (x, y). By the result we just obtained in Example 4.9, there are points p and q such that $\|p - a\| \le \|x - a\|$, $\|q - a\| \le \|x - a\|$, and:

$$f(\mathbf{x}) - f(\mathbf{a}) = \frac{\partial f}{\partial x}(\mathbf{p}) \cdot (x - c) + \frac{\partial f}{\partial y}(\mathbf{q}) \cdot (y - d).$$

Also:

$$Df(\mathbf{a}) \cdot (\mathbf{x} - \mathbf{a}) = \left[\frac{\partial f}{\partial x}(\mathbf{a}) \quad \frac{\partial f}{\partial y}(\mathbf{a})\right] \begin{bmatrix} x - c \\ y - d \end{bmatrix}$$
$$= \frac{\partial f}{\partial x}(\mathbf{a}) \cdot (x - c) + \frac{\partial f}{\partial y}(\mathbf{a}) \cdot (y - d).$$

Hence:

$$f(\mathbf{x}) - f(\mathbf{a}) - Df(\mathbf{a}) \cdot (\mathbf{x} - \mathbf{a}) = (\frac{\partial f}{\partial x}(\mathbf{p}) - \frac{\partial f}{\partial x}(\mathbf{a})) \cdot (x - c) + (\frac{\partial f}{\partial y}(\mathbf{q}) - \frac{\partial f}{\partial y}(\mathbf{a})) \cdot (y - d).$$

so:

$$\frac{|f(\mathbf{x}) - f(\mathbf{a}) - Df(\mathbf{a}) \cdot (\mathbf{x} - \mathbf{a})|}{\|\mathbf{x} - \mathbf{a}\|}$$

$$= \frac{|(\frac{\partial f}{\partial x}(\mathbf{p}) - \frac{\partial f}{\partial x}(\mathbf{a})) \cdot (x - c) + (\frac{\partial f}{\partial y}(\mathbf{q}) - \frac{\partial f}{\partial y}(\mathbf{a})) \cdot (y - d)|}{\|\mathbf{x} - \mathbf{a}\|}$$

$$\leq \left| \frac{\partial f}{\partial x}(\mathbf{p}) - \frac{\partial f}{\partial x}(\mathbf{a}) \right| \cdot \frac{|x - c|}{\|\mathbf{x} - \mathbf{a}\|} + \left| \frac{\partial f}{\partial y}(\mathbf{q}) - \frac{\partial f}{\partial y}(\mathbf{a}) \right| \cdot \frac{|y - d|}{\|\mathbf{x} - \mathbf{a}\|},$$

where the last step uses the triangle inequality. Note that $\frac{|x-c|}{\|\mathbf{x}-\mathbf{a}\|} = \frac{\sqrt{(x-c)^2}}{\sqrt{(x-c)^2+(y-d)^2}} \leq 1.$
Similarly, $\frac{|y-d|}{\|\mathbf{x}-\mathbf{a}\|} \leq 1.$ Therefore:

$$\frac{|f(\mathbf{x}) - f(\mathbf{a}) - Df(\mathbf{a}) \cdot (\mathbf{x} - \mathbf{a})|}{\|\mathbf{x} - \mathbf{a}\|} \leq \left| \frac{\partial f}{\partial x}(\mathbf{p}) - \frac{\partial f}{\partial x}(\mathbf{a}) \right| + \left| \frac{\partial f}{\partial y}(\mathbf{q}) - \frac{\partial f}{\partial y}(\mathbf{a}) \right|.$$

At last, we bring in the assumption that f has continuous partial derivatives. Since $\|\mathbf{p} - \mathbf{a}\| \leq \|\mathbf{x} - \mathbf{a}\|$ and $\|\mathbf{q} - \mathbf{a}\| \leq \|\mathbf{x} - \mathbf{a}\|$, it follows that, as x approaches a, so do p and q. By the continuity of the partial derivatives, this means that $\frac{\partial f}{\partial x}(\mathbf{p})$ approaches $\frac{\partial f}{\partial x}(\mathbf{a})$ and $\frac{\partial f}{\partial y}(\mathbf{q})$ approaches $\frac{\partial f}{\partial y}(\mathbf{a})$.

Hence, in the limit as x goes to a, the right side of (4.12) approaches 0, and therefore so must the left. By definition, f is differentiable at a.

Example 4.11. In Example 4.2, we showed that the function $f \colon \mathbb{R}^2 \to$ R, $f(x, y) = x^2 + y^2$, is differentiable at the point a = (1, 2) by using the definition of differentiability. It is much simpler to note that $\frac{\partial f}{\partial x} = 2x$ and $\frac{\partial f}{\partial y} = 2y$ are both continuous on $\mathbb{R}^2$. Hence, by the C^1 test, f is differentiable at every point of $\mathbb{R}^2$.

Example 4.12. The other example we looked at carefully is the function:

$$f(x, y) = \begin{cases} \frac{xy}{x^2+y^2} & \text{if } (x, y) \neq (0, 0), \\ 0 & \text{if } (x, y) = (0, 0) \end{cases}$$

of (4.7). We saw that f is not differentiable at the origin. On the other hand, at points other than the origin, the partial derivatives can be found using the quotient rule. This gives:

$$\frac{\partial f}{\partial x} = \frac{(x^2 + y^2) \cdot y - xy \cdot 2x}{(x^2 + y^2)^2} = \frac{y^3 - x^2 y}{(x^2 + y^2)^2} = \frac{y(y^2 - x^2)}{(x^2 + y^2)^2}$$

and $\quad \frac{\partial f}{\partial y} = \frac{(x^2 + y^2) \cdot x - xy \cdot 2y}{(x^2 + y^2)^2} = \frac{x(x^2 - y^2)}{(x^2 + y^2)^2}.$

As algebraic combinations of x and y, both of these partial derivatives are continuous on the open set $\mathbb{R}^2 - \{(0, 0)\}$. As a result, the C^1 test says that f is differentiable at all points other than the origin.

4.5. THE LITTLE CHAIN RULE

In one-variable calculus, the derivative measures the rate of change of a function. When there is more than one variable, the derivative is no longer just a number, but it still has something to say about rates of change.

Let U be an open set in $\mathbb{R}^n$, and let $f: U \to R$ be a real-valued function defined on U. Suppose that $\alpha : I \to U$ is a path in U defined on an interval I in R. The composition $f \circ \alpha : I \to R$ is a real-valued function of one variable that describes how f behaves along the path. We examine the rate of change of this composition.

Let t_0 be a point of I, and let a = α(t0). If f is differentiable at a, then its first-order approximation applies: $f(x) \approx f(a) + \nabla f(a) \cdot (x - a)$, or:

$$f(x) - f(a) \approx \nabla f(a) \cdot (x - a) \text{ when x is near a.}$$

Here, we use the gradient form of the derivative.

Assume in addition that α is differentiable at t_0. Then α is continuous at t_0, so, when t is near t_0, α(t) is near $\alpha(t_0)$. Substituting x = α(t) and a = $\alpha(t_0)$ into the previous approximation gives:

$$f(\alpha(t)) - f(\alpha(t_0)) \approx \nabla f(\alpha(t_0)) \cdot (\alpha(t) - \alpha(t_0)) \quad \text{when } t \text{ is near } t_0.$$

Thus $\frac{f(\alpha(t)) - f(\alpha(t_0))}{t - t_0} \approx \nabla f(\alpha(t_0)) \cdot \frac{\alpha(t) - \alpha(t_0)}{t - t_0}$. In the limit as t approaches t0, the left side approaches the one-variable derivative (f ∘ α) (t0) and the right approaches ∇f $(\alpha(t_0)) \cdot \alpha$ (t_0). Writing t in place of t_0, this gives the following result.

Theorem 4.13 (Little Chain Rule). Let U be an open set in $\mathbb{R}^n$, $\alpha : I \to U$ a path in U defined on an open interval I in R, and $f: U \to R$ a real-valued function. If α is differentiable at t and f is differentiable at α(t), then f ∘ α is differentiable at t and:

$$(f \circ \alpha)'(t) = \nabla f(\alpha(t)) \cdot \alpha'(t).$$

In words, the derivative of the composition is "gradient dot velocity." While we outlined the idea why this is true, a complete proof of the result as stated entails a more careful treatment of the approximations involved. The details appear in the exercises (see Exercise 5.6).

4.6 DIRECTIONAL DERIVATIVES

In this section and the next, we present some interpretations of the gradient that follow from the Little Chain Rule. The first concerns the rate of change of a function of n variables in a given direction. Let a be a point of an open set U in $\mathbb{R}^n$. To represent a direction going away from a, we choose a unit vector u pointing in that direction. We travel in that direction on the line through a and parallel to u, which is parametrized by α(t) = a + tu. The parametrization has velocity α'(t) = u and hence constant speed $\|u\| = 1$.

If $f : U \to R$ is a real-valued function, then the derivative $(f \circ \alpha)'(t)$ is the rate of change of f along this line. In particular, since $\alpha(0) = a$, we think of $(f \circ \alpha)'(0)$ as the rate of change at a in the direction of u.

Definition. Given the input described above, $(f \circ \alpha)'(0)$ is called the **directional derivative of** f at a in the direction of the unit vector u, denoted by $(D_u f)(a)$. If v is any nonzero vector in $\mathbb{R}^n$, we define the directional derivative in the direction of v to be $(D_u f)(a)$, where $u = \frac{1}{\|v\|} v$ is the unit vector in the direction of v.

Directional derivatives are easy to compute, for by the Little Chain Rule:

$$(f \circ \alpha)'(0) = \nabla f(\alpha(0)) \cdot \alpha'(0) = \nabla f(a) \cdot u.$$

In other words:

Proposition 4.14. If f is differentiable at a and u is a unit vector, then the directional derivative is given by:

$$(D_u f)(a) = \nabla f(a) \cdot u.$$

As a result, $(D_u f)(a) = \|\nabla f(a)\| \, \|u\| \cos \theta = \|\nabla f(a)\| \cos \theta$, where θ is the angle between $\nabla f(a)$ and the direction vector u. Keeping a fixed but varying the direction, we see that $(D_u f)(a)$ is maximized when $\theta = 0$, i.e., when u points in the same direction as $\nabla f(a)$, and that, in this direction, $(D_u f)(a) = \|\nabla f(a)\|$.

Corollary 4.15. The maximum directional derivative of f at a:

- ⊙ occurs in the direction of $\nabla f(a)$ and
- ⊙ has a value of $\|\nabla f(a)\|$.

Sometimes, weather reports mention the "temperature gradient." Usually, this does not refer to the gradient vector that we have been studying but rather to one of the aspects of the gradient in the corollary, that is, the magnitude and/or direction of the maximum rate of temperature increase.

Example 4.16. Let $f : R^3 \to R$ be $f(x, y, z) = x^2 + y^2 + z^2$, and let a = (1, 2, 3).

Find the directional derivative of f at a in the direction of v = (2, –1, 1).

Find the direction and rate of maximum increase at a.

First, $\nabla f = (2x, 2y, 2z)$, so $\nabla f(a) = (2, 4, 6)$. Next, the unit vector in the direction of v is $u = \frac{1}{\|v\|} v = \frac{1}{\sqrt{6}}(2, -1, 1)$. Thus:

$$(D_u f)(a) = \nabla f(a) \cdot u$$
$$= (2, 4, 6) \cdot \frac{1}{\sqrt{6}}(2, -1, 1) = \frac{1}{\sqrt{6}}(4 - 4 + 6) = \frac{6}{\sqrt{6}} = \sqrt{6}.$$

The direction of maximum increase is the direction of $\nabla f(a) = (2, 4, 6)$, that is, the direction of the unit vector $u = \frac{1}{\sqrt{4+16+36}}(2,4,6) = \frac{1}{\sqrt{56}}(2,4,6) = \frac{1}{\sqrt{14}}(1,2,3)$. The rate of maximum increase is $\|\nabla f(a)\| = \sqrt{56} = 2\sqrt{14}$.

4.7 ∇F AS NORMAL VECTOR

Let $f : U \to R$ be a differentiable real-valued function of n variables defined on an open set U in $\mathbb{R}^n$, and let S be the level set of f corresponding to $f = c$. Let a be a point of S, and let v be a vector tangent to S at a. We take this to mean that v can be realized as the velocity vector of a path in S, that is, there is a differentiable path $\alpha : I \to \mathbb{R}^n$ that:

- lies in S, i.e., $\alpha(t) \in$ S for all t,

- passes through a, say $a = \alpha(t_0)$ for some t_0 in I, and

- has velocity v at a, i.e., $\alpha'(t_0) = v$.

This is illustrated in Figure 4.4. Then $f(\alpha(t)) = c$ for all t, so $(f \circ \alpha)'(t) = c' = 0$. By the Little Chain Rule, this can be written $\nabla f(\alpha(t)) \cdot \alpha'(t) = 0$. In particular, when $t = t_0$, we obtain $\nabla f(a) \cdot v = 0$. This is true for all vectors v tangent to S at a, which has the following consequence.

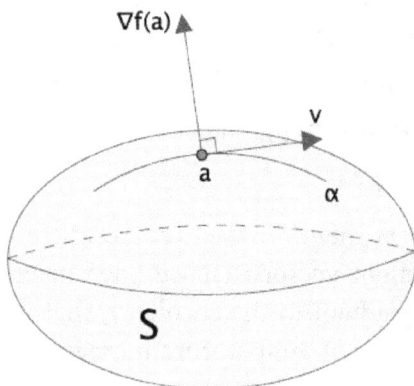

Figure 4.4: A level set S of f and a vector v tangent to it

Proposition 4.17. If a is a point on a level set S of f, then $\nabla f(a)$ is a normal vector to S at a.

Example 4.18. Find an equation for the plane tangent to the graph $z = x^2 + y^2$ at the point $(1, 2, 5)$.

As with the equation of any plane, it suffices to find a point p on the plane and a normal vector n. As a point, we can use the point of tangency, $p = (1, 2, 5)$. To find a normal vector, we rewrite the graph as a level set. That is, rewrite $z = x^2 + y^2$ as $x^2 + y^2 - z = 0$. Then the graph is the level set of the function of three variables F (x, y,

z) = $x^2 + y^2 - z$ corresponding to c = 0. In particular, ∇F (p) is a normal vector to the level set, that is, to the graph, at p, so it is normal to the tangent plane as well. See Figure 4.5.

The rest is calculation: $\nabla F = (2x, 2y, -1)$, so n = ∇F (p) = (2, 4, -1) is our normal vector. Substituting into the equation $n \cdot x = n \cdot p$ for a plane gives:

$$(2, 4, -1) \cdot (x, y, z) = (2, 4, -1) \cdot (1, 2, 5) = 2 + 8 - 5 = 5,$$
$$\text{or} \quad 2x + 4y - z = 5.$$

Note that we just studied the graph of the function $f(x, y) = x^2 + y^2$ of two variables, and we considered the point p = (1, 2, 5) = (a, f(a)) on the graph, where a = (1, 2). In the remarks following Example 4.2, we saw that the first-order approximation of f at a is given by $\ell(x, y) = 2x + 4y - 5$ (see equation (4.6)). The graph of the approximation ℓ is given by z = 2x + 4y - 5, which is exactly the tangent plane we just found.

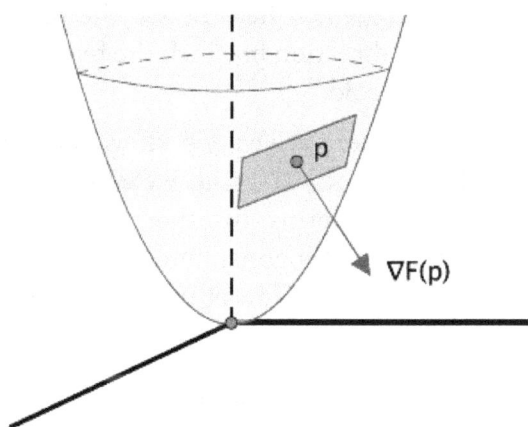

Figure 4.5: The tangent plane to z = $x^2 + y^2$ at p = (1, 2, 5)

In other words, geometrically, the approximation we get by using the tangent plane at (a, f(a)) to approximate the graph coincides with the first-order approximation at a. This extends a familiar idea from first-year calculus, where tangent lines are used to obtain the first-order approximation of a differentiable function. In fact, it is the idea we used to motivate our multivariable definition of differentiability in the first place.

Example 4.19. Suppose that T (x, y) represents the temperature at the point (x, y) in an open set U of the plane. There are two natural families of curves associated with this function.

The first is the curves along which the temperature is constant. They are called **isotherms** and are the level sets of T : all points on a single isotherm have the same temperature. See Figure 4.6.

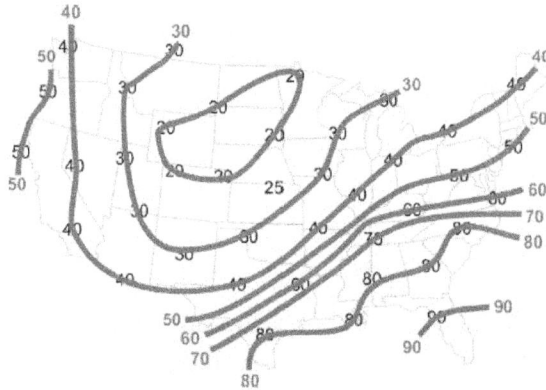

Figure 4.6: Isotherms are level curves of temperature. Heat travels along curves that are always orthogonal to the isotherms.

The second is the curves along which heat flows. The principle here is that, at every point, heat travels in the direction in which the temperature is decreasing as rapidly as possible, from hot to cold.

From our work on directional derivatives, we know that the maximum rate of decrease occurs in the direction of $-\nabla T$. And we've just seen that ∇T is always normal to the level sets, the isotherms. In other words, the isotherms and the curves of heat flow are mutually orthogonal families of curves. If you have a map of the isotherms, you can visualize the flow of heat: always move at right angles to the isotherms.

4.8. HIGHER-ORDER PARTIAL DERIVATIVES

Once they start partial differentiating, most people can't stop.

Example 4.20. Let $f(x, y) = x^3 - 3x^2y^4 + e^{xy}$. Then:

$$\frac{\partial f}{\partial x} = 3x^2 - 6xy^4 + ye^{xy}$$

$$\text{and} \quad \frac{\partial f}{\partial y} = -12x^2y^3 + xe^{xy}.$$

Both of these are again real-valued functions of x and y, so they have partial derivatives of their own. In the case of $\frac{\partial f}{\partial x}$:

$$\frac{\partial^2 f}{\partial x^2} = \frac{\partial}{\partial x}\left(\frac{\partial f}{\partial x}\right) = 6x - 6y^4 + y^2 e^{xy}$$

$$\text{and} \quad \frac{\partial^2 f}{\partial y \, \partial x} = \frac{\partial}{\partial y}\left(\frac{\partial f}{\partial x}\right) = -24xy^3 + e^{xy} + xye^{xy}.$$

Similarly, for $\frac{\partial f}{\partial y}$:

$$\frac{\partial^2 f}{\partial x \, \partial y} = \frac{\partial}{\partial x}\left(\frac{\partial f}{\partial y}\right) = -24xy^3 + e^{xy} + xye^{xy}$$

$$\text{and} \quad \frac{\partial^2 f}{\partial y^2} = \frac{\partial}{\partial y}\left(\frac{\partial f}{\partial y}\right) = -36x^2y^2 + x^2 e^{xy}.$$

These are the second-order partial derivatives. They too have partial derivatives, and the process could continue indefinitely to obtain partial derivatives of higher and higher order.

Perhaps the most striking part of the preceding calculation is that the two partial derivatives $\frac{\partial^2 f}{\partial y \, \partial x}$ and $\frac{\partial^2 f}{\partial x \, \partial y}$ turnsed out to be equal. They are called the mixed partials. To see whether they might be equal in general, we returns to the definition of a partial derivative as a limit of difference quotients and try to describe the second-order partials in terms of the original function f. In fact, what we are about to discuss is a method one might use to numerically approximate the second-order partials.

Let a = (c, d). First, consider $\frac{\partial^2 f}{\partial x \, \partial y} = \frac{\partial}{\partial x}\left(\frac{\partial f}{\partial y}\right)$. The function being differentiated is $\frac{\partial f}{\partial y}$. It is being differentiated with respect to x, so we let x vary and hold y constant:

$$\frac{\partial^2 f}{\partial x \, \partial y}(c, d) = \left(\frac{\partial}{\partial x}\left(\frac{\partial f}{\partial y}\right)\right)(c, d) \approx \frac{\frac{\partial f}{\partial y}(c + h, d) - \frac{\partial f}{\partial y}(c, d)}{h}.$$

The approximation is meant to hold when h is small. The terms in the numerator are both partial derivatives of f with respect to y, so in each case we differentiate f, letting y vary and holding x constant:

$$\frac{\partial^2 f}{\partial x \, \partial y}(c, d) \approx \frac{\frac{f(c+h,d+k)-f(c+h,d)}{k} - \frac{f(c,d+k)-f(c,d)}{k}}{h}$$

$$\approx \frac{f(c + h, d + k) - f(c + h, d) - f(c, d + k) + f(c, d)}{hk}$$

when h and k are small.

This expression is symmetric enough in h and k that perhaps you are willing to believe that you would get the same thing when the order of differentiation is reversed. But if you are nervous about it, here is the parallel calculation for $\frac{\partial^2 f}{\partial y \, \partial x}$:

$$\frac{\partial^2 f}{\partial y\,\partial x}(c,d) = \left(\frac{\partial}{\partial y}\left(\frac{\partial f}{\partial x}\right)\right)(c,d)$$

$$\approx \frac{\frac{\partial f}{\partial x}(c,d+k) - \frac{\partial f}{\partial x}(c,d)}{k}$$

$$\approx \frac{\frac{f(c+h,d+k)-f(c,d+k)}{h} - \frac{f(c+h,d)-f(c,d)}{h}}{k}$$

$$\approx \frac{f(c+h,d+k) - f(c,d+k) - f(c+h,d) + f(c,d)}{hk}. \tag{4.14}$$

Indeed, comparing the approximations shows that they are the same, so:

$$\frac{\partial^2 f}{\partial x\,\partial y}(c,d) \approx \frac{\partial^2 f}{\partial y\,\partial x}(c,d).$$

A proper proof that the two sides are actually equal requires tightening up the approximations that are involved. We leave the details for the exercises (Exercise 8.7), but the main idea is to use the mean value theorem to show that the quotient $Q = \frac{f(c+h,d+k)-f(c+h,d)-f(c,d+k)+f(c,d)}{hk}$ common to approximations (4.13) and (4.14) is equal on the nose to the values of the mixed partials at nearby points, that is:

$$\frac{\partial^2 f}{\partial x\,\partial y}(\mathbf{p}) = Q = \frac{\partial^2 f}{\partial y\,\partial x}(\mathbf{q}), \tag{4.15}$$

where p and q are points that are close to a but depend on h and k. As h and k go to 0, p and q approach a. If the mixed partials were known to be continuous at a, then, in the limit, equation (4.15) becomes:

$$\frac{\partial^2 f}{\partial x\,\partial y}(\mathbf{a}) = \frac{\partial^2 f}{\partial y\,\partial x}(\mathbf{a}).$$

Thus the mixed partials are equal provided one invokes a requirement of continuity. In fact, there are functions for which the mixed partials are not equal, so some sort of restriction of this type is unavoidable.

For functions of more than two variables, the second-order mixed partials treat all but two variables as fixed, so we remain essentially in the two-variable case. In other words, if the mixed partials are equal for functions of two variables, they are also equal for more than two. We summarize the discussion with the following result.

Theorem 4.21 (Equality of mixed partials). Let $f: U \to \mathbb{R}$ be a real-valued function defined on an open set U in $\mathbb{R}^n$, all of whose first and second-order partial derivatives are continuous on U. (Such a function is said to be of class C^2.) Then:

$$\frac{\partial^2 f}{\partial x_i \, \partial x_j} = \frac{\partial^2 f}{\partial x_j \, \partial x_i}$$

at all points of U and for all i, j.

Example 4.22. Find $\frac{\partial^4 f}{\partial x \, \partial y^2 \, \partial z}$ if $f(x, y, z) = (x^2 + y^3 + \sin(xz))(\cos x - xz^5)$.

The order of differentiation doesn't matter, so we may as well choose one that simplifies the bookkeeping. Differentiating with respect to x or z would involve the product rule, and who wants that, so let's differentiate first with respect to y: $\frac{\partial f}{\partial y} = 3y^2(\cos x - xz^5)$. The next choices don't make much difference, but let's try differentiating with respect to z: $\frac{\partial^2 f}{\partial z \, \partial y} = -15xy^2 z^4$. Then, with respect x: $\frac{\partial^3 f}{\partial x \, \partial z \, \partial y} = -15y^2 z^4$. And lastly with respect to y again:

$$\frac{\partial^4 f}{\partial y \, \partial x \, \partial z \, \partial y} = -30yz^4.$$

By the equality of mixed partials, this is the same as the originally requested $\frac{\partial^4 f}{\partial x \, \partial y^2 \, \partial z}$.

4.9 SMOOTH FUNCTIONS

The C^1 test states that functions with continuous partial derivatives are differentiable. The converse is false, however. There are plenty of functions that are differentiable but don't have continuous partials, though actually writing one down takes some determination. In the one-variable case, f (x) = $x^2 \sin(1/x)$ is an example. If f (0) is defined to be 0, then f is differentiable at a = 0 but f' is not continuous there. Similarly, there are functions whose first-order partials are continuous but whose second-order partials are not. There is an example in the exercises (Exercise 8.6), though again it is somewhat contrived to make the point.

Most of the differentiable functions that we work with have continuous partial derivatives of all possible orders. Such functions are said to be **of class C^∞**. We refer to them simply as **smooth**. We shall usually take them as the default and state our results involving derivatives for smooth functions. We may not actually need all those higher-order partial derivatives to draw the conclusion we are trying to reach, but keeping track of the degree of differentiability that is really necessary in every case is not something we want to dwell on.

In addition, so far we have defined continuity and differentiability only for functions defined on open sets. This is so that every point in the domain can be approached within the domain from all possible directions, making the corresponding limiting processes easier to handle. Sometimes, however, we are interested in domains that are not open. If D is a subset of $\mathbb{R}^n$, not necessarily open, we say that a function f : D → R is differentiable at a point a of D if it agrees with a differentiable function near a, that is, there is an open ball B containing a and a function g : B → R such that $g(x) = f(x)$ for all x in B ∩ D and g is differentiable at a in the sense previously defined. We make the analogous modification for smoothness. For continuity, we keep the definition as before but restrict our attention to points where f is defined: f is continuous at a if, given any open ball B(f (a), ϵ) about f (a), there exists an open ball B(a, δ) about a such that f (B(a, δ) ∩ D) ⊂ B(f (a), ϵ). We won't dwell on these points either. In most concrete examples, f is defined by a well-behaved formula that applies to an open set larger than the domain in which we happen to be interested, so there isn't a serious problem.

4.10. MAX/MIN: CRITICAL POINTS

In first-year calculus, one of the featured applications of the derivative is finding the largest and/or smallest values that a function attains and where it attains them. Whenever one wants to do something in an optimal way—for instance, the most productive or the least costly—and the objective can be modeled by a smooth function, calculus is likely to be of great use. We discuss this for functions of more than one variable. This is a big subject that overflows with applications. Our treatment is not comprehensive. Instead, we have selected a couple of topics that illustrate some of the similarities and differences that pertain to moving to the multivariable case.

Definition. Let $f : U → R$ be a function defined on an open set U in $\mathbb{R}^n$. f is said to have a local maximum at a point a of U if there is an open ball B(a, r) centered at a such that f (a) ≥ f (x) for all x in B(a, r). It is said to have a **local minimum** at a if instead f (a) ≤ f (x) for all x in B(a, r). See Figure 4.7. It has a **global maximum** or **global minimum** at a if the associated inequalities are true for all x in U , not just in B(a, r).

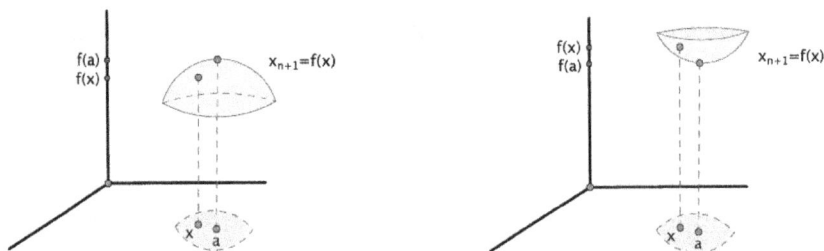

Figure 4.7: Local maximum (left) and local minimum (right) at a

If f is smooth and has a local maximum or local minimum at a, then it has a local maximum or local minimum in every direction, in particular, in each of the coordinate directions. The derivatives in these directions are the partial derivatives.

They are essentially one-variable derivatives, so from first-year calculus, at a local maximum or minimum, all of them must be 0, that is, $\frac{\partial f}{\partial x_j}(a) = 0$ for j = 1, 2, . . . , n.

Definition. A point a is called a critical point of f if $\frac{\partial f}{\partial x_j}(a) = 0$ for j = 1, 2, . . . , n. Equivalently, $Df(a) = [0\ 0 \ldots 0]$, or $\nabla f(a) = 0$.

The preceding discussion proves the following.

Proposition 4.23. Let U be an open set in $\mathbb{R}^n$, and let $f: U \to R$ be a smooth function. If f has a local maximum or a local minimum at a, then a is a critical point of f.

Hence, if one is looking for local maxima or minima of a smooth function on an open set, the critical points are the only possible candidates. This is just like the one-variable case. Conversely, as in the one-variable case, not every critical point is necessarily a local max or min, though there are new ways in which this failure can occur. A good example to keep in mind are functions whose graphs are saddles, such as $f(x, y) = x^2 - y^2$ from Example 3.3 in Chapter 3. At a saddle point, the function has a local maximum in one direction and a local minimum in another, so there is no open ball about the point in which it is exclusively one or the other.

Even so, in first-year calculus, a given critical point can often be classified as a local maximum or minimum using the second derivative. There is an analogue of this for more than one variable that is similar in spirit while reflecting the greater number of possibilities. We discuss the situation for two variables.

Let U be an open set in $\mathbb{R}^2$, and let $f: U \to R$ be a smooth function. If $a \in U$, consider the matrix of second-order partial derivatives:

$$H(a) = \begin{bmatrix} \frac{\partial^2 f}{\partial x^2}(a) & \frac{\partial^2 f}{\partial y \partial x}(a) \\ \frac{\partial^2 f}{\partial x \partial y}(a) & \frac{\partial^2 f}{\partial y^2}(a) \end{bmatrix}.$$

It is called the **Hessian matrix** of f at a. By the equality of mixed partials, the two off-diagonal terms, $\frac{\partial^2 f}{\partial x \partial y}(a)$ and $\frac{\partial^2 f}{\partial y \partial x}(a)$, are actually equal.

Theorem 4.24 (Second derivative test for functions of two variables). Let a be a critical point of f such that det $H(a) \neq 0$. Under this assumption, a is called a **nondegenerate** critical point.

1. If det H(a) > 0 and:

    a. if $\frac{\partial^2 f}{\partial x^2}(a) > 0$, then f has a local minimum at a,

    b. if $\frac{\partial^2 f}{\partial x^2}(a) < 0$, then f has a local maximum at a.

3. If det H(a) < 0, then f has a saddle point at a.
4. If det H(a) = 0, the test is inconclusive.

We discuss the ideas behind the test in the next section, but, before that, we present some examples. The first involves three simple functions for which we already know the answers, but it is meant to illustrate how the test works and confirm that it gives the right results. In a way, these three functions are prototypes of the general situation.

Example 4.25. Find all critical points of the following functions and, if possible, classify each as a local maximum, local minimum, or saddle point.

a. $f(x, y) = x^2 + y^2$
b. $f(x, y) = -x_2 - y^2$
c. $f(x, y) = x^2 - y^2$

Intuitively, it's clear that the first two of these attain their minimum and maximum values, respectively, at the origin, while, as noted earlier, the third has a saddle point there. The graphs of the functions are shown in Figure 4.8.

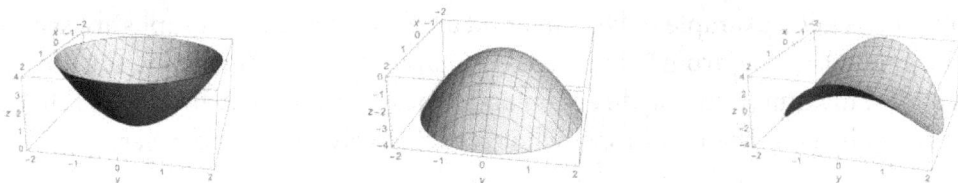

**Figure 4.8: Prototypical nondegenerate critical points at (0, 0): f (x, y) = x² + y²
(left), f (x, y) = -x² - y² (middle), f (x, y) = x² - y² (right)**

The second derivative test reaches these conclusions as follows.

a. To find the critical points, we set $\frac{\partial f}{\partial x} = 2x = 0$ and $\frac{\partial f}{\partial y} = 2y = 0.$ The only solution is x = 0 and y = 0. Hence the only critical point is (0, 0). To classify its type, we look at the second-order partials: $\frac{\partial^2 f}{\partial x^2} = 2, \frac{\partial^2 f}{\partial x \partial y} = \frac{\partial^2 f}{\partial y \partial x} = 0,$ and $\frac{\partial^2 f}{\partial y^2} = 2.$ Therefore:

$$H(0,0) = \begin{bmatrix} 2 & 0 \\ 0 & 2 \end{bmatrix}.$$

Since det H(0, 0) = 4 > 0 and $\frac{\partial^2 f}{\partial x^2}(0,0) = 2 > 0,$ the second derivative test implies that f has a local minimum at (0, 0), as expected.

b. By the same calculation, (0, 0) is the only critical point, only this time:

$$H(0,0) = \begin{bmatrix} -2 & 0 \\ 0 & -2 \end{bmatrix}.$$

Hence det H(0, 0) = 4 > 0 and $\frac{\partial^2 f}{\partial x^2}(0,0) = -2 < 0,$ so f has a local maximum at (0, 0).

c. Again, (0, 0) is the only critical point, but:

$$H(0,0) = \begin{bmatrix} 2 & 0 \\ 0 & -2 \end{bmatrix}.$$

Since det H(0, 0) = -4 < 0, f has a saddle point at (0, 0).

Example 4.26. We repeat for the function $f(x, y) = x^3 + 8y^3 - 3xy$. A portion of the graph of f is shown in Figure 4.9.

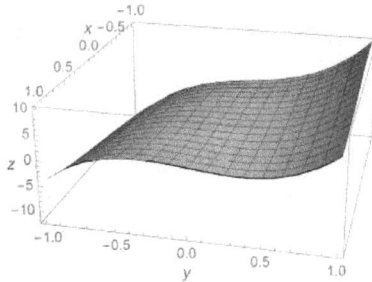

Figure 4.9: The graph of $f(x, y) = x^3 + 8y^3 - 3xy$ for $-1 \le x \le 1$, $-1 \le y \le 1$

First, to find the critical points, we set:

$$\begin{cases} \frac{\partial f}{\partial x} = 3x^2 - 3y = 0 \\ \frac{\partial f}{\partial y} = 24y^2 - 3x = 0. \end{cases}$$

The first equation implies that y = x2, and substituting this into the second gives 24x⁴ - 3x = 0, or 3x(8x³ - 1) = 0. Hence x = 0, in which case y = 0² = 0, or $x^3 = \frac{1}{8}$, i.e., $x = \frac{1}{2}$, in which case $y = \left(\frac{1}{2}\right)^2 = \frac{1}{4}$. As a result, f has two critical points, (0, 0) and $(\frac{1}{2}, \frac{1}{4})$.

To classify their types, we consider the Hessian:

$$H(x, y) = \begin{bmatrix} 6x & -3 \\ -3 & 48y \end{bmatrix}.$$

For instance, at the critical point (0, 0), det $H(0,0) = \det \begin{bmatrix} 0 & -3 \\ -3 & 0 \end{bmatrix} = -9.$ Since this is negative, (0, 0) is a saddle point.

Similarly, det $H(\frac{1}{2}, \frac{1}{4}) = \det \begin{bmatrix} 3 & -3 \\ -3 & 12 \end{bmatrix} = 36 - 9 = 27.$ This is positive and $\frac{\partial^2 f}{\partial x^2}(0,0) = 3$ is positive (it's the upper left entry of the Hessian), so, by the second derivative test, $(\frac{1}{2}, \frac{1}{4})$ is a local minimum.

As a check on the plausibility of these conclusions, see Figure 4.10 for what the graph of f looks like near each of the critical points.

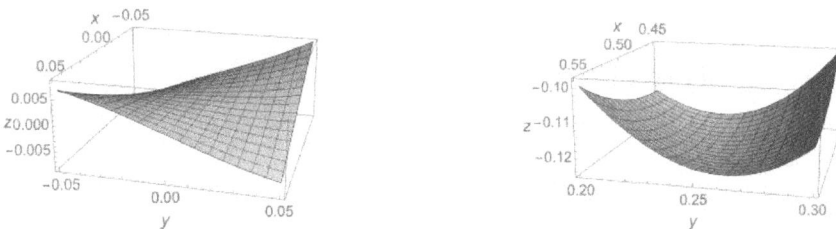

Figure 4.10: The graph of $f(x, y) = x^3 + 8y^3 - 3xy$ near (0, 0) (left) and near $(\frac{1}{2}, \frac{1}{4})$ (right)

4.11. CLASSIFYING NON-DEGENERATE CRITICAL POINTS

We now say a few words about why the second derivative test for functions of two variables works (Theorem 4.24 of the preceding section). We won't give a complete proof, but, since the test may appear to come out of nowhere, a word or two of explanation is warranted.

We begin by reviewing the corresponding situation for functions of one variable. Let $f : I \to R$ be a smooth function defined on an open interval I in R. The critical points of f are the points a where f'(a) = 0. At such points, the usual first-order approximation $f(x) \approx f(a) + f'(a) \cdot (x - a)$ degenerates to $f(x) \approx f(a)$, which gives us little information about the behavior of $f(x)$ when x is near a.

As a refinement, we add an additional term by considering the second-order Taylor approximation at a:

$$f(x) \approx f(a) + f'(a) \cdot (x - a) + \frac{f''(a)}{2} \cdot (x - a)^2. \tag{4.16}$$

If a is a critical point, this reduces to $f(x) \approx f(a) + \frac{f''(a)}{2} \cdot (x-a)^2$, a parabola whose general shape is determined by the sign of the coefficient $\frac{f''(a)}{2}$. If $f''(a) > 0$, then the approximating parabola is concave up and f has a local minimum at a. If $f''(a) < 0$, the parabola is concave down and f has a local maximum. If f''(a) = 0, the approximation degenerates again. Further information is required to determine the type of critical point in this case.

4.11.1 The second-order approximation

Let $f : U \to R$ be a smooth function of two variables defined on an open set U in $\mathbb{R}^2$, and let a be a point of U. The situation here is that the second-order approximation of $f(x)$ for x near a is:

$$f(\mathbf{x}) \approx f(\mathbf{a}) + Df(\mathbf{a}) \cdot (\mathbf{x} - \mathbf{a}) + \frac{1}{2}(\mathbf{x} - \mathbf{a})^t \cdot H(\mathbf{a}) \cdot (\mathbf{x} - \mathbf{a}). \tag{4.17}$$

The various dots are matrix products, where, writing x = (x, y) and a = (c, d):

⊙
$$H(\mathbf{a}) = \begin{bmatrix} \frac{\partial^2 f}{\partial x^2}(\mathbf{a}) & \frac{\partial^2 f}{\partial x\,\partial y}(\mathbf{a}) \\ \frac{\partial^2 f}{\partial x\,\partial y}(\mathbf{a}) & \frac{\partial^2 f}{\partial y^2}(\mathbf{a}) \end{bmatrix}$$

is the Hessian matrix at a,

⊙ $\mathbf{x} - \mathbf{a} = \begin{bmatrix} x - c \\ y - d \end{bmatrix}$, and

⊙ (x - a)^t = [x - c y - d], the transpose of x - a.

To justify the approximation, let x be a point near a, and let v = x − a. Consider the real- valued function g defined by g(t) = f(a + tv). Then g(0) = f(a) and g(1) = f(a + v) = f(x), so finding an approximation for f(x) is the same as approximating g(1). As a real-valued function of one variable, g(t) has a second-order Taylor approximation (4.16) at 0, which for g(1) gives:

$$f(\mathbf{x}) = g(1) \approx g(0) + g'(0) \cdot (1 - 0) + \frac{g''(0)}{2} \cdot (1 - 0)^2$$

$$\approx f(\mathbf{a}) + g'(0) + \frac{g''(0)}{2}. \tag{4.18}$$

To find the derivatives of g, note that, by construction, g is the composition g(t) = f (α(t)), where α(t) = a + tv. Hence, by the Little Chain Rule:

$$g'(t) = \nabla f(\alpha(t)) \cdot \alpha'(t) = \nabla f(\mathbf{a} + t\mathbf{v}) \cdot \mathbf{v}. \tag{4.19}$$

In particular, g'(0) = ∇f(a) · v, or in matrix form, g'(0) = Df(a) · v. Substituting this into (4.18) and recalling that v = x − a gives:

$$f(\mathbf{x}) \approx f(\mathbf{a}) + Df(\mathbf{a}) \cdot (\mathbf{x} - \mathbf{a}) + \frac{g''(0)}{2}. \tag{4.20}$$

So far, we have succeeded only in reconstructing the first-order approximation of f at a. The really new information comes from determining g''(0). From equation (4.19), note that g' is also a composition, namely, g'(t) = u(α(t)), where u(x) = ∇f(x)· v and α(t) = a + tv as before.

Therefore, by the same Little Chain Rule calculation:

$$g''(0) = (u \circ \alpha)'(0) = \nabla u(\mathbf{a}) \cdot \mathbf{v}. \tag{4.21}$$

To find ∇u, we let v = (h, k) and write out $u(x) = \nabla f(x) \cdot \mathbf{v} = (\frac{\partial f}{\partial x}, \frac{\partial f}{\partial y}) \cdot (h, k) = h\frac{\partial f}{\partial x} + k\frac{\partial f}{\partial y}$. Then $\nabla u = (h\frac{\partial^2 f}{\partial x^2} + k\frac{\partial^2 f}{\partial x \partial y}, h\frac{\partial^2 f}{\partial y \partial x} + k\frac{\partial^2 f}{\partial y^2})$. From equation (4.21) and the equality of mixed partials, we obtain:

$$g''(0) = \left(h\frac{\partial^2 f}{\partial x^2}(\mathbf{a}) + k\frac{\partial^2 f}{\partial x \partial y}(\mathbf{a}), \ h\frac{\partial^2 f}{\partial y \partial x}(\mathbf{a}) + k\frac{\partial^2 f}{\partial y^2}(\mathbf{a}) \right) \cdot (h, k)$$

$$= \frac{\partial^2 f}{\partial x^2}(\mathbf{a}) h^2 + 2\frac{\partial^2 f}{\partial x \partial y}(\mathbf{a}) hk + \frac{\partial^2 f}{\partial y^2}(\mathbf{a}) k^2. \tag{4.22}$$

To simplify the notation somewhat, let $A = \frac{\partial^2 f}{\partial x^2}(\mathbf{a})$, $B = \frac{\partial^2 f}{\partial x \partial y}(\mathbf{a})$, and $C = \frac{\partial^2 f}{\partial y^2}(\mathbf{a})$. We do some algebraic rearrangement in order to get the quadratic term of the approximation into the desired form (4.17):

$$g''(0) = Ah^2 + 2Bhk + Ck^2 \tag{4.23}$$
$$= (Ah^2 + Bhk) + (Bhk + Ck^2)$$
$$= \begin{bmatrix} h & k \end{bmatrix} \begin{bmatrix} Ah + Bk \\ Bh + Ck \end{bmatrix}$$
$$= \begin{bmatrix} h & k \end{bmatrix} \begin{bmatrix} A & B \\ B & C \end{bmatrix} \begin{bmatrix} h \\ k \end{bmatrix}$$
$$= (\mathbf{x} - \mathbf{a})^t \cdot H(\mathbf{a}) \cdot (\mathbf{x} - \mathbf{a}).$$

Substituting this into (4.20) gives the second-order approximation $f(\mathbf{x}) \approx f(\mathbf{a}) + Df$ (a)· (x - a),+ $\frac{1}{2}$(x - a)^t· H(a)· (x - a) as stated in (4.17). This accomplishes what we set out to do.

We remark that one can continue in this way to obtain higher-order approximations of f by using higher-order Taylor approximations of g. The coefficients involve the higher-order derivatives g(n)(0) of g. The idea is that, by the Little Chain Rule, each additional derivative of g corresponds to apply the "operator" $\mathbf{v} \cdot \nabla = (h, k) \cdot (\frac{\partial}{\partial x}, \frac{\partial}{\partial y}) = h\frac{\partial}{\partial x} + k\frac{\partial}{\partial y}$. Thus $g^{(n)}(0) = ((h\frac{\partial}{\partial x} + k\frac{\partial}{\partial y})^n f)(\mathbf{a})$, where $(h\frac{\partial}{\partial x} + k\frac{\partial}{\partial y})^n$ denotes the n-fold composition $(h\frac{\partial}{\partial x} + k\frac{\partial}{\partial y}) \circ (h\frac{\partial}{\partial x} + k\frac{\partial}{\partial y}) \circ \cdots \circ (h\frac{\partial}{\partial x} + k\frac{\partial}{\partial y})$.

We leave for the curious and ambitious reader the work of checking that this formula is consistent with our results for the first two derivatives and of working out a general formula for the nth-order approximation of f . For example, see Exercise 11.3 for the third-order approximation.

4.11.2. Sums and differences of squares

Still assuming that f is a smooth function of two variables, if a is a critical point, then Df (a) = [0 0] , so the second-order approximation (4.17) becomes:

$$f(\mathbf{x}) = f(\mathbf{a} + \mathbf{v}) \approx f(\mathbf{a}) + \frac{1}{2}(\mathbf{x} - \mathbf{a})^t \cdot H(\mathbf{a}) \cdot (\mathbf{x} - \mathbf{a})$$
$$\approx f(\mathbf{a}) + \frac{1}{2}\mathbf{v}^t \cdot H(\mathbf{a}) \cdot \mathbf{v},$$

where v = x - a. The approximation holds when v is near 0.

We assume that det H(a)/= 0 and determine the nature of the critical point a. To do this, it is simpler to write the quadratic term of the approximation using the notation of (4.23) with v = (h, k) so that:

$$f(\mathbf{a} + \mathbf{v}) \approx f(\mathbf{a}) + \frac{1}{2}(Ah^2 + 2Bhk + Ck^2) \tag{4.24}$$

and $H(\mathbf{a}) = \begin{bmatrix} A & B \\ B & C \end{bmatrix}$.

We consider the case A≠ 0. Similar arguments apply if A = 0 (see Exercise 11.5). Completing the square on the expression in parentheses in (4.24) gives:

$$
\begin{aligned}
Ah^2 + 2Bhk + Ck^2 &= A(h^2 + \tfrac{2B}{A}hk) + Ck^2 \\
&= A(h^2 + \tfrac{2B}{A}hk + \tfrac{B^2}{A^2}k^2) - \tfrac{B^2}{A}k^2 + Ck^2 \\
&= A(h + \tfrac{B}{A}k)^2 + (C - \tfrac{B^2}{A})k^2 \\
&= A(h + \tfrac{B}{A}k)^2 + \tfrac{AC-B^2}{A}k^2 \\
&= A\tilde{h}^2 + \tfrac{AC-B^2}{A}k^2,
\end{aligned}
$$

where $\tilde{h} = h + \tfrac{B}{A}k$. Then (4.24) becomes:

$$
f(\mathbf{a}+\mathbf{v}) \approx f(\mathbf{a}) + \frac{1}{2}\left(A\tilde{h}^2 + \tfrac{AC-B^2}{A}k^2\right).
$$

The important point is that the expression in parentheses is now a combination of squares, and we know from Example 4.25 what to expect of expressions of this type. For instance, if both coefficients A and $\tfrac{AC-B^2}{A}$ are positive, then f has a local minimum at v = 0, like $x^2 + y^2$. If both coefficients are negative, then f has a local maximum, like $-x^2 - y^2$. If the coefficients have opposite signs, then f has a saddle point, like $x^2 - y^2$. Note that th)is last case arises precisely when the product of the coefficients is negative, that is, when $\left|A\left(\tfrac{AC-B^2}{A}\right) = AC - B^2 < 0.\right.$

Recall that $A = \tfrac{\partial^2 f}{\partial x^2}(a)$, and observe that the expression $AC - B^2$ is the determinant of the Hessian: $\det H(\mathbf{a}) = \det\begin{bmatrix} A & B \\ B & C \end{bmatrix} = AC - B^2$. Collecting the information in the previous paragraph about the signs of the coefficients A and $\tfrac{AC-B^2}{A}$ then gives the second-derivative test as stated in Theorem 4.24.

The key idea of the argument was to complete the square to turns f into one of the prototypes of being a sum and/or difference of squares, at least up to second order. This is not as ad hoc as it might seem. A theorem in linear algebra states that, if M is a symmetric matrix in the sense that $M^t = M$, then it is always possible to complete the square and convert the expression $v^t \cdot M \cdot v$ into a sum and/or difference of squares. This is one interpretation of a powerful result known as the spectral theorem. Thanks to the equality of mixed partials, the Hessian matrix H(a) is symmetric, so the spectral theorem applies. In fact, the spectral theorem is true for n by n symmetric matrices for all n. Using similar ideas, this leads to a corresponding classification of nondegenerate critical points for real-valued functions of n variables when n > 2.

4.12 MAX/MIN: LAGRANGE MULTIPLIERS

We now consider the problem of maximizing or minimizing a function subject to a "constraint." The constraint is often what gives an optimization problem substance.

For instance, trying to find the maximum value of a function like $f(x, y, z) = x + 2y + 3z$ is silly: the function can be made arbitrarily large by going far away from the origin where x, y, and z are all large. But if (x, y, z) is confined to a limited region of space, then there's something to think about. We illustrate with a specific example.

Example 4.27. Let S be the surface described by $\frac{x^2}{4} + \frac{y^2}{9} + z^2 = 1$ in R^3. It is an ellipsoid, as sketched in Figure 4.11. If $f(x, y, z) = x + 2y + 3z$, at what points of S does f attain its maximum and minimum values, and what are those values?

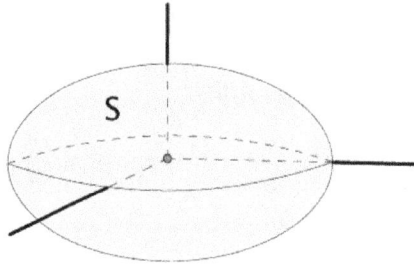

Figure 4.11: The ellipsoid $\frac{x^2}{4} + \frac{y^2}{9} + z^2 = 1$

Here, we want to maximize and minimize $f = x + 2y + 3z$ subject to the constraint $\frac{x^2}{4} + \frac{y^2}{9} + z^2 = 1$. One approach is to convert this to a two-variable problem, for instance, writing $z = \pm\sqrt{1 - \frac{x^2}{4} - \frac{y^2}{9}}$ on S, so $f = x + 2y + 3\sqrt{1 - \frac{x^2}{4} - \frac{y^2}{9}}$ or $f = x + 2y - 3\sqrt{1 - \frac{x^2}{4} - \frac{y^2}{9}}$. These are functions of two variables, so we could find the critical points, i.e., set the partial derivatives with respect to x and y equal to 0, and proceed as in the previous two sections.

Instead, we try a new approach based on organizing the values of f on S according to the level sets of f. Each level set $f = x + 2y + 3z = c$ is a plane that intersects S typically in a curve, if at all. As c varies, we get a family of level curves on S with f constant on each. See Figure 4.12.

Let's say we want to maximize f on S. Then, starting at some point, we move on S in the direction of increasing c. We continue doing this until we reach a point a where we can't go any farther. Let cmax denote the value of f at this point a. Observe that at a:

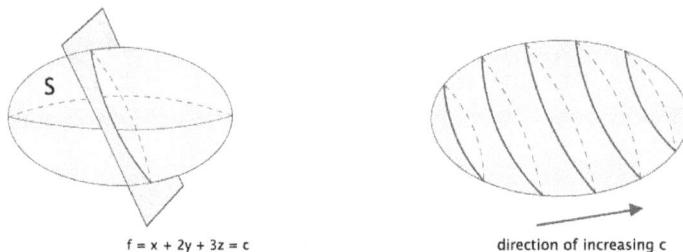

f = x + 2y + 3z = c   direction of increasing c

Figure 4.12: Intersecting S with a level set f = c (left) and a family of intersections (right)

- The level set $f = c_{max}$ must be tangent to S. Otherwise we would be able to cross the level curve at a and make the value of f larger while remaining on S.

- $\nabla f(a)$ is a normal vector to the level set $f = c_{max}$. This is because gradients are always normal to level sets (Proposition 4.17).

- Similarly, the surface S is itself a level set, namely, the set where:

$$g(x, y, z) = \frac{x^2}{4} + \frac{y^2}{9} + z^2 = 1.$$

g(x, y, z) = Hence $\nabla g(a)$ is a normal vector to S.

The situation is shown in Figure 4.13.

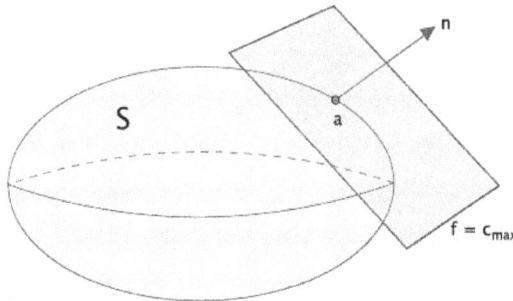

Figure 4.13: The level set f = c_{max} and the ellipsoid S are tangent at a.

Since the level set $f = c_{max}$ and the surface S are tangent at a, their normal vectors are scalar multiples of each other. In other words, there is a scalar λ such that $\nabla f(a)$ = $\lambda \nabla g(a)$. This is the idea behind the following general principle.

Proposition 4.28 (Method of Lagrange multipliers). Let U be an open set in $\mathbb{R}^n$, and suppose that you want to maximize or minimize a smooth function $f : U \to R$ subject to the constraint g(x) = c for some smooth function g : U $\to$ R and some constant c. If f has a local maximum or local minimum at a point a and if $\nabla g(a) \neq 0$, then there is a scalar λ such that:

$$\nabla f(\mathbf{a}) = \lambda \nabla g(\mathbf{a}).$$

The scalar λ is called a Lagrange multiplier.

We returns to Example 4.27 and find the maximum and minimum of f = x + 2y + 3z subject to the constraint $g = \frac{x^2}{4} + \frac{y^2}{9} + z^2 = 1$. According to the method of Lagrange multipliers, the solutions occur at points where $\nabla f = \lambda \nabla g$, that is, where $(1, 2, 3) = \lambda(\frac{1}{2}x, \frac{2}{9}y, 2z)$. This gives a system of equations:

$$\begin{cases} 1 = \dfrac{1}{2}\lambda x \\ 2 = \dfrac{2}{9}\lambda y \\ 3 = 2\lambda z. \end{cases}$$

There are four unknowns and only three equations which may seem like not enough information, but we must remember that the constraint provides a fourth equation that can be included as part of the system. In this case, the equations in the system imply that $\lambda = 0$, so we can solve for each of x, y, and z in terms of λ, then substitute into the constraint: $x = \frac{2}{\lambda}$, $y = \frac{9}{\lambda}$, $z = \frac{3}{2\lambda}$, and:

$$\frac{\left(\frac{2}{\lambda}\right)^2}{4} + \frac{\left(\frac{9}{\lambda}\right)^2}{9} + \left(\frac{3}{2\lambda}\right)^2 = 1.$$

This simplifies to $\frac{1}{\lambda^2} + \frac{9}{\lambda^2} + \frac{9}{4\lambda^2} = 1$, or $\frac{4+36+9}{4\lambda^2} = 1$. Hence $4\lambda^2 = 49$, so $\lambda = \pm\frac{7}{2}$. If $\lambda = \frac{7}{2}$, then

$$x = \frac{2}{\lambda} = \frac{2}{\frac{7}{2}} = \frac{4}{7}, \qquad y = \frac{9}{\frac{7}{2}} = \frac{18}{7}, \qquad z = \frac{3}{2 \cdot \frac{7}{2}} = \frac{3}{7}.$$

and $f = x + 2y + 3z = \frac{4}{7} + \frac{36}{7} + \frac{9}{7} = 7$. If $\lambda = -\frac{7}{2}$, it's the same thing with minus signs. Thus, on S, f attains a maximum value of 7 at $(\frac{4}{7}, \frac{18}{7}, \frac{3}{7})$ and a minimum value of -7 at $(-\frac{4}{7}, -\frac{18}{7}, -\frac{3}{7})$.

Example 4.29. Suppose that, when a manufacturer spends x dollars on labor, y dollars on equipment, and z dollars on research, the total number of units that it produces is:

$$f(x, y, z) = 60x^{1/6}y^{1/3}z^{1/2}.$$

This is an example of what is called a Cobb-Douglas production function. If the manufacturer has a total budget of B dollars, how should it distribute its spending among the three categories in order to maximize production?

Here, we want to maximize the function f above subject to the budget constraint g = x+y +z =B. Following the method of Lagrange multipliers, we set $\nabla f = \lambda \nabla g$:

$$(10x^{-5/6}y^{1/3}z^{1/2}, 20x^{1/6}y^{-2/3}z^{1/2}, 30x^{1/6}y^{1/3}z^{-1/2}) = \lambda(1, 1, 1).$$

In other words:

$$\begin{cases} 10x^{-5/6}y^{1/3}z^{1/2} = \lambda & (4.25) \\ 20x^{1/6}y^{-2/3}z^{1/2} = \lambda & (4.26) \\ 30x^{1/6}y^{1/3}z^{-1/2} = \lambda. & (4.27) \end{cases}$$

Both (4.25) and (4.26) equal λ, and setting them equal gives:

$$10x^{-5/6}y^{1/3}z^{1/2} = 20x^{1/6}y^{-2/3}z^{1/2}. \tag{4.28}$$

If z = 0, then f = 0, which is clearly not the maximum value. Thus we may assume z ≠ 0 so that the factors of $z^{1/2}$ in equation (4.28) can be canceled, leaving $10x^{-5/6}y^{1/3}$ = $20x^{1/6}y^{-2/3}$, or y = 2x. Similarly, setting (4.25) and (4.27) equal and assuming y ≠ 0 gives $10x^{-5/6}z^{1/2} = 30x^{1/6}z^{-1/2}$, or z = 3x.

Thus the budget constraint becomes x + y + z = x + 2x + 3x = 6x = B, so $x = \frac{B}{6}$.. We conclude that the optimal level of production is achieved by spending $x = \frac{B}{6}$. dollars on labor, $y = 2x = \frac{B}{3}$ dollars on equipment, and $z = 3x = \frac{B}{2}$ .dollars on research.

It's worth stepping back to review this calculation, as there is more here than meets the eye. A partial derivative can be viewed as a one-variable rate of change. For instance, in the previous example, $\frac{\partial f}{\partial x}$,is the rate at which the output changes per dollar spent on labor. This is called the marginal productivity with respect to labor. There are similar interpretations of $\frac{\partial f}{\partial y}$ and $\frac{\partial f}{\partial z}$ with respect to equipment and research, respectively.

The Lagrange multiplier condition ∇f = $\lambda \nabla g$ with g = x + y + z says that $\left(\frac{\partial f}{\partial x}, \frac{\partial f}{\partial y}, \frac{\partial f}{\partial z}\right)$ = $\lambda(1, 1, 1)$ = $(\lambda, \lambda, \lambda)$, or:

$$\frac{\partial f}{\partial x} = \frac{\partial f}{\partial y} = \frac{\partial f}{\partial z} = \lambda. \tag{4.29}$$

In other words, the marginal productivities with respect to labor, equipment, and research are equal. This part of the analysis applies to any production function f , not just the Cobb-Douglas model of the example. Thus, at the optimal level of production, an extra dollar spent on labor, equipment, or research all increase production by the same amount. This makes sense since, if there were an advantage to increasing the amount spent on one of them, then the manufacturer could increase production by spending more on that one and less on the others. At the optimal level, no such advantage can exist.

Morcover, according to (4.29), the common value of these marginal productivities is the La- grange multiplier λ. In the example, by (4.25):

$$\lambda = \frac{10y^{1/3}z^{1/2}}{x^{5/6}}$$

$$= \frac{10\left(\frac{B}{3}\right)^{1/3}\left(\frac{B}{2}\right)^{1/2}}{\left(\frac{B}{6}\right)^{5/6}}$$

$$= 10\left(\frac{1}{3}\right)^{1/3}\left(\frac{1}{2}\right)^{1/2}B^{5/6} \cdot \frac{6^{5/6}}{B^{5/6}}$$

$$= 60\left(\frac{1}{3}\right)^{1/3}\left(\frac{1}{2}\right)^{1/2}\left(\frac{1}{6}\right)^{1/6}$$

$$\approx 21.82.$$

So, at the optimal level, an additional dollar spent on any combination of labor, equipment, or research increases production by approximately 21.82 units. The seemingly extraneous Lagrange mutliplier has told us something interesting.

4.13 EXERCISES FOR CHAPTER 4

Section 1 The first-order approximation

1.1. Let $f(x, y) = x^2 - y^2$, and let a = (2, 1).

    a. Find $\frac{\partial f}{\partial x}(a)$ and $\frac{\partial f}{\partial y}(a)$.

    b. Find $Df(a)$ and $\nabla f(a)$.

    c. Find the first-order approximation $\ell(x, y)$ of $f(x, y)$ at a. (You may assume that f is differentiable at a.)

    d. Compare the values of f (2.01, 1.01) and ℓ(2.01, 1.01).

1.2. Let $f(x, y, z) = x^2y + y^2z + z^2x$, and let a = (1, -1, 1).

    a. Find $\partial \frac{\partial f}{\partial x}(a)$, $\frac{\partial f}{\partial y}(a)$, and $\frac{\partial f}{\partial z}(a)$.

    b. Find $Df(a)$ and $\nabla f(a)$.

    c. Find the first-order approximation $\ell(x, y, z)$ of $f(x, y, z)$ at a. (You may assume that f is differentiable at a.)

    d. Compare the values of $f(1.05, -1.1, 0.95)$ and $\ell(1.05, -1.1, 0.95)$.

In Exercises 1.3–1.8, find (a) the partial derivatives $\frac{\partial f}{\partial x}$ and $\frac{\partial f}{\partial y}$ and (b) the matrix Df (x, y).

1.3. $f(x, y) = x^3 - 2x^2y + 3xy^2 - 4y^3$

1.4. $f(x, y) = \sin x \sin 2y$

1.5. $f(x, y) = xe^{xy}$

1.6. $f(x, y) = \frac{x^2-y^2}{x^2+y^2}$

1.7. $f(x, y) = \frac{1}{x^2+y^2}$

1.8. $f(x, y) = x^y$

In Exercises 1.9–1.13, find (a) the partial derivatives $\frac{\partial f}{\partial x}$, $\frac{\partial f}{\partial y}$, and $\frac{\partial f}{\partial z}$ and (b) the gradient $\nabla f(x, y, z)$.

1.9. $f(x, y, z) = x + 2y + 3z + 4$

1.10. $f(x, y, z) = xy + xz + yz - xyz$

1.11. $f(x, y, z) = e^{-x2-y2-z2} \sin(x + 2y)$

1.12. $f(x, y, z) = \frac{x+y}{y+z}$

1.13. $f(x, y, z) = \ln(x^2 + y^2)$

1.14. Let:

$$f(x, y) = x^{y^{x^{y^{x^y}}}} \sin(xy) + \frac{x^3 y + x^4}{y^2 + xy + 1} \arctan\left(\frac{\ln x}{x^5 + y^6}\right).$$

1.15 Let $f(x, y) = x + 2y$. Use the definition of differentiability to prove that f is differentiable at every point $a = (c, d)$ of $\mathbb{R}^2$.

1.16 Use the definition of differentiability to show that the function $f(x, y) = x^2 + y^2$ is differentiable at every point $a = (c, d)$ of $\mathbb{R}^2$.

1.17 Let $f: \mathbb{R}^2 \to \mathbb{R}$ be the function:

$$f(x, y) = \sqrt{x^4 + y^4}.$$

a. Find formulas for the partial derivatives $\frac{\partial f}{\partial x}$ and $\frac{\partial f}{\partial y}$ at all points (x, y) other than the origin.
b. Find the values of the partial derivatives $\frac{\partial f}{\partial x}(0, 0)$ and $\frac{\partial f}{\partial y}(0, 0)$. (Note that the formulas you found in part (a) probably do not apply at $(0, 0)$.)
c. Use the definition of differentiability to determine whether f is differentiable at $(0, 0)$. (Hint: Polar coordinates might be useful.)

1.18 Consider the function $f: \mathbb{R}^2 \to \mathbb{R}$ defined by:

$$f(x, y) = \begin{cases} \frac{x^3 + 2y^3}{x^2 + y^2} & \text{if } (x, y) \neq (0, 0). \\ 0 & \text{if } (x, y) = (0, 0). \end{cases}$$

Find the values of the partial derivatives $\frac{\partial f}{\partial x}(0, 0)$ and $\frac{\partial f}{\partial y}(0, 0)$. (This shouldn't require much calculation.)

Use the definition of differentiability to determine whether f is differentiable at $(0, 0)$.

1.19 Let f and g be differentiable real-valued functions defined on an open set U in $\mathbb{R}^n$. Prove that:

$$\nabla(fg) = f\nabla g + g\nabla f.$$

(Hint: Partial derivatives are one-variable derivatives.)

1.20 Let $f, g : U \to \mathbb{R}$ be real-valued functions defined on an open set U in $\mathbb{R}^n$, and let a be a point of U. Use the definition of differentiability to prove the following statements.

a. If f and g are differentiable at a, so is $f + g$. (It's reasonably clear that $D(f + g)(a) = Df(a) + Dg(a)$, so the problem is to show that the limit in the definition of differentiability for $f + g$ is 0.)
b. If f is differentiable at a, so is cf for any scalar c.

Section 2 Conditions for differentiability

2.1 Let $f : \mathbb{R}^2 \to R$ be the function defined by:

$$f(x, y) = \begin{cases} x^2 - y^2 & \text{if } (x, y) \neq (0, 0), \\ \pi & \text{if } (x, y) = (0, 0). \end{cases}$$

Is f differentiable at $(0, 0)$? If so, find $Df(0, 0)$.

2.2 Is the function $f(x, y) = x^{2/3} + y^{2/3}$ differentiable at $(0, 0)$? If so, find $Df(0, 0)$.

2.3 This exercise gives the proof of Proposition 4.4, that differentiability implies continuity. Let U be an open set in $\mathbb{R}^n$, $f : U \to R$ a real-valued function, and a a point of U . Define a function $Q : U \to R$ by:

$$Q(x) = \begin{cases} \dfrac{f(x) - f(a) - \nabla f(a) \cdot (x - a)}{\|x - a\|} & \text{if } x \neq a, \\ 0 & \text{if } x = a. \end{cases}$$

Note that the quotient is the one that appears in the definition of differentiability using the gradient form of the derivative.

    a. If f is differentiable at a, show that Q is continuous at a.
    b. If f is differentiable at a, show that there exists a $\delta > 0$ such that, if $x \in B(a, \delta)$, then $|Q(x)| < 1$.
    c. If f is differentiable at a, prove that f is continuous at a. (Hint: Use the triangle inequality and the Cauchy-Schwarz inequality to show that:

$$|f(x) - f(a)| \leq |Q(x)| \|x - a\| + \|\nabla f(a)\| \|x - a\|.)$$

Section 3 The mean value theorem

3.1. Let $B = B(a, r)$ be an open ball in $\mathbb{R}^2$ centered at the point $a = (c, d)$, and let $f : B \to R$ be a differentiable function such that:

$$Df(x) = \begin{bmatrix} 0 & 0 \end{bmatrix} \text{ for all x in } B.$$

Prove that $f(x) = f(a)$ for all x in B. In other words, if the derivative of f is zero everywhere, then f is a constant function.

Section 4 The C¹ test

4.1 We showed in Example 4.7 that the function $f(x, y) = \sqrt{x^2 + y^2}$ is not differentiable at $(0, 0)$. Is f differentiable at points other than the origin? If so, which points?

In Exercises 4.2–4.6, use the various criteria for differentiability discussed so far to determine the points at which the function is differentiable and the points at which it is not. Your reasons may be brief as long as they are clear and precise. It should not be necessary to use the definition of differentiability.

4.2 $f: \mathbb{R}^2 \to R, f(x, y) = \sin(x + y)$

4.3 $f: \mathbb{R}^2 \to \mathbb{R}, f(x, y) = \begin{cases} \frac{x^2 - y^2}{x^2 + y^2} & \text{if } (x, y) \neq (0, 0), \\ 1 & \text{if } (x, y) = (0, 0) \end{cases}$

4.4. $f: \mathbb{R}^2 \to R, f(x, y) = |x| + |y|$

4.5. $f: R^3 \to R, f(x, y, z) = x^4 + 2x^2yz - y^3z^3$

4.6 $f: \mathbb{R}^4 \to \mathbb{R}, f(w, x, y, z) = \det \begin{bmatrix} w & x \\ y & z \end{bmatrix}$

4.7. Let $f(x, y) = \sqrt[3]{x^3 + 8y^3}$.

Find formulas for the partial derivatives $\frac{\partial f}{\partial x}$ and $\frac{\partial f}{\partial y}$ .at all points (x, y) where $x^3 + 8y^3 \neq 0$.

Do $\frac{\partial f}{\partial x}(0, 0)$ and $\frac{\partial f}{\partial y}(0, 0)$ exist? If so, what are their values? If not, why not?

Is f differentiable at (0, 0)?

More generally, at which points of $\mathbb{R}^2$ is f differentiable? (Hint: If a =(c, d) is a point other than the origin where $c^3 + 8d^3 = 0$, note that $\sqrt[3]{x^3 + 8d^3} = \sqrt[3]{x^3 - c^3} = \sqrt[3]{(x - c)(x^2 + cx + c^2)} = \sqrt[3]{x - c} \cdot \sqrt[3]{x^2 + cx + c^2}$ where $x^2 + cx + c^2 \neq 0$ when x = c.)

Section 5 The Little Chain Rule

5.1 Let $f(x, y) = x^2 + y^2$, and let $\alpha(t) = (t^2, t^3)$. Calculate $(f \circ \alpha)'(1)$ in two different ways:

  a. by using the Little Chain Rule,
  b. by substituting for x and y in terms of t in the formula for f to obtain (f °α)(t) directly and differentiating the result.

5.2 Let $f(x, y, z) = xyz$, and let $\alpha(t) = (\cos t, \sin t, t)$. Calculate $(f \circ \alpha)'(\frac{\pi}{6})$ .in two different ways:

  a. by using the Little Chain Rule,
  b. by substituting for x, y, and z in terms of t in the formula for f to obtain (f ° α)(t) directly and differentiating the result.

5.3 Let $f(x, y, z)$ be a differentiable real-valued function of three variables, and let $\alpha(t) = (x(t), y(t), z(t))$ be a differentiable path in R3. If $w = f(\alpha(t))$, use the Little Chain Rule to find a formula for $\frac{dw}{dt}$ in terms of the partial derivatives of f and the derivatives with respect to t of x, y, and z.

5.4 The Little Chain Rule often lurks in the background of a type of first-year calculus problem known as "related rates." For instance, as an (admittedly artificial) example, suppose that the length ℓ and width w of a rectangular region in the plane are changing as functions of time t, so that the area A = ℓw also changes.

a. Find a formula for $\frac{dA}{dt}$ in terms of ℓ, w, and their derivatives by differentiating the formula $A = \ell w$ directly with respect to t.

b. Obtain the same result using the Little Chain Rule. (Hint: Let $\alpha(t) = (\ell(t), w(t))$.)

c. Suppose that, at the instant that the length of the region is 100 inches and its width is 40 inches, the length is increasing at a rate of 2 inches/second and the width is increasing at a rate of 3 inches/second. How fast is the area changing?

5.5 Let a be a point of $\mathbb{R}^n$, and let B = B(a, r) be an open ball centered at a. Let $f : B \to$ R be a differentiable function. If b is any point of B, show that there exists a point c on the line segment connecting a and b such that:

$$f(\mathbf{b}) - f(\mathbf{a}) = \nabla f(\mathbf{c}) \cdot (\mathbf{b} - \mathbf{a}).$$

Note that this is a generalization of the mean value theorem to real-valued functions of more than one variable. (Hint: Explain why the line segment can be parametrized by $\alpha(t) = a + t(b - a)$, $0 \leq t \leq 1$. Then, note that the composition $f \circ \alpha$ is a real-valued function of one variable, so the one-variable mean value theorem applies.)

5.6 In this exercise, we prove the Little Chain Rule (Theorem 4.13). Let U be an open set in $\mathbb{R}^n$, $\alpha : I \to U$ a path in U defined on an open interval I, and $f : U \to R$ a real-valued function. Let t0 be a point of I. Assume that α is differentiable at t0 and f is differentiable at $\alpha(t_0)$. Let $a = \alpha(t_0)$, and let $Q : U \to R$ be the function defined in Exercise 2.3:

$$Q(\mathbf{x}) = \begin{cases} \frac{f(\mathbf{x}) - f(\mathbf{a}) - \nabla f(\mathbf{a}) \cdot (\mathbf{x} - \mathbf{a})}{\|\mathbf{x} - \mathbf{a}\|} & \text{if } \mathbf{x} \neq \mathbf{a}, \\ 0 & \text{if } \mathbf{x} = \mathbf{a}. \end{cases}$$

a. Prove that $\lim_{t \to t_0} \left| \frac{f(\alpha(t)) - f(\alpha(t_0)) - \nabla f(\alpha(t_0)) \cdot (\alpha(t) - \alpha(t_0))}{t - t_0} \right| = 0$. (Hint: f (x)-f(a)-∇f(a)·(x - a) = ||x - a|| Q(x). Or see Exercise 8.2 in Chapter 3.)

b. Prove the Little Chain Rule: $(f \circ \alpha)'(t0) = \nabla f(\alpha(t_0)) \cdot \alpha'(t_0)$.

5.7. Let $f : R^3 \to R$ be a differentiable function with the property that $\nabla f(x)$ points in the same direction as x for all nonzero x in R3. If a > 0, prove that f is constant on the sphere $x^2 + y^2 + z^2 = a^2$. (Hint: If p and q are any two points on the sphere, there is a differentiable path α on the sphere from p to q.)

Section 6 Directional derivatives

In Exercises 6.1–6.4, find the directional derivative of f at the point a in the direction of the vector v.

6.1. $f(x, y) = e^{-2x}y^3$, $a = (0, 1)$, $v = (4, 1)$

6.2. $f(x, y) = x^2 + y^2$, $a = (1, 2)$, $v = (-2, 1)$

6.3. $f(x, y, z) = x^3 - 2x^2yz + xz - 3$, $a = (1, 0, -1)$, $v = (1, -1, 2)$

6.4. $f(x, y, z) = \sin x \sin y \cos z$, $a = (0, \frac{\pi}{2}, \pi)$, $v = (\pi, 2\pi, 2\pi)$

6.5. For a certain differentiable real-valued function $f : \mathbb{R}^2 \to R$, the maximum directional derivative at the point a = (0, 0) has a value of 6 and occurs in the direction from a to the point b = (4, 1). Find $\nabla f(a)$.

6.6 Let $f(x, y) = x^2 - y^2$.

    a. At the point $a = (\sqrt{2}, 1)$, find a unit vector u that points in the direction in which f is increasing most rapidly. What is the rate of increase in this direction?

    b. Describe the set of all points a = (x, y) in $\mathbb{R}^2$ such that f increases most rapidly at a in the direction that points directly towards the origin.

6.7 A certain function has the form:

$$f(x, y) = rx^2 + sy^2,$$

where r and s are constant. Find values of r and s so that the maximum directional derivative of f at the point a = (1, 2) has a value of 10 and occurs in the direction from a to the point b = (2, 3).

6.8 Let $f(x, y) = xy$, and let a = (5, -5).

    a. Find all directions in which the directional derivative of f at a is equal to 1.

    b. Is there a direction in which the directional derivative at a is equal to 10?

6.9 The temperature in a certain region of space, in degrees Celsius, is modeled by the function T $(x, y, z) = 20e^{-x2-2y2-4z2}$, where x, y, z are measured in meters. At the point a = (2, -1, 3):

    a. In what direction is the temperature increasing most rapidly?

    b. In what direction is it decreasing most rapidly?

    c. If you travel in the direction described in part (a) at a speed of 10 meters/second, how fast is the observed temperature changing at a in degrees Celsius per second?

6.10. Let $c = (c_1, c_2, \ldots, c_n)$ be a vector in $\mathbb{R}^n$, and define $f : \mathbb{R}^n \to R$ by $f(x) = c \cdot x$.

    a. Find the derivative $Df(x)$.

    b. If u is a unit vector in $\mathbb{R}^n$, find a formula for the directional derivative $(D_u f)$ (x).

6.11. You discover that your happiness is a function of your location in the plane. At the point (x, y), your happiness is given by the formula:

$$H(x, y) = x^3 e^2 y \text{ happs,}$$

where "happ" is a unit of happiness. For instance, at the point $(1, 0)$, you are $H(1, 0) = 1$ happ happy.

    a. At the point $(1, 0)$, in which direction is your happiness increasing most rapidly? What is the value of the directional derivative in this direction?

    b. Still at the point $(1, 0)$, in which direction is your happiness decreasing most rapidly? What is the value of the directional derivative in this direction?

    c. Suppose that, starting at $(1, 0)$, you follow a curve so that you are always traveling in the direction in which your happiness increases most rapidly. Find an equation for the curve, and sketch the curve. An equation describing the relationship between x and y along the curve is fine. No further parametrization is necessary. (Hint: What can you say about the slope at each point of the curve?)

    d. Suppose instead that, starting at $(1, 0)$, you travel along a curve such that the directional derivative in the tangent direction is always equal to zero. Find an equation describing this curve, and sketch the curve.

6.12. Does there exist a differentiable function $f: U \to R$ defined on an open subset U of $\mathbb{R}^n$ with the property that, for some point a in U, the directional derivatives at a satisfy $(D_u f)(a) > 0$ for all unit vectors u in $\mathbb{R}^n$, i.e., the directional derivative at a is positive in every direction? Either find such a function and point a, or explain why none exists.

Section 7 ∇f as normal vector

7.1. Find an equation of the tangent plane to the surface $z = x^2 - y^2$ at the point a = $(1, 2, -3)$.

7.2. Find an equation of the tangent plane to the surface $x^2 + y^2 - z^2 = 1$ at the point a = $(1, -1, 1)$.

7.3. The surface in R^3 given by was featured in an article in the February 14, 2019 issue of The New York Times. See Figure 4.14. Nowhere in the article was the tangent plane at the point $(0, 1, 1)$ mentioned. Scoop the Times by finding an equation of the tangent plane.

$$(9x^2 + y^2 + z^2 - 1)^3 - y^2 z^3 - \frac{2}{5} x^2 z^3 = 0$$

7.4. Let S be the ellipsoid $x^2 + \frac{y^2}{2} + \frac{z^2}{4} = 4$. Let a = $(1, 2, 2)$, and let n be the unit normal vector to S at a that points outward from S. If $f(x, y, z) = xyz$, find the directional derivative $(D_n f)(a)$.

7.5. The saddle surface $y^2 - x^2 - z = 0$ and the sphere $x^2 + y^2 + z^2 = 14$ intersect in a curve C that passes through the point a = $(1, 2, 3)$. Find a parametrization of the line tangent to C at a.

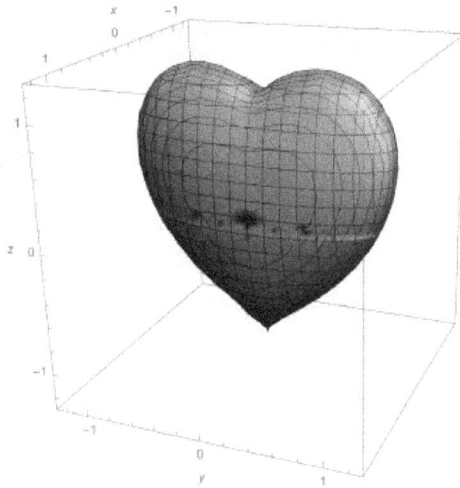

Figure 4.14: The surface $(9x^2 + y^2 + z^2 - 1)^3 - y^2 z^3 - \frac{2}{5}x^2 z^3 = 0$

7.6. Find all values of c such that, at every point of intersection of the spheres the respective tangent planes are perpendicular to one another.

$$(x - c)^2 + y^2 + z^2 = 3 \quad \text{and} \quad x^2 + (y - 1)^2 + z^2 = 1.$$

Section 8 Higher-order partial derivatives

In Exercises 8.1–8.4, find all four second-order partial derivatives of the given function f.

8.1. $f(x, y) = x^4 - 2x^3 y + 3x^2 y2 - 4xy^3 + 5y^4$

8.2. $f(x, y) = \sin x \cos y$

8.3. $f(x, y) = e^{-x2-y2}$

8.4. $f(x, y) = \dfrac{y}{x+y}$

8.5. Let $f(x, y) = x^4 y^3$. Find i and j so that the (i + j)th-order partial derivative $\dfrac{\partial^{i+j} f}{\partial x^i \partial y^j}(0,0)$ is nonzero. What is the value of this partial derivative?

8.6. Consider the function $f: \mathbb{R}^2 \to$ R defined by:

$$f(x, y) = \begin{cases} \frac{x^3 y}{x^2 + y^2} & \text{if } (x, y) \neq (0,0), \\ 0 & \text{if } (x, y) = (0,0). \end{cases}$$

    a. Find formulas for the partial derivatives $\frac{\partial f}{\partial x}(x, y)$ and $\frac{\partial f}{\partial y}(x, y)$ if (x, y)≠(0, 0).

b. Find the values of $\frac{\partial f}{\partial x}(0,0)$ and $\frac{\partial f}{\partial y}(0,0)$.

c. Use your answers to parts (a) and (b) to evaluate the second-order partial derivatives:

$$\frac{\partial^2 f}{\partial y\, \partial x}(0,0) = \left(\frac{\partial}{\partial y}\left(\frac{\partial f}{\partial x}\right)\right)(0,0) \quad \text{and} \quad \frac{\partial^2 f}{\partial x\, \partial y}(0,0) = \left(\frac{\partial}{\partial x}\left(\frac{\partial f}{\partial y}\right)\right)(0,0)$$

at the origin (and only at the origin). (WaℝⁿWarning: Do not assume that the second-order partial derivatives are continuous.)

d. Is f of class C^2 on $\mathbb{R}^2$? Is it of class C^1?

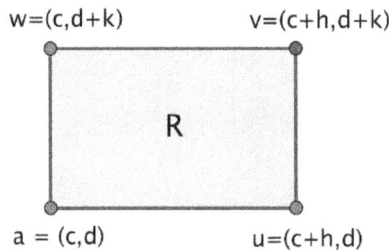

Figure 4.15: The rectangle with vertices a, u, v, w

8.7. This exercise gives the details of a full proof of the equality of mixed partials (Theorem 4.21). Let a = (c, d) be a point of $\mathbb{R}^2$, and let $f : B \to R$ be a real-valued function defined on an open ball B = B(a, r) centered at a. Assume that the first and second-order partial derivatives of f are continuous on B. Let R be the rectangle whose vertices are a = (c, d), u = (c + h, d), v = (c + h, d + k), and w = (c, d + k), where h and k are nonzero real numbers small enough that R ⊂ B. See Figure 4.15.

Let:

$$\Delta = f(\mathbf{v}) - f(\mathbf{u}) - f(\mathbf{w}) + f(\mathbf{a})$$
$$= f(c + h, d + k) - f(c + h, d) - f(c, d + k) + f(c, d).$$

Show that $\Delta = \left(\frac{\partial f}{\partial y}(c + h, y_1) - \frac{\partial f}{\partial y}(c, y_1)\right)k$ for some number y1 between d and d + k. (Hint: Note that Δ = g(d + k) – g(d), where g is the function defined by g(y) = f(c + h, y) – f(c, y). Apply the mean value theorem.)

Show that there is a point p = (x$_1$,y$_1$) in R such that $\Delta = \frac{\partial^2 f}{\partial x\, \partial y}(\mathbf{p})\, hk$. (Hint: Apply the mean value theorem again, this time to the difference in the result of part (a).)

Similarly, show that there is a point q = (x$_2$, y$_2$) in R such that $\Delta = \frac{\partial^2 f}{\partial y\, \partial x}(\mathbf{q})\, hk$. (Hint: Consider j(x) = f(x, d + k) – f(x, d).)

Deduce that the mixed partials at a are equal: $\frac{\partial^2 f}{\partial x\, \partial y}(\mathbf{a}) = \frac{\partial^2 f}{\partial y\, \partial x}(\mathbf{a})$. (Hint: What happens to R as h, k → 0?)

Section 10 Max/min: critical points

In Exercises 10.1–10.6, find all critical points of f, and, if possible, classify their type.

10.1. $f(x, y) = 2 - 10x - 3x^2 - 20y + 8xy + 3y^2$

10.2. $f(x, y) = 5 - 2x - 2x^2 - 2xy + 2y - y^2$

10.3. $f(x, y) = x^4 + 2y^4 + 3x^2y^2 + 4x^2 + 5y^2$

10.4. $f(x, y) = x^3 + x^2 + 4xy - 2y^2 + 9$

10.5. $f(x, y) = 2x^3 + y^2 - 6xy$

10.6. $f(x, y) = 4x^3 + y^3 + (1 - x - y)3$

10.7. Let k be a real number, and let $f(x, y) = x^3 + y^3 - kxy$. Find all critical points of f, and classify each one as a local maximum, local minimum, or neither. (The answers may depend on the value of k.)

10.8. Let $f(x, y) = 3x^4 - 4x^2y + y2$.

    a. Make a sketch in $\mathbb{R}^2$ indicating the set of points (x, y) where $f(x, y) > 0$ and the set where $f(x, y) < 0$. (Hint: Factor f.)

    b. Explain why f does not have a local maximum or minimum at $(0, 0)$.

    c. On the other hand, show that f does have a local minimum at $(0, 0)$ when restricted to any line $y = mx$.

10.9   a. Given a point a in $\mathbb{R}^n$, define $f : \mathbb{R}^n \to \mathbb{R}$ by $f(x) = \|x - a\|^2$. Show that $\nabla f(x) = 2(x - a)$.

    b. Let $p_1, p_2, \ldots, p_m$ be m distinct points in $\mathbb{R}^n$, and let $F(x) = \sum_{k=1}^{m} \|x - p_k\|^2$. Assuming that F has a global minimum, show that it must occur when:

$$x = \frac{1}{m} \sum_{k=1}^{m} p_k \quad \text{(the "centroid")}.$$

10.10. Let $(x_1, y_1), (x_2, y_2), \ldots, (x_n, y_n)$ be a collection of n distinct points in $\mathbb{R}^2$, where $n \geq 2$. We think of the points as "data." The line $y = mx + b$ that best fits the data **in the sense of least squares** is defined to be the one that minimizes the quantity:

$$E = \sum_{i=1}^{n} (mx_i + b - y_i)^2.$$

Note that using a line $y = mx + b$ to predict the data value $y = y_i$ when $x = x_i$ gives an error of magnitude $|mx_i + b - y_i|$. Consequently E is called the total squared error. It depends on the line, which we identify here by its slope m and y-intercept b. Thus E is a function of the two variables m and b, and we can apply the methods we have been studying to find where it assumes its minimum.

Calculate the partial derivatives $\frac{\partial E}{\partial m}$ and $\frac{\partial E}{\partial b}$., and show that the critical points (m, b) of E are the solutions of the system of equations:

$$\begin{cases} \left(\sum_{i=1}^{n} x_i^2\right) m + \left(\sum_{i=1}^{n} x_i\right) b &= \sum_{i=1}^{n} x_i y_i \\ \left(\sum_{i=1}^{n} x_i\right) m + nb &= \sum_{i=1}^{n} y_i. \end{cases} \tag{4.30}$$

In Exercises 10.11–10.12, find the best fitting line in the sense of least squares for the given data points by solving the system (4.30). Then, plot the points and sketch the line.

10.11. $(1, 1), (2, 3), (3, 3)$

10.12. $(1, 2), (2, 1), (3, 4), (4, 3)$

10.13. In this exercise, we use the second derivative test to verify that, for the best fitting line in the sense of least squares, the critical point (m, b) given by the system (4.30) is a local minimum of the total squared error E.

    a. If $x_1, x_2, \ldots, x_n$ are real numbers, show that $\left(\sum_{i=1}^{n} x_i\right)^2 \leq n\left(\sum_{i=1}^{n} x_i^2\right)$ and that equality holds if and only if $x_1 = x_2 = \cdots = x_n$. (Hint: Consider the dot product $x \cdot v$, where $x = (x_1, x_2, \ldots, x_n)$ and $v = (1, 1, \ldots, 1)$.)

    b. Let $(x_1, y_1), (x_2, y_2), \ldots, (x_n, y_n)$ be given data points. Show that the Hessian of the total squared error E is given by:

$$H(m, b) = \begin{bmatrix} 2\left(\sum_{i=1}^{n} x_i^2\right) & 2\left(\sum_{i=1}^{n} x_i\right) \\ 2\left(\sum_{i=1}^{n} x_i\right) & 2n \end{bmatrix}.$$

    c. Assuming that the data points don't all have the same x-coordinate, show that the critical point of E given by (4.30) is a local minimum. (In fact, it is a global minimum, as follows once one factors in that E is a quadratic polynomial in m and b, though we won't go through the details to justify this.)

Section 11 Classifying nondegenerate critical points

11.1 Find the second-order approximation of $f(x, y) = \cos(x + y)$ at a = (0, 0).

11.2 Find the second-order approximation of $f(x, y)$ = x2 + y2 at a = (1, 2).

11.3 Let U be an open set in $\mathbb{R}^2$, and let $f : U \to$ R be a smooth real-valued function defined on U . Let a be a point of U , and let x = a + v be a nearby point, where v = (h, k).

Due to the equality of mixed partials, the powers of the operator $v \cdot \nabla = h\frac{\partial}{\partial x} + k\frac{\partial}{\partial y}$ under composition expand in the same way as ordinary binomial expressions. For instance:

$$\left(h\frac{\partial}{\partial x} + k\frac{\partial}{\partial y} \right)^2 f = \left(\left(h\frac{\partial}{\partial x} + k\frac{\partial}{\partial y} \right) \circ \left(h\frac{\partial}{\partial x} + k\frac{\partial}{\partial y} \right) \right) f$$

$$= \left(h^2\frac{\partial}{\partial x} \circ \frac{\partial}{\partial x} + hk\frac{\partial}{\partial x} \circ \frac{\partial}{\partial y} + hk\frac{\partial}{\partial y} \circ \frac{\partial}{\partial x} + k^2\frac{\partial}{\partial y} \circ \frac{\partial}{\partial y} \right) f$$

$$= \left(h^2\frac{\partial^2}{\partial x^2} + 2hk\frac{\partial^2}{\partial x\,\partial y} + k^2\frac{\partial^2}{\partial y^2} \right) f$$

$$= h^2\frac{\partial^2 f}{\partial x^2} + 2hk\frac{\partial^2 f}{\partial x\,\partial y} + k^2\frac{\partial^2 f}{\partial y^2}.$$

Note that this agrees with the expression derived in (4.22) as part of the quadratic term in the second-order approximation of f.

Find the corresponding expansion of $\left(h\frac{\partial}{\partial x} + k\frac{\partial}{\partial y} \right)^3 f$.

Find a formula for the third-order approximation of f at a. (Recall from first-year calculus that the third-order Taylor approximation of a function of one variable is $f(x) = f(a+h) \approx f(a) + f'(a)h + \frac{f''(a)}{2}h^2 + \frac{f'''(a)}{3!}h^3.$)

11.4 Let $f(x, y) = ax^2 + by^2$, where a and b are nonzero constants. For which values of a and b is the origin a local maximum? For which is it a local minimum? For which is it a saddle point?

11.5 Complete the description of the behavior of the second-order approximation near a critical point a by considering the case that det H(a) $\neq$ 0 and A = 0, where $H(\mathbf{a}) = \begin{bmatrix} A & B \\ B & C \end{bmatrix}$.

    a. Show that det H(a) < 0.

    b. Show that the quadratic term $Ah^2 + 2Bhk + Ck^2 = 2Bhk + Ck^2$ in the second-order approximation can be written as a difference of two perfect squares. (Hence, based on our prototypical models, we predict a to be a saddle point. Hint: In the case that C = 0, too, consider the expansions of $(a \pm b)^2$.)

Section 12 Max/min: Lagrange multipliers

Consider the problem of finding the maximum and minimum values of the function $f(x, y) = xy$ subject to the constraint x + 2y = 1.

    a. Explain why a minimum value does not exist.

    b. On the other hand, you may assume that a maximum does exist. Use Lagrange multi- pliers to find the maximum value and the point at which it is attained.

12.2   a. Find the points on the curve $x^2 - xy + y^2 = 4$ at which the function $f(x, y) = x^2 + y^2$ attains its maximum and minimum values. You may assume that these maximum and minimum values exist.

b. Use your answer to part (a) to help sketch the curve $x^2\,xy+y^2=4$. (Hint: Note that $f(x,y)=x^2+y^2$ represents the square of the distance from the origin.)

12.3 At what points on the unit sphere $x^2+y^2+z^2=1$ does the function $f(x,y,z)=2x-3y+4z$ attain its maximum and minimum values?

12.4 Find the minimum value of the function $f(x,y,z)=x^2+y^2+z^2$ subject to the constraint $2x-3y+4z=1$. At what point is the minimum attained?

12.5 Find the positive values of x, y, and z for which the function $f(x,y,z)=xy^2z^3$ attains its maximum value subject to the constraint $3x+4y+5z=12$. You may assume that a maximum exists.

12.6   a. Use the method of Lagrange multipliers to find the positive values of x, y, and z such that $xyz=1$ and $x+y+z$ is as small as possible. What is the minimum value? You may assume that a minimum exists.

b. Use part (a) to show that, for all positive real numbers a, b, c:

$$\frac{a+b+c}{3} \geq \sqrt[3]{abc}.$$

This is called the inequality of the arithmetic and geometric means. (Hint: Let $k=\sqrt[3]{abc}$, and consider a/k, b/k, and c/k.)

Let A be a 3 by 3 symmetric matrix, $A=\begin{bmatrix} a & b & c \\ b & d & e \\ c & e & f \end{bmatrix}$, and consider the quadratic polynomial:

$$Q(\mathbf{x})=\mathbf{x}^t A\mathbf{x}=\begin{bmatrix} x & y & z \end{bmatrix}\begin{bmatrix} a & b & c \\ b & d & e \\ c & e & f \end{bmatrix}\begin{bmatrix} x \\ y \\ z \end{bmatrix}.$$

where $\mathbf{x}=\begin{bmatrix} x \\ y \\ z \end{bmatrix}$.

Use Lagrange multipliers to show that, when Q is restricted to the unit sphere $x^2+y^2+z^2=1$, any point x at which it attains its maximum or minimum value satisfies $Ax=\lambda x$ for some scalar λ and that the maximum or minimum value is the corresponding value of λ. (In the language of linear algebra, x is called an eigenvector of A and λ is called the corresponding eigenvalue.) In fact, maximum and minimum values do exist, so the conclusion is not an empty one.

12.8 You discover that your happiness is a function of your daily routine. After a great deal of soul-searching, you decide to focus on three activities to which you will devote your entire day: if x, y, and z are the number of hours a day that you spend eating, sleeping, and studying multivariable calculus, respectively, then, based on data provided by the federal gov$\mathbb{R}^n$ment, your happiness is given by the function:

$$h(x, y, z) = 1000 - 23x^2 - 23y^2 - 46y - z^2 \text{ happs.}^6$$

There is a mix of eating, sleeping, and studying that leads to the greatest possible daily happiness (you may assume this). What is it? Answer in two ways, as follows.

   a. Substitute for z in terms of x and y in the formula for h, and find the critical points of the resulting function of x and y.
   b. Use Lagrange multipliers.

After conducting extensive market research, a manufacturer of monkey saddles discovers that, if it produces a saddle consisting of x ounces of leather, y ounces of copper, and z ounces of premium bananas, the monkey rider experiences a total of:

S = 2xy + 3xz + 4yz units of satisfaction.

Leather costs $1 per ounce, copper $2 per ounce, and premium bananas $3 per ounce, and the manufacturer is willing to spend at most $1000 per saddle. What combination of ingredients yields the most satisfied monkey? You may assume that a maximum exists.

5
Chapter

REAL-VALUED FUNCTIONS: INTEGRATION

5.1 VOLUME AND ITERATED INTEGRALS

We now study the integral of real-valued functions of more than one variable. Most of our time is devoted to functions of two variables. In this section, we introduce the integral informally by thinking about how to visualize it. This uses an idea likely to be familiar from first-year calculus. It won't be until the next section that we define what the integral actually is.

Let D be a subset of $\mathbb{R}^2$. At this point, we won't be too careful about putting any conditions on D, but it's best to think of it as some sort of two-dimensional blob, as opposed to a set of discrete points or a curve. Let $f: A \to R$ be a real-valued function whose domain A is a subset of $\mathbb{R}^2$ that contains D. For the moment, assume that f (x) ≥ 0 for all x in D.

Let W be the region in R^3 that lies below the graph z = $f(x, y)$ and above D, as shown in Figure 5.1. That is:

$$W = \{(x, y, z) \in \mathbb{R}^3 : (x, y) \in D \text{ and } 0 \leq z \leq f(x, y)\}.$$

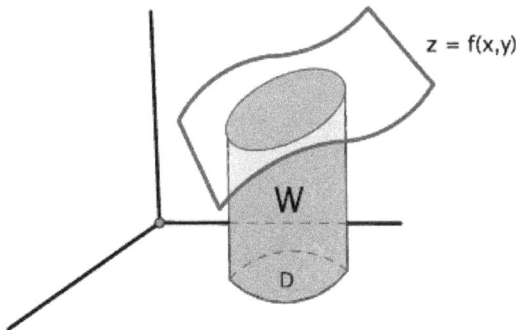

Figure 5.1: The region W under the graph z = f (x, y) and above D

Consider the volume of W , which we denote variously by $\iint_D f(x,y)\,dA$, $\iint_D f(x,y)\,dx\,dy$, or just $\iint_D f\,dA$. To calculate it, we apply the following principle from first-year calculus:

Volume = Integral of cross-sectional area.

For instance, in first-year calculus, the cross-sections may turns out to be disks, washers, or, in a slight variant, cylinders.

Example 5.1. Find the volume of the region under the plane z = 4 - x + 2y and above the square 1 ≤ x ≤ 3, 2 ≤ y ≤ 4, in the xy-plane. See Figure 5.2.

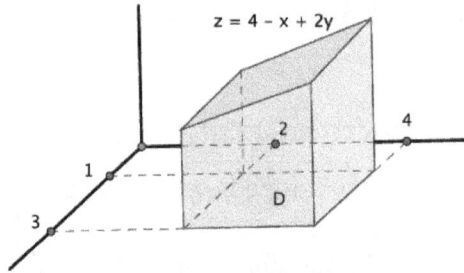

Figure 5.2: The region under z = 4 - x + 2y and above the square 1 ≤ x ≤ 3, 2 ≤ y ≤ 4

Here, $f(x, y) = 4 - x + 2y$, and, if D denotes the specified square, we are looking for $\iint_D (4 - x + 2y)\, dA$. As above, let W denote the region in question, below the plane and above the square. We consider first cross-sections perpendicular to the x-axis, that is, intersections of W with planes of the form x = constant. There is such a cross-section for each value of x from x = 1 to x = 3, so:

$$\text{Volume} = \int_1^3 A(x)\, dx, \tag{5.1}$$

where A(x) is the area of the cross-section at x. A typical cross-section is shown in Figure 5.3. The base is a line segment from y = 2 to y = 4, and the top is a curve lying in the graph z = 4 - x + 2y. (In fact, the cross-sections in this example are trapezoids, but let's not worry about that.) In other words, A(x) is an area under a curve, so it too is an integral: $A(x) = \int_2^4 (4 - x + 2y)\, dy$, where, for the given cross-section, x is fixed. Integrating the cross-sectional area as in equation (5.1) and evaluating gives:

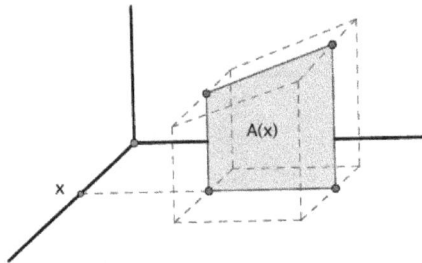

Figure 5.3: A cross-section perpendicular to the x-axis

$$\text{Volume} = \int_1^3 \left(\int_2^4 (4 - x + 2y) \, dy \right) dx$$

$$= \int_1^3 (4y - xy + y^2) \Big|_{y=2}^{y=4} dx$$

$$= \int_1^3 ((16 - 4x + 16) - (8 - 2x + 4)) \, dx$$

$$= \int_1^3 (20 - 2x) \, dx$$

$$= 20x - x^2 \Big|_1^3$$

$$= (60 - 9) - (20 - 1)$$

$$= 32.$$

Alternatively, we could have used cross-sections perpendicular to the y-axis, in which case the cross-sections occur from y = 2 to y = 4 (Figure 5.4):

$$\text{Volume} = \int_2^4 A(y) \, dy. \qquad (5.2)$$

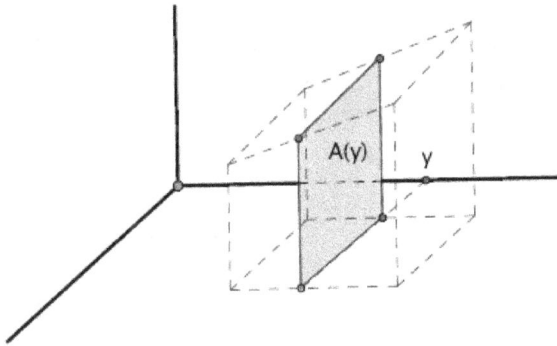

Figure 5.4: A cross-section perpendicular to the y-axis

Each cross-section lies below the graph and above a line segment that goes from x = 1 to x = 3, so $A(y) = \int_1^3 (4 - x + 2y) \, dx$. This time, the integral of the cross-sectional area (5.2) gives:

$$\text{Volume} = \int_2^4 \left(\int_1^3 (4 - x + 2y) \, dx \right) dy$$

$$= \int_2^4 \left(4x - \frac{1}{2}x^2 + 2yx \right) \Big|_{x=1}^{x=3} dy$$

$$= \int_2^4 \left(\left(12 - \frac{9}{2} + 6y \right) - \left(4 - \frac{1}{2} + 2y \right) \right) dy$$

$$= \int_2^4 (4 + 4y) \, dy$$

$$= 4y + 2y^2 \big|_2^4$$

$$= 48 - 16$$

$$= 32.$$

According to this approach, the volume is the integral of an integral, or an **iterated integral**. Evaluating the integral consists of a sequence of "partial antidifferentiations" with respect to one variable at a time, all other variables being held constant. Once you integrate with respect to a particular variable, that variable disappears from the remainder of the calculation.

Example 5.2. Find the volume of the solid that lies inside the cylinders $x^2 + z^2 = 1$ and $y^2 + z^2 = 1$ and above the xy-plane.

The cylinders have axes along the y and x-axis, respectively, and intersect one another at right angles, as in Figure 5.5. Perhaps it's simplest to focus on the quarter of the solid that lies in the "first octant," i.e., the region of R3 in which x, y, and z are all nonnegative. The base of this portion is the unit square $0 \leq x \leq 1, 0 \leq y \leq 1$ (Figure 5.6, left). Half of the solid lies under $x^2 + z^2 = 1$ (Figure 5.6, right), running in the y-direction above the triangle in the xy-plane bounded by the x-axis, the line x = 1, and the line y = x. Call this triangle D1. The other half runs in the x-direction, under y2 + z2 = 1 and above a complementary triangle D2 within the unit square.

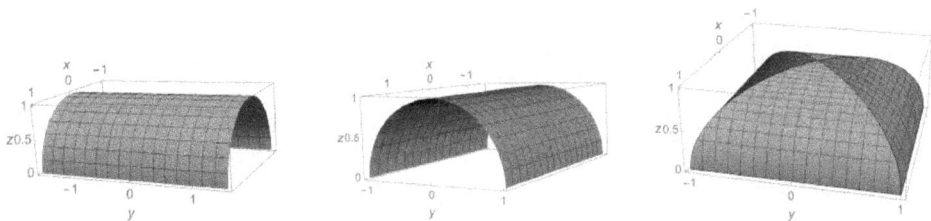

Figure 5.5: The cylinders $x^2 + z^2 = 1$ (left), $y^2 + z^2 = 1$ (middle), and the region contained in both (right)

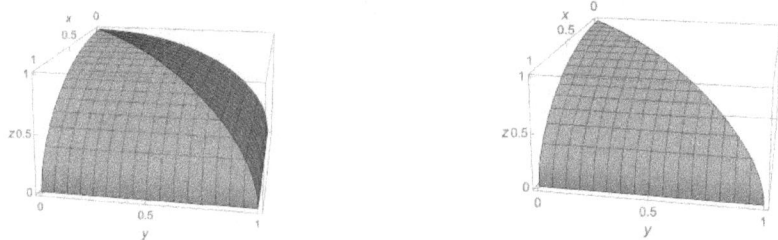

Figure 5.6: The portion in the first octant (left) and the portion under only x² + z² = 1 (right)

The volume of the original solid is 8 times the volume of the portion over D1. This lies below $x^2 + z^2 = 1$, i.e., below the graph $z = \sqrt{1 - x^2}$. Hence:

$$\text{Volume} = 8 \iint_{D_1} \sqrt{1 - x^2}\, dA.$$

To calculate the volume over D_1, let's try cross-sections perpendicular to the x-axis. These exist from x = 0 to x = 1:

$$\text{Volume} = 8 \int_0^1 A(x)\, dx.$$

The base of each cross-section is a line segment perpendicular to the x-axis within D_1. The lower endpoint is always at y = 0, but the upper endpoint depends on x. Indeed, the upper endpoint lies on the line y = x. See Figure 5.7.

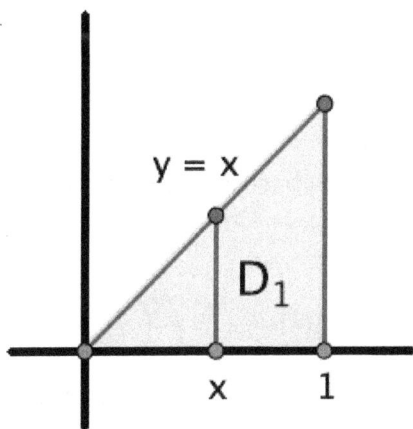

Figure 5.7: The base of the cross-section at x

Hence $A(x) = \int_0^x \sqrt{1 - x^2}\, dy$, so:

$$\text{Volume} = 8 \int_0^1 \left(\int_0^x \sqrt{1 - x^2} \, dy \right) dx$$

$$= 8 \int_0^1 \left(\sqrt{1 - x^2} \cdot y \, \Big|_{y=0}^{y=x} \right) dx$$

$$= 8 \int_0^1 x\sqrt{1 - x^2} \, dx \qquad (\text{let } u = 1 - x^2, du = -2x \, dx)$$

$$= 8 \left(-\frac{1}{2} \cdot \frac{2}{3}(1 - x^2)^{3/2} \Big|_0^1 \right)$$

$$= -\frac{8}{3}(0^{3/2} - 1^{3/2})$$

$$= \frac{8}{3}.$$

Alternatively, suppose we had used cross-sections perpendicular to the y-axis. The cross-sections exist from y = 0 to y = 1, and, for each such y, the base of the cross-section is a line segment in the x-direction that goes from x = y to x = 1, as illustrated in Figure 5.8.

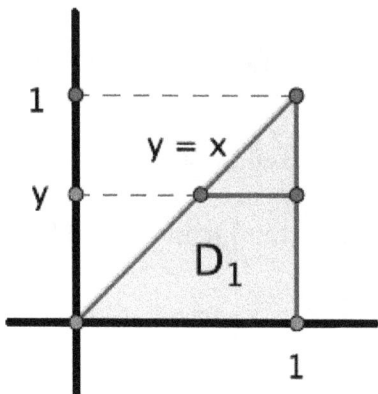

Figure 5.8: The base of the cross-section at y

Consequently:

$$\text{Volume} = 8 \int_0^1 A(y) \, dy = 8 \int_0^1 \left(\int_y^1 \sqrt{1 - x^2} \, dx \right) dy.$$

While not impossible, it is a little less obvious how to do the first partial antidifferentiation with respect to x by hand. In other words, our original order of antidifferentiation might be preferable.

Example 5.3. Consider the iterated integral $\int_0^2 \left(\int_{x^2}^4 x^3 e^{y^3} \, dy \right) dx$.

    a. Sketch the domain of integration D in the xy-plane.

    b. Evaluate the integral.

    c. The domain of integration can be reconstructed from the endpoints of the integrals. The outermost integral says that x goes from x = 0 to x = 2. Geometrically, we are integrating areas of cross-sections perpendicular to the x-axis. Then, the inner integral says that, for each x, y goes from y = x2 to y = 4. Figure 5.9 shows the relevant input. Hence D is described by the conditions:

$$D = \{(x, y) \in \mathbb{R}^2 : 0 \le x \le 2, x^2 \le y \le 4\}.$$

This is the region in the first quadrant bounded by the parabola y = x2, the line y = 4, and the y-axis.

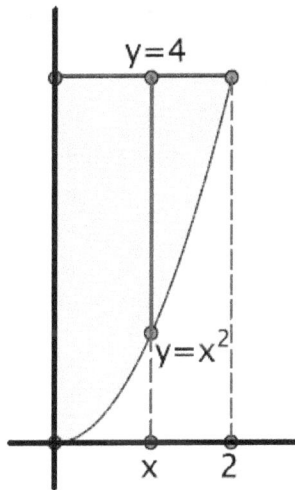

Figure 5.9: The domain of integration: cross-sections perpendicular to the x-axis

To evaluate the integral as presented, we would antidifferentiate first with respect to y, treating x as constant:

$$\int_0^2 \left(\int_{x^2}^4 x^3 e^{y^3} \, dy \right) dx = \int_0^2 \left(x^3 \int_{x^2}^4 e^{y^3} \, dy \right) dx.$$

The innermost antiderivative looks hard. So, having nothing better to do, we try switching the order of antidifferentiation. Changing to cross-sections perpendicular to the y-axis, we see from the description of D that y goes from y = 0 to y = 4, and, for each y, x goes from x = 0 to $x = \sqrt{y}$.

The thinking behind the switched order is illustrated in Figure 5.10.

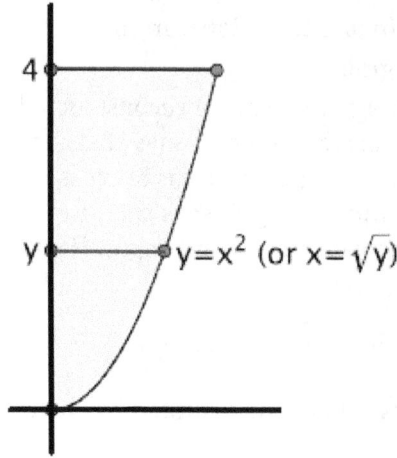

Figure 5.10: Reversing the order of antidifferentiation

Therefore:

$$\int_0^2 \left(\int_{x^2}^4 x^3 e^{y^3} \, dy \right) dx = \int_0^4 \left(\int_0^{\sqrt{y}} x^3 e^{y^3} \, dx \right) dy$$

$$= \int_0^4 \left(\frac{1}{4} x^4 e^{y^3} \Big|_{x=0}^{x=\sqrt{y}} \right) dy$$

$$= \int_0^4 (\frac{1}{4} y^2 e^{y^3} - 0) \, dy \qquad (\text{let } u = y^3, du = 3y^2 \, dy)$$

$$= \frac{1}{4} \cdot \frac{1}{3} e^{y^3} \Big|_0^4$$

$$= \frac{1}{12} (e^{64} - 1).$$

5.2 THE DOUBLE INTEGRAL

We now define the integral $\iint_D f(x,y) \, dA$ of a real-valued function f of two variables whose domain is a subset D of $\mathbb{R}^2$. We assume the following:

- ⊙ D is a **bounded** subset of $\mathbb{R}^2$: this means that there is a rectangle R in $\mathbb{R}^2$ such that $D \subset R$.

- ⊙ f is a **bounded** function: this means that there is a scalar M such that $|f(x)| \leq M$ for all x in D.

To define the integral, we take as a guiding principle that it is a limit of weighted sums. This is often what gives the integral its power.

We begin by reviewing the one-variable case. Let $f: [a, b] \to R$ be defined on a closed interval [a, b]. We subdivide the interval into n subintervals of widths Δx_1, Δx_2, , Δx_n and choose a point $\sum_i f(p_i) \Delta x_i$. nterval for i = 1, 2, . . . , n. See Figure 5.11. We then form the sum

This is called a **Riemann sum**, and we refer to the pi as **sample points**. We think of the Riemann sum as the sum of the values f (pi) at the sample points weighted by the width of the subintervals (though it may be equally appropriate on occasion to think of it as the widths of the subintervals weighted by the function values). The integral is defined as a limit of the Riemann sums:

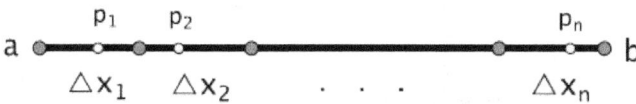

Figure 5.11: Subdividing [a, b] into n subintervals of widths Δx_1, Δx_2, . . . , Δx_n with sample points p_1, p_2, . . . , p_n

$$\int_a^b f(x)\, dx = \lim_{\Delta x_i \to 0} \sum_i f(p_i) \Delta x_i.$$

The limit is a different type from the ones we encountered before. Here, it means that, given any $\epsilon > 0$, there exists a $\delta > 0$ such that, for all Riemann sums $\sum_i f(p_i) \Delta x_i$ based on subdivisions of [a, b] with $\Delta x_i < \delta$ for all i, we have $|\sum_i f(p_i) \Delta x_i - \int_a^b f(x)\, dx| < \epsilon$. This is extremely cumbersome to work with.

Moving on to functions of two variables $f: D \to R$, where $D \subset \mathbb{R}^2$, we do our best to mimic the one-variable case and proceed in two steps.

Step 1. Integrals over rectangles.

We first introduce some notation that makes it easier to specify the kind of rectangles we have in mind.

Definition. If X and Y are sets, then the Cartesian product X × Y is defined to be the set of all ordered pairs:

$$X \times Y = \{(x, y) : x \in X, y \in Y\}.$$

For instance, the product of two closed intervals R = [a, b] × [c, d] = {(x, y) : x ∈ [a, b], y ∈ [c, d]} = {(x, y) : a ≤ x ≤ b, c ≤ y ≤ d} is a rectangle in $\mathbb{R}^2$ whose sides are parallel to the coordinate axes (Figure 5.12). We define how to integrate a function f (x, y) over such a rectangle R.

We chop up R into a grid of subrectangles with sides parallel to the axes and of dimensions Δx_i by Δy_j. A simple example is shown in Figure 5.13. We choose a sample

Multivariable Calculus

point p_{ij} in each subrectangle, and consider the Riemann sum $\sum_{i,j} f(p_{ij}) \triangle x_i \triangle y_j$. This is a weighted sum in which the function values at the sample points are weighted by the areas of the subrectangles.

Now, let $\triangle x_i$ and $\triangle y_j$ go to 0.

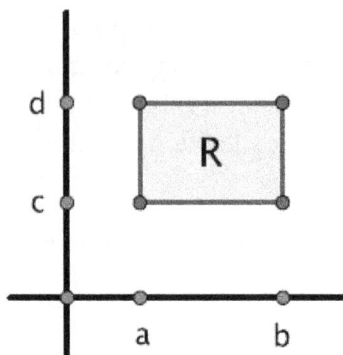

Figure 5.12: The rectangle R = [a, b] × [c, d] in $\mathbb{R}^2$

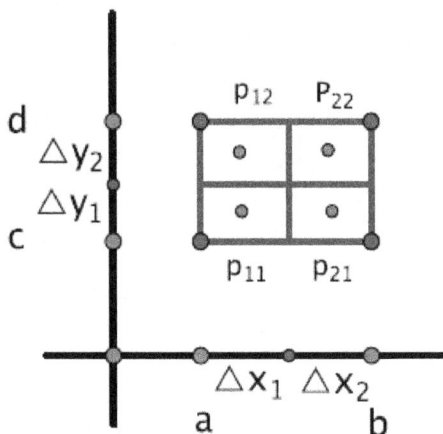

Figure 5.13: A subdivision of a rectangle into 4 subrectangles

Definition. Let R = [a, b] × [c, d] be a rectangle in $\mathbb{R}^2$. The integral of a bounded function $f: R \to R$ is defined by:

$$\iint_R f(x,y)\, dA = \lim_{\substack{\triangle x_i \to 0 \\ \triangle y_j \to 0}} \sum_{i,j} f(p_{ij}) \triangle x_i \triangle y_j.$$

assuming that the limit exists. When it does, we say that f is **integrable** on R.

There are other versions of integrals, too, and the one defined here is known technically as the **Riemann integral**. Integrals of functions of two variables are called **double integrals**.

We begin with a standard somewhat artificial example that is constructed to make a point.

Example 5.4. Let R = [0, 1] × [0, 1], and define $f: R \to R$ by:

$$f(x, y) = \begin{cases} 0 & \text{if } x \text{ and } y \text{ are both rational,} \\ 1 & \text{otherwise.} \end{cases}$$

For instance, $f(\frac{1}{2}, \frac{3}{5}) = 0$, while $f(\frac{1}{2}, \frac{\sqrt{2}}{2}) = 1$ and $f(\frac{\pi}{6}, \frac{\sqrt{2}}{2}) = 1$. The graph z = f (x, y) consists of two 1 by 1 squares, one at height z = 0 and one at height z = 1, both with lots of missing points.

To evaluate the integral $\iint_R f(x, y)\, dA$ using the definition, consider a subdivision of R into subrectangles. Regardless of the subdivision, each subrectangle contains points where $f = 0$ as well as points where $f = 1$. Thus, depending on the choice of sample points, some Riemann sums have the form $\sum_{i,j} 0 \cdot \triangle x_i \triangle y_j = 0$, and some have the form $\sum_{i,j} 1 \cdot \triangle x_i \triangle y_j = $ Area of R = 1. Still others have values in between. This is true no matter what $\triangle x_i$ and $\triangle y_j$ are, so $\sum_{i,j} f(\mathbf{p}_{ij}) \triangle x_i \triangle y_j$ does not approach a limit as $\triangle x_i$ and $\triangle y_j$ go to 0. Consequently f is not integrable on R.

This raises the question of whether it's possible to tell when a function is integrable. For the preceding function, in some sense the problem is that it is too discontinuous.

Theorem 5.5. Let $f: R \to R$ be a bounded function defined on a rectangle R = [a, b] × [c, d]. If f is continuous on R, except possibly at a finite number of points or on a finite number of smooth curves, then f is integrable on R.

This is actually a special case of a more general theorem that characterizes precisely when a function is Riemann integrable. Unfortunately, the proof uses concepts that are best left for a course in real analysis, so we do not go into it further. One consequence, however, is the useful fact that continuous functions are integrable. This is worth remembering. It covers most of the functions that we work with.

A second question is that, once a function is known to be integrable, is there a practical way to evaluate the integral? Here again, we only indicate why the answer is plausible without attempting anything like a correct proof. The intuition is that, after you subdivide a rectangle and choose sample points, you can organize the Riemann sum $\sum_{i,j} f(\mathbf{p}_{ij}) \triangle x_i \triangle y_j$ either:

- ⊙ column-by-column: $\sum_i (\sum_j f(\mathbf{p}_{ij}) \triangle y_j) \triangle x_i$, or
- ⊙ row-by-row: $\sum_j (\sum_i f(\mathbf{p}_{ij}) \triangle x_i) \triangle y_j$.

These are Riemann sums of Riemann sums. The first sum is an approximation of the iterated integral $\int_a^b \left(\int_c^d f(x,y)\, dy \right) dx$, while the second sum approximates $\int_c^d \left(\int_a^b f(x,y)\, dx \right) dy$. Refer to Figure 5.14. As Δx_i and Δy_j go to 0, these approximations become exact, and, in the limit, the iterated integrals coincide with the double integral.

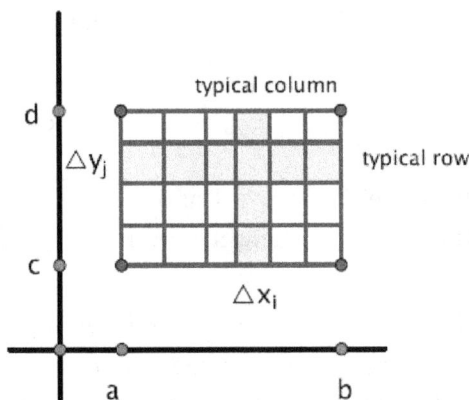

Figure 5.14: Organizing Riemann sums row-by-row or column-by-column. For instance, a typical column sum is $\sum_j f(\mathrm{p}_{ij}) \Delta y_j$, where i is fixed.

Theorem 5.6 (Fubini's theorem). If f is integrable on R = [a, b] × [c, d], then $\iint_R f(x,y)\, dA$ is equal to both of the iterated integrals:

$$\iint_R f(x,y)\, dA = \int_a^b \left(\int_c^d f(x,y)\, dy \right) dx = \int_c^d \left(\int_a^b f(x,y)\, dx \right) dy.$$

In particular, the iterated integrals that we computed in Section 5.1 were actually examples of double integrals.

Step 2. Integrals over bounded sets in general.

Let $f : D \to R$ be a bounded function defined on a bounded subset D of $\mathbb{R}^2$. By definition, D is contained in a rectangle R. By taking an even larger rectangle, if necessary, we may assume that R has the form [a, b] × [c, d]. Define a function $\widetilde{f} : R \to \mathbb{R}$ by:

$$\widetilde{f}(x,y) = \begin{cases} f(x,y) & \text{if } (x,y) \in D. \\ 0 & \text{otherwise.} \end{cases}$$

An example of a graph of $\widetilde{f}$ is shown in Figure 5.15.

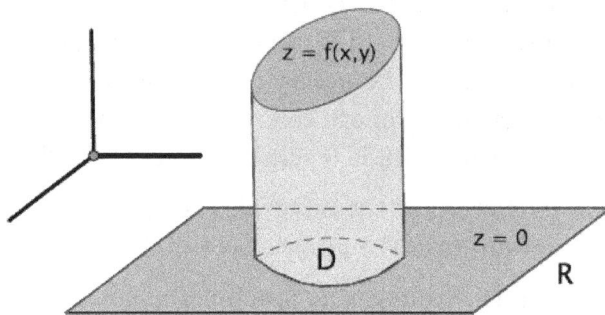

Figure 5.15: The general case: the graph $z = \tilde{f}(x, y)$

Definition. We sa y that $f \colon D \to R$ is integrable on D if $\tilde{f}$ is integrable on R. If so, we define $\iint_D f(x, y)\, dA = \iint_R \tilde{f}(x, y)\, dA$.

Thus the integral of f over D is defined as a limit of Riemann sums for $\tilde{f}$, $\sum_{i,j} \tilde{f}(\mathbf{p}_{ij}) \triangle x_i \triangle y_j$, based on subdivisions of the larger rectangle R into subrectangles. Since $\tilde{f} = 0$ away from D, however, only those subrectangles that intersect D contribute to the Riemann sum. Hence we might right:

$$\iint_D f(x, y)\, dA = \lim_{\substack{\triangle x_i \to 0 \\ \triangle y_j \to 0}} \sum_{\substack{\text{subrectangles} \\ \text{that intersect } D}} f(\mathbf{p}_{ij}) \triangle x_i \triangle y_j.$$

The subdivision is indicated in Figure 5.16.

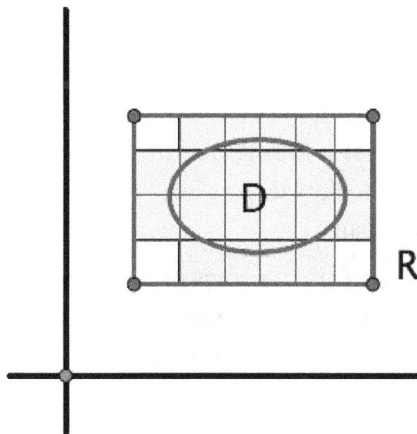

Figure 5.16: The general case: subdividing a rectangle containing D into subrectangles. Only those subrectangles that intersect D may contribute to a Riemann sum for $\tilde{f}$ **.**

As Figure 5.15 indicates, even if f is continuous on D, the function f might well be discontinuous on the boundary of D. Nevertheless, by Theorem 5.5, f will still be integrable on R as long as the boundary of D consists of finitely many smooth curves. In other words, the integral can tolerate a certain amount of bad behavior on the boundary. Having this latitude is something we take advantage of later on.

Here are some basic properties of the integral. They are quite reasonable and can be proven using the definition of the integral as a limit of sums.

Proposition 5.7. Let f and g be integrable functions on a bounded set D in $\mathbb{R}^2$. Then:

$$\iint_D (f + g)\, dA = \iint_D f\, dA + \iint_D g\, dA.$$

$\iint_D cf\, dA = c \iint_D f\, dA$ for any scalar c.

If f (x, y) ≤ g(x, y) for all (x, y) in D, then $\iint_D f\, dA \leq \iint_D g\, dA$.

5.3 INTERPRETATIONS OF THE DOUBLE INTEGRAL

We have defined the integral of a bounded real-valued function f (x, y) of two variables over a bounded subset D of $\mathbb{R}^2$ as:

$$\iint_D f(x,y)\, dA = \lim_{\substack{\Delta x_i \to 0 \\ \Delta y_j \to 0}} \sum_{\substack{\text{subrectangles} \\ \text{that intersect } D}} f(\mathbf{p}_{ij})\, \triangle x_i\, \triangle y_j,$$

where the subrectangles come from subdividing a rectangle R that contains D. Now, we go about trying to understand why this might be useful.

First, the integral is a limit of sums. Therefore, if a typical summand f (pij) Δxi Δyj represents "something," then $\iint_D f(x,y)\, dA$ represents the total "something." Also, we have noted before that a Riemann sum is the sum of the sample values f (pij) weighted by the area of the subrectangles. We consider a few situations where such a weighted sum might arise.

Example 5.8. The first is when f (x, y) represents some sort of density, or quantity per unit area, at the point (x, y), for instance, mass density or population density. Then $f (p_{ij})\, \Delta x_i\, \Delta y_j$ is approximately the quantity on a typical subrectangle, and:

$$\iint_D (\text{density})\, dA = \text{total quantity in } D.$$

for instance, the total mass or total population.

Example 5.9. Suppose that f is the constant function f (x, y) = 1 for all (x, y) in D. Then $f (p_{ij})\, \Delta x_i\, \Delta y_j$ = 1 · Δxi Δyj is the area of a subrectangle, so:

$$\iint_D 1\, dA = \text{Area}\,(D).$$

Example 5.10. Let's returns to the graph of a function $z = f(x, y)$, where $f(x, y) \geq 0$ for all (x, y) in D. Then $f(p_{ij})$ is the height of the graph above the sample point p_{ij}, and $f(p_{ij})\,\Delta x_i\,\Delta y_j$ is approximately the volume below the graph and above a typical subrectangle (Figure 5.17). Thus:

$$\iint_D f(x, y)\, dA = \text{volume below the graph of } f \text{ and above } D.$$

As a result, volume can be regarded either as an iterated integral, as in Section 5.1, or as a double integral, i.e., a limit of Riemann sums. That these two approaches give the same thing is another way of understanding Fubini's theorem.

Example 5.11. Average value of a function.

To average a list of numbers, say 2, 4, 7, 4, 4, we find the sum and divide by the number of entries:

$$\text{average} = \frac{2 + 4 + 7 + 4 + 4}{5} = \frac{21}{5} = 4.2.$$

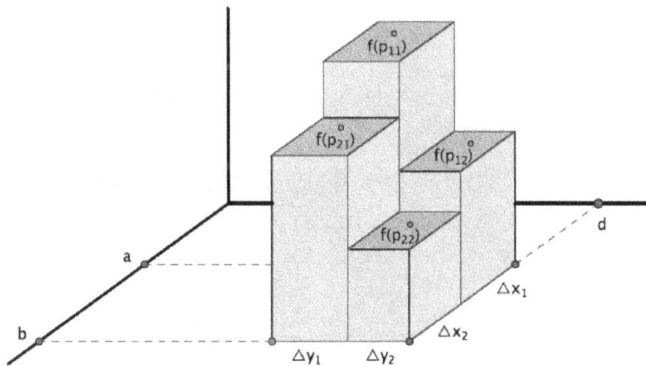

Figure 5.17: Visualizing a Riemann sum as approximating the volume under a graph This could also be written as:

This could also be written as:

$$2 \cdot \frac{1}{5} + 4 \cdot \frac{1}{5} + 7 \cdot \frac{1}{5} + 4 \cdot \frac{1}{5} + 4 \cdot \frac{1}{5} \quad \text{or} \quad 2 \cdot \frac{1}{5} + 4 \cdot \frac{3}{5} + 7 \cdot \frac{1}{5}.$$

In other words, it's a sum in which each value is weighted by its relative frequency.

Similarly, to find the average value of a function $f(x, y)$ over a region D, we take a sampling of points pij in D and consider the sum in which each function value $f(p_{ij})$ is weighted by the fraction of D that it represents. If the sampling comes from a subdivision into subrectangles and we think of each sample point as representing its subrectangle, this gives:

$$\text{Average value} \approx \sum f(\mathbf{p}_{ij}) \frac{\triangle x_i \, \triangle y_j}{\text{Area}(D)} = \frac{1}{\text{Area}(D)} \sum f(\mathbf{p}_{ij}) \triangle x_i \, \triangle y_j.$$

Taking the limit as $\triangle x_i$ and $\triangle y_j$ go to 0 leads to the following definition.

Definition.

The **average value** of a bounded function $f(x, y)$ on a bounded subset D of $\mathbb{R}^2$, denoted by $\bar{f}$, is defined to be:

$$\bar{f} = \frac{1}{\text{Area}(D)} \iint_D f \, dA.$$

For example, the average x and y-coordinates on D are:

$$\bar{x} = \frac{1}{\text{Area}(D)} \iint_D x \, dA \qquad \text{and} \qquad \bar{y} = \frac{1}{\text{Area}(D)} \iint_D y \, dA.$$

The point $(\bar{x}, \bar{y})$ is called the centroid of D.

Example 5.12. Let D be the triangular region in $\mathbb{R}^2$ with vertices (0, 0), (2, 2), and (-2, 2).

    a. Find the average value of $f(x, y) = y^2 - x^2$ on D. The graph of f is in Figure 5.18.
    b. Find the centroid of D.

1. Here, $\bar{f} = \frac{1}{\text{Area}(D)} \iint_D (y^2 - x^2) \, dA,$ where the region D is shown in Figure 5.19. We could find the area of D using Area $(D) = \iint_D 1 \, dA$ as in Example 5.9, but, since D is a triangle, we can use

$$\text{Area}(D) = \frac{1}{2}(\text{base})(\text{height}) = \frac{1}{2} \cdot 4 \cdot 2 = 4.$$

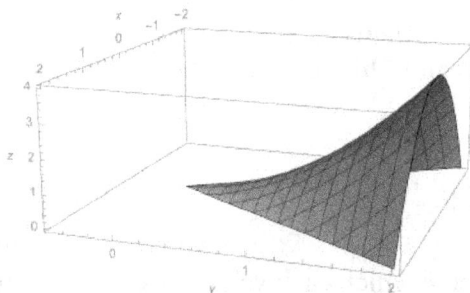

Figure 5.18: The graph of f (x, y) = y² - x² over the triangle D

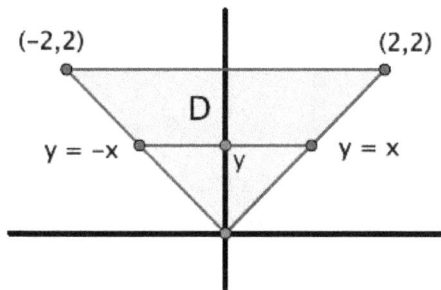

Figure 5.19: The triangle D

To set up the double integral $\iint_D (y^2 - x^2)\, dA$ as an iterated integral, note that D is bounded on the left by y = -x, on the right by y = x, and on top by y = 2. We use cross-sections perpendicular to the y-axis. These cross-sections exist from y = 0 to y = 2, and, for each y, x goes from x = -y to x = y. Hence:

$$\iint_D (y^2 - x^2)\, dA = \int_0^2 \left(\int_{-y}^{y} (y^2 - x^2)\, dx \right) dy$$

$$= \int_0^2 (y^2 x - \frac{1}{3} x^3) \Big|_{x=-y}^{x=y} dy$$

$$= \int_0^2 ((y^3 - \frac{1}{3} y^3) - (-y^3 + \frac{1}{3} y^3))\, dy$$

$$= \int_0^2 \frac{4}{3} y^3\, dy$$

$$= \frac{1}{3} y^4 \Big|_0^2$$

$$= \frac{16}{3}.$$

Therefore $\overline{f} = \frac{1}{4} \cdot \frac{16}{3} = \frac{4}{3}$.

As a side note, suppose we had tried to calculate the double integral using cross-sections perpendicular to the x-axis instead. These cross-sections exist from x = -2 to x = 2, but the y endpoints of a given cross-section are different for the left and right halves of the triangle. When x is between -2 and 0, the cross-section at x goes from y = -x to y = 2, whereas when x is between 0 and 2, it goes from y = x to y = 2 (Figure 5.20). Thus, to describe D this way, the integral must be split into two pieces:

$$\iint_D (y^2 - x^2)\, dA = \int_{-2}^{0} \left(\int_{-x}^{2} (y^2 - x^2)\, dy \right) dx + \int_0^2 \left(\int_x^2 (y^2 - x^2)\, dy \right) dx.$$

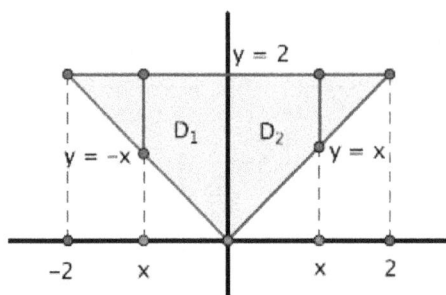

Figure 5.20: Reversing the order of integration to integrate over D

Actually, we could try to take advantage of symmetry properties of D and f to avoid having to evaluate both halves. We take up this idea more fully when we discuss the change of variables theorem in Chapter 7.

2. Because of the symmetry of D in the y-axis, it's intuitively clear that the average x-coordinate is 0: x = 0. Again, we justify this more carefully later (see Example 7.7 in Chapter 7).

For the average y-coordinate, by definition

$$\bar{y} = \frac{1}{\text{Area}\,(D)} \iint_D y \, dA = \tfrac{1}{4} \iint_D y \, dA.$$

We describe D using the same limits of integration as in part (a) with cross-sections perpendicular to the y-axis:

$$
\begin{aligned}
\iint_D y \, dA &= \int_0^2 \left(\int_{-y}^y y \, dx \right) dy \\
&= \int_0^2 \left(y x \Big|_{x=-y}^{x=y} \right) dy \\
&= \int_0^2 (y^2 - (-y^2)) \, dy \\
&= \int_0^2 2y^2 \, dy \\
&= \frac{2}{3} y^3 \Big|_0^2 \\
&= \frac{16}{3}.
\end{aligned}
$$

Thus $\bar{y} = \tfrac{1}{4} \cdot \tfrac{16}{3} = \tfrac{4}{3}$, , and the centroid is the point $(\bar{x}, \bar{y}) = (0, \tfrac{4}{3})$.

5.4 PARAMETRIZATIONS OF SURFACES

Thus far, we have studied surfaces that arise as:

- ⊙ graphs $z = f(x, y)$ (or, analogously, $y = g(x, z)$ and $x = h(y, z)$) or
- ⊙ level sets $f(x, y, z) = c$.

We next describe surfaces by how they are traced out. This is like the way we described curves using parametrizations, except that now, since surfaces are two-dimensional, we need two parameters. In other words, surfaces will be described by functions $\sigma : D \to \mathbb{R}^n$, where $D \subset \mathbb{R}^2$, as illustrated in Figure 5.21. We want the domain of the parametrization to be two-dimensional, so we assume that D is an open set U in $\mathbb{R}^2$ together with all of its boundary points. The technical term is that D is the closure of U . Actually, the theory carries over if some of the boundary points are missing, but, in our examples, D will often be a rectangle or a closed disk in $\mathbb{R}^2$, in which case all boundary points are included.

The coordinates in D are the parameters. There are often natural choices for what to call them depending on the surface, though generically we call them s and t. For every (s, t) in D, $\sigma(s, t)$ is a point in $\mathbb{R}^n$, so we may write $\sigma(s, t) = (x_1(s, t), x_2(s, t), \ldots , x_n(s, t))$. We shall focus almost exclusively on surfaces in $\mathbb{R}^3$, in which case $\sigma(s, t) = (x(s, t), y(s, t), z(s, t))$.

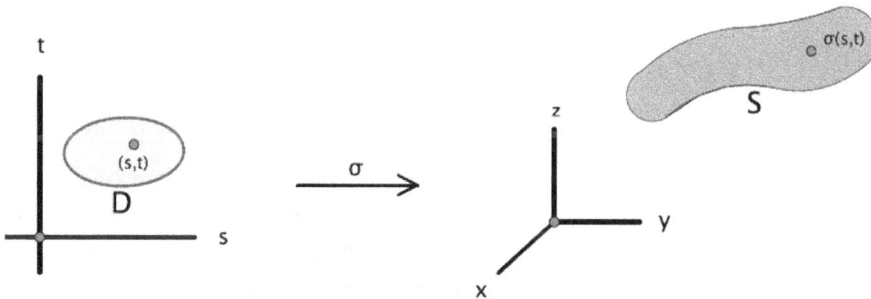

Figure 5.21: A parametrization σ of a surface S in $\mathbb{R}^3$

The reason for introducing the topic now is that this is a good time to talk about how to integrate real-valued functions over surfaces. Because surfaces are two-dimensional objects, in principle, these integrals should be extensions of the double integrals we are in the midst of studying. For example, the interpretations of limits of weighted sums that apply to integrals over regions in the plane should then carry over to integrals over surfaces as well.

In the remainder of this section, we look at several examples of parametrizations of surfaces and, in particular, some common geometric parameters. Integrals we leave for the next section.

Example 5.13. Consider the portion of the cone $z = \sqrt{x^2 + y^2}$ in R³ where $0 \le z \le 2$. See Figure 5.22. The surface is already described in terms of x and y, so we use (x, y) as parameters with $z = \sqrt{x^2 + y^2}$. That is, let:

$$\sigma(x, y) = (x, y, \sqrt{x^2 + y^2}).$$

The domain D of σ is the set of all (x, y) such that $0 \le \sqrt{x^2 + y^2} \le 2$, or x² + y² ≤ 4. This is a disk of radius 2 in the xy-plane. As (x, y) ranges over D, σ(x, y) sweeps out the given portion of the cone.

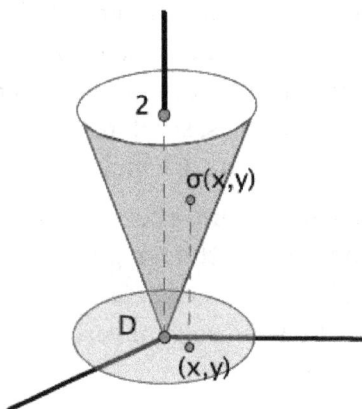

Figure 5.22: A parametrization of a cone using x and y as parameters

In this example, we used the x and y coordinates of the xyz-coordinate system as parameters for the surface. This is always an option for surfaces that are graphs z = f (x, y) of functions of two variables. We next introduce some other coordinate systems that can be used to locate points in the plane or in three-dimensional space. These coordinate systems provide the most convenient way to approach a variety of situations, though, for the time being, we just show how two of the coordinates are natural parameters for certain common surfaces.

5.4.1 Polar coordinates (r, θ) in $\mathbb{R}^2$

We referred to polar coordinates in Chapter 3, but perhaps it is best to introduce them formally. To plot a point in $\mathbb{R}^2$ with polar coordinates (r, θ), go out r units along the positive x-axis, then rotate counterclockwise about the origin by the angle θ (Figure 5.23). This brings you to the point (r cos θ, r sin θ) in rectangular coordinates. Thus the conversions from polar to rectangular coordinates are:

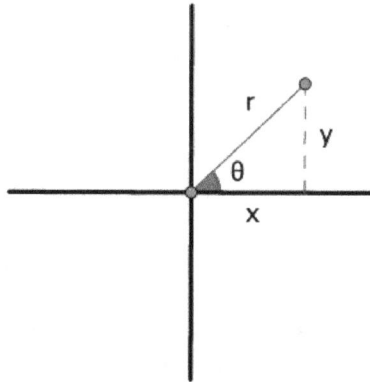

Figure 5.23: Polar coordinates (r, θ)

$$\begin{cases} x = r\cos\theta \\ y = r\sin\theta. \end{cases}$$

Also, by the Pythagorean theorem, $r = \sqrt{x^2 + y^2}$. . All of $\mathbb{R}^2$ can be described by choosing r and θ in the ranges r ≥ 0, 0 ≤ θ ≤ 2π. Actually, θ = 0 and θ = 2π represent the same points, along the positive x-axis, but it is convenient to allow this degree of redundancy.

For example, the disk D of radius 2 about the origin in Example 5.13 can be described in polar coordinates by the conditions 0 ≤ r ≤ 2, 0 ≤ θ ≤ 2π, that is, (r, θ) ∈ [0, 2] × [0, 2π]. We think of imposing polar coordinates on D as a transformation T : [0, 2] × [0, 2π] → $\mathbb{R}^2$ from the rθ-plane to the xy-plane, where T (r, θ) = (r cos θ, r sin θ). Horizontal segments 0 ≤ r ≤ 2, θ = constant, in the rθ-plane get mapped to radial segments emanating from the origin in the xy-plane, and vertical segments r = constant, 0 ≤ θ ≤ 2π, get mapped to circles centered at the origin. This is illustrated in Figure 5.24.

Example 5.14. We returns to the cone $z = \sqrt{x^2 + y^2}$, $0 \le z \le 2$, from Example 5.13. In polar coordinates, the equation of the cone is z = r, and combining this with the polar description of the disk D gives a parametrization σ of the cone with r and θ as parameters:

$$\tilde{\sigma}: [0, 2] \times [0, 2\pi] \to \mathbb{R}^3, \quad \tilde{\sigma}(r, \theta) = (r\cos\theta, r\sin\theta, r).$$

See Figure 5.25. The actual cone is the same one as before, but one advantage of this parametrization is that, for the purposes of integration, rectangles are nicer domains than disks.

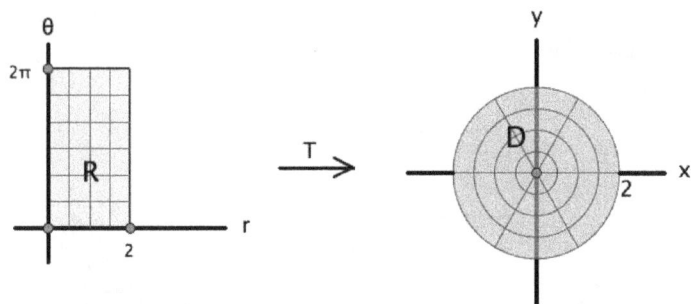

Figure 5.24: The polar coordinate transformation from the rθ-plane to the xy-plane

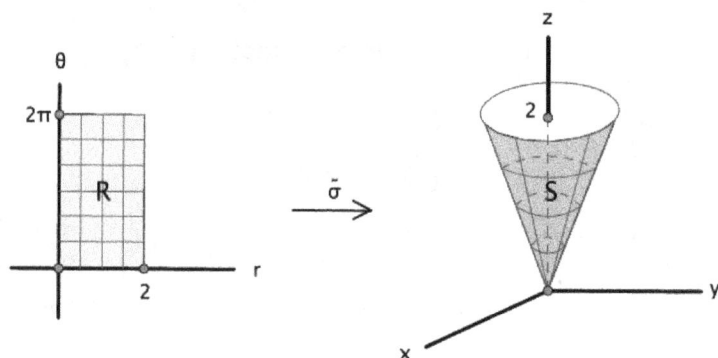

Figure 5.25: A parametrization of a cone using polar coordinates r and θ as parameters

5.4.2 Cylindrical coordinates (r, θ, z) in R³

Cylindrical coordinates are polar coordinates in the xy-plane together with the usual z-coordinate. To plot a point with cylindrical coordinates (r, θ, z), go out r units along the positive x-axis, rotate counterclockwise about the positive z-axis by the angle θ, and then go vertically z units. These coordinates are shown in Figure 5.26. This brings you to the point (r cos θ, r sin θ, z) in xyz-space.

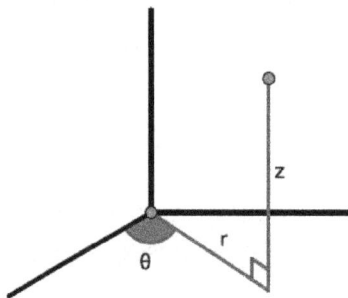

Figure 5.26: Cylindrical coordinates (r, θ, z)

Thus the conversions from cylindrical to rectangular coordinates are:

$$\begin{cases} x = r\cos\theta \\ y = r\sin\theta \\ z = z. \end{cases}$$

Again $r = \sqrt{x^2 + y^2}.$ All of R^3 is covered by choosing r ≥ 0, 0 ≤ θ ≤ 2π, -∞ < z < ∞.

Example 5.15 (Circular cylinders). We parametrize the circular cylinder of radius a and height h given by the conditions:

$x^2 + y^2 = a^2,$ 0 ≤ z ≤ h.

The axis of the cylinder is the z-axis.

In cylindrical coordinates, along the cylinder, r is fixed at the radius a, but θ and z can vary independently. The surface is traced out by setting r = a and letting θ and z range over 0 ≤ θ ≤ 2π, 0 ≤ z ≤ h. In other words, we can parametrize the cylinder using θ and z as parameters. With r = a fixed, the conversions from cylindrical to rectangular coordinates give the parametrization:

$$\sigma: [0, 2\pi] \times [0, h] \to \mathbb{R}^3, \quad \sigma(\theta, z) = (a\cos\theta, a\sin\theta, z).$$

See Figure 5.27.

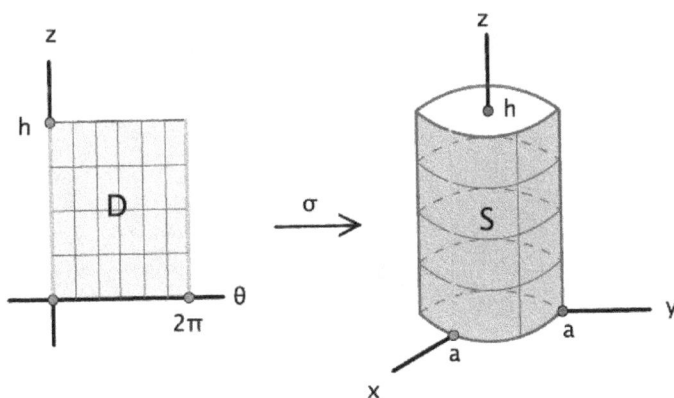

Figure 5.27: A parametrization of a cylinder using cylindrical coordinates θ and z as parameters

It is instructive to think about how σ transforms the rectangle D = [0, 2π] × [0, h] to become the cylinder S. In polar/cylindrical coordinates, θ = 0 and θ = 2π represent the same thing in xyz-space. Thus σ maps the points (0, z) and (2π, z) in the θz-parameter plane to the same point in R3. Apart from this, distinct points in D get mapped to distinct points in xyz-space. One way to construct a cylinder is to take a rectangle like D and glue together, or identify, points (0, z) on the left edge with points $(2\pi, z)$ on the right. It is clear geometrically that the result is a cylinder, and σ gives a formula that carries it out.

5.4.3 Spherical coordinates (ρ, ϕ, θ) in R³

The three spherical coordinates of a point in R³ are described geometrically as follows.

- ρ is the distance from the origin to the point.

- ϕ is the angle from the positive z-axis to the point. It is like an angle of latitude, except measured down from the positive z-axis rather than up or down from the equatorial plane.

- θ is the usual polar/cylindrical angle.

To plot a point in R3 with spherical coordinates (ρ, ϕ, θ), first go out ρ units along the positive z- axis. This brings you to the point $(0, 0, \rho)$ in rectangular coordinates. Then, rotate counterclockwise about the positive y-axis by the angle ϕ, bringing you to the point $(\rho \sin \phi, 0, \rho \cos \phi)$ in the xz- plane, a distance ρ from the origin and $r = \rho \sin \phi$ from the z-axis. Finally, rotate counterclockwise about the positive z-axis by the angle θ. This lands at the point $(\rho \sin \phi \cos \theta, \rho \sin \phi \sin \theta, \rho \cos \phi)$. The coordinates are labeled in Figure 5.28.

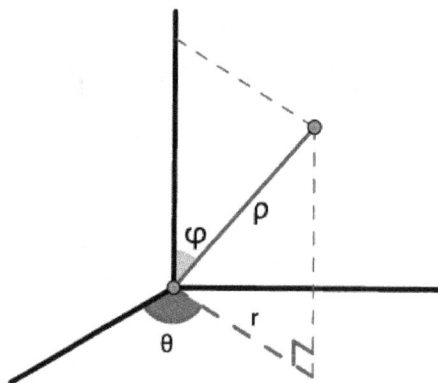

Figure 5.28: Spherical coordinates (ρ, ϕ, θ)

Thus the conversions from spherical to rectangular coordinates are:

$$\begin{cases} x = \rho \sin \phi \cos \theta \\ y = \rho \sin \phi \sin \theta \\ z = \rho \cos \phi. \end{cases}$$

From the Pythagorean theorem, the distance from the origin is $\rho = \sqrt{x^2 + y^2 + z^2}$. All of R³ is covered by spherical coordinates in the intervals $\rho \geq 0$, $0 \leq \phi \leq \pi$, $0 \leq$

$\theta \leq 2\pi$. (Note that the angular interval $0 \leq \phi \leq \pi$ ranges from the direction of the positive z-axis to the direction of the negative z-axis and the interval $0 \leq \theta \leq 2\pi$ sweeps all the way around the z-axis. This is why values of ϕ in the interval $\pi < \phi \leq 2\pi$ are not needed—they would duplicate points already covered.)

Example 5.16 (Spheres). We parametrize the sphere of radius a centered at the origin:

$$x^2 + y^2 + z^2 = a^2.$$

In spherical coordinates, the sphere is traced out by keeping the value of ρ fixed at the radius a, while ϕ and θ vary independently. More precisely, set $\rho = a$, and let ϕ and θ range over $0 \leq \phi \leq \pi, 0 \leq \theta \leq 2\pi$. Hence we use ϕ and θ as parameters, and, from the spherical to rectangular conversions with $\rho = a$ fixed, obtain the parametrization:

$$\sigma: [0, \pi] \times [0, 2\pi] \to \mathbb{R}^3, \quad \sigma(\phi, \theta) = (a \sin \phi \cos \theta, a \sin \phi \sin \theta, a \cos \phi).$$

This is illustrated in Figure 5.29.

Again the parametrization gives precise instructions for how to transform the rectangle $D = [0, \pi] \times [0, 2\pi]$ into a sphere. Here, all points where $\phi = 0$ get mapped to the north pole. These are the points along the left edge of D. Similarly, the right edge of D, where $\phi = \pi$, gets mapped to the south pole. Finally, points where $\theta = 0$ and $\theta = 2\pi$, that is, along the top and bottom edges, are identified in pairs along a meridian on the sphere in the xz-plane. Thus, to construct a sphere from a rectangle, collapse each of the left and right sides to points and zip together the remaining two sides to close up the surface.

Example 5.17. We returns one last time to the cone $z = \sqrt{x^2 + y^2}, 0 \leq z \leq 2$, which we have parametrized twice already, once using rectangular coordinates (x, y) and once using polar/cylindrical coordinates (r, θ) as parameters. It can be parametrized using spherical coordinates as well.

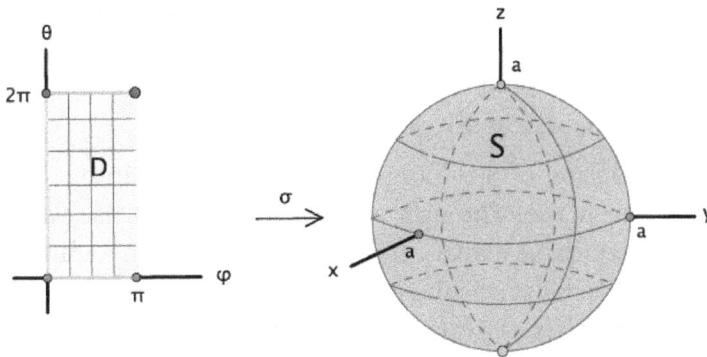

Figure 5.29: A parametrization of a sphere using spherical coordinates ϕ and θ as parameters

Here, along the cone, the angle ϕ down from the positive z-axis is fixed at $\frac{\pi}{4}$. To see this, note, for instance, that the cross-section with the yz-plane i$\sqrt{}$s z = |y|. On the other hand, ρ and θ can vary independently. For example, ρ ranges from 0 to $2\sqrt{2}$. (To determine the upper endpoint, one can either use trigonometry or solve for ρ using z = ρ cos ϕ with z = 2 and $\phi = \frac{\pi}{4}$.) Also, θ rotates all the way around from 0 to 2π. From the spherical conversions to rectangular coordinates, we obtain a parametrization:

$$\hat{\sigma}: [0, 2\sqrt{2}] \times [0, 2\pi] \to \mathbb{R}^3,$$

$$\hat{\sigma}(\rho, \theta) = \left(\rho \sin \frac{\pi}{4} \cos \theta, \rho \sin \frac{\pi}{4} \sin \theta, \rho \cos \frac{\pi}{4}\right) = \left(\frac{\sqrt{2}}{2}\rho \cos \theta, \frac{\sqrt{2}}{2}\rho \sin \theta, \frac{\sqrt{2}}{2}\rho\right).$$

We leave as an exercise the instructions that σ gives for turnsing a rectangle into a cone. See Figure 5.30.

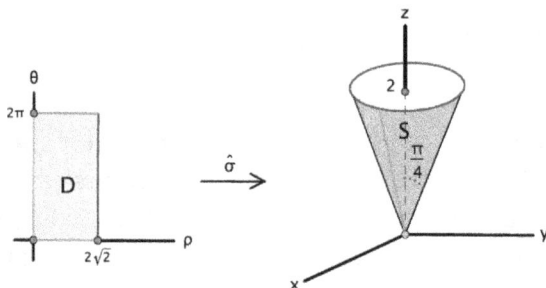

Figure 5.30: A parametrization of a cone using spherical coordinates ρ and θ as parameters

As the cone illustrates, it may be possible to parametrize a given surface in several different ways.

5.5 INTEGRALS WITH RESPECT TO SURFACE AREA

We now discuss how to integrate a real-valued function $f: S \to R$ defined on a surface S in $\mathbb{R}^3$. To formulate a reasonable definition, we follow the usual procedure for integrating: chop up S into a two-dimensional partition of small pieces of surface area ΔS_{ij}, choose a sample point p_{ij} in each piece, form a sum $\sum_{i,j} f(\mathbf{p}_{ij}) \triangle S_{ij}$, and take the limit as the size of the pieces goes to zero.

We use a parametrization of S to convert the calculation into a double integral over a region in the plane. The approach is similar to what we did when we defined integrals over curves with respect to arclength, though that was a long time ago (Section 2.3 to be precise). Thus let $\sigma : D \to \mathbb{R}^3$ be a parametrization of S, where $D \subset \mathbb{R}^2$. We write $\sigma(s, t) = (x(s, t), y(s, t), z(s, t))$. We subdivide D (or, really, a rectangle that contains D) into small subrectangles of dimensions Δs_i by Δt_j, and choose a sample point a_{ij} in each subrectangle. Intuitively, the expectation is that σ

transforms each subrectangle into a small "curvy quadrilateral" in S. This is one of the small pieces of S alluded to earlier. We denote its area by ΔS_{ij} and take $p_{ij} = \sigma(a_{ij})$ as the corresponding sample point. This is depicted in Figure 5.31.

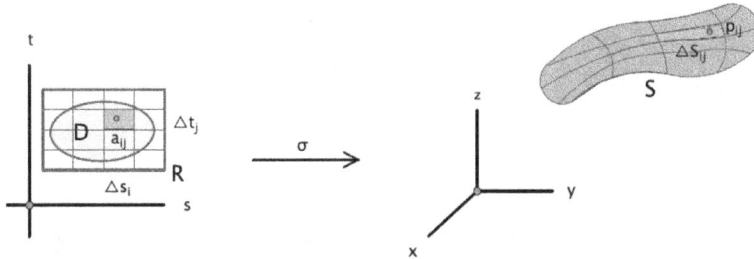

Figure 5.31: The integral with respect to surface area

Let $\frac{\partial \sigma}{\partial s}$ .and $\frac{\partial \sigma}{\partial t}$ be the vectors given by:

$$\frac{\partial \sigma}{\partial s} = \left(\frac{\partial x}{\partial s}, \frac{\partial y}{\partial s}, \frac{\partial z}{\partial s} \right) \quad \text{and} \quad \frac{\partial \sigma}{\partial t} = \left(\frac{\partial x}{\partial t}, \frac{\partial y}{\partial t}, \frac{\partial z}{\partial t} \right).$$

We think of these vectors as velocities with respect to s and t, respectively. Then σ transforms the Δs_i side of a typical subrectangle of D approximately to the vector $\partial \sigma \frac{\partial \sigma}{\partial s} \Delta s_i$ and the Δt_j side approximately to $\frac{\partial \sigma}{\partial t} \Delta t_j$. This gives an approximation of the area of the curvy quadrilaterals:

$\Delta S_{ij} \approx$ area of the parallelogram determined by $\frac{\partial \sigma}{\partial s} \Delta s_i$ and $\frac{\partial \sigma}{\partial t} \Delta t_j$

t (by an old property about the cross product and areas, page 36)

$\approx \left\| \frac{\partial \sigma}{\partial s} \Delta s_i \times \frac{\partial \sigma}{\partial t} \Delta t_j \right\|$ (by an old property about the cross product and areas, page 36)

$$\approx \left\| \frac{\partial \sigma}{\partial s} \times \frac{\partial \sigma}{\partial t} \right\| \Delta s_i \, \Delta t_j.$$

See Figure 5.32.

Hence $\sum_{i,j} f(p_{ij}) \Delta S_{ij} \approx \sum_{i,j} f(\sigma(a_{ij})) \left\| \frac{\partial \sigma}{\partial s} \times \frac{\partial \sigma}{\partial t} \right\| \Delta s_i \, \Delta t_j$. In the limit as Δs_i and Δt_j go to zero, this becomes the double integral $\iint_D f(\sigma(s,t)) \left\| \frac{\partial \sigma}{\partial s} \times \frac{\partial \sigma}{\partial t} \right\| ds \, dt$. With this as motivation, we adopt the following definition.

Definition. Let S be a surface in R^3, and let $f : S \to R$ be a conti n uous real-valued function.

Then the **integral of f with respect to surface area,** denoted by $\iint_S f \, dS$ is defined to be:

$$\iint_S f \, dS = \iint_D f(\sigma(s,t)) \left\| \frac{\partial \sigma}{\partial s} \times \frac{\partial \sigma}{\partial t} \right\| ds \, dt,$$

where $\sigma : D \to R^3$ is a parametrization of S.

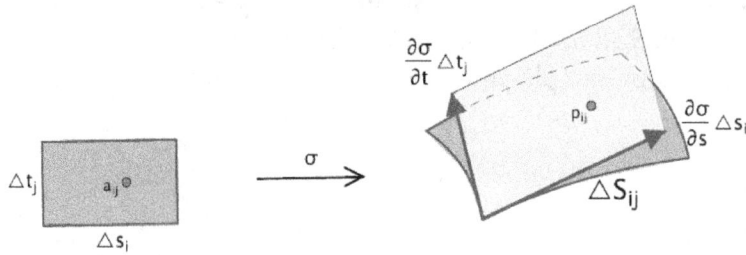

Figure 5.32: A blowup of the previous figure: σ sends a small Δs_i by Δt_j subrectangle to a curvy quadrilateral whose area ΔS_{ij} is approximately $\left\| \frac{\partial \sigma}{\partial s} \times \frac{\partial \sigma}{\partial t} \right\| \Delta s_i \, \Delta t_j$.

We should acknowledge a point that was also glossed over in connection with integrals with respect to arclength. Namely, we motivated the surface integral using Riemann sums $\sum_{i,j} f(\mathbf{p}_{ij}) \Delta S_{ij}$ based on the surface, but the actual definition relies on a parametrization σ. Perhaps a better notation for the integral as defined would be $\iint_\sigma f \, dS$. The issue is whether the choice of parametrization affects the value of the integral. Once we have a few more tools available, we shall see that there is no need for concern: using the definition of the integral with two different parametrizations of the same surface gives the same value. The effect of parametrizations on surface integrals is discussed at the end of Chapter 10.

Example 5.18. If $f(x, y, z) = 1$ for all (x, y, z) in S, then $\sum_{i,j} f(\mathbf{p}_{ij}) \Delta S_{ij} = \sum_{i,j} 1 \cdot \Delta S_{ij}$, the sum of the small pieces of surface area. This is the total area of S, that is:

$$\text{Surface area of } S = \iint_S 1 \, dS. \tag{5.3}$$

Example 5.19. Let S be the sphere $x^2 + y^2 + z^2 = a^2$ of radius a.

1. Find the surface area of S.
2. Find $\iint_S z^2 \, dS$.
3. Find $\iint_S (x^2 + y^2 + z^2) \, dS$.

We use the parametrization of a sphere derived in Example 5.16 with spherical coordinates (ϕ, θ) as parameters and $\rho = a$ fixed:

$$\sigma : [0, \pi] \times [0, 2\pi] \to \mathbb{R}^3, \quad \sigma(\phi, \theta) = (a \sin \phi \cos \theta, a \sin \phi \sin \theta, a \cos \phi).$$

Let D denote the domain of the parametrization, D = $[0, \pi] \times [0, 2\pi]$.

    a. Applying the previous example, we integrate the function $f = 1$. Then, using the definition of the integral:

$$\text{Area}(S) = \iint_S 1 \, dS = \iint_D 1 \cdot \left\| \frac{\partial \sigma}{\partial \phi} \times \frac{\partial \sigma}{\partial \theta} \right\| d\phi \, d\theta.$$

We calculate $\frac{\partial\sigma}{\partial\phi}$ and $\frac{\partial\sigma}{\partial\theta}$ and place them in the rows of the determinant for a cross product:

$$\frac{\partial\sigma}{\partial\phi}\times\frac{\partial\sigma}{\partial\theta}=\det\begin{bmatrix} \mathbf{i} & \mathbf{j} & \mathbf{k} \\ a\cos\phi\cos\theta & a\cos\phi\sin\theta & -a\sin\phi \\ -a\sin\phi\sin\theta & a\sin\phi\cos\theta & 0 \end{bmatrix}$$

$$=\left(a^2\sin^2\phi\cos\theta,\ a^2\sin^2\phi\sin\theta,\ a^2\cos\phi\sin\phi(\cos^2\theta+\sin^2\theta)\right)$$

$$=a^2\sin\phi\left(\sin\phi\cos\theta,\ \sin\phi\sin\theta,\ \cos\phi\right),$$

so:

$$\left\|\frac{\partial\sigma}{\partial\phi}\times\frac{\partial\sigma}{\partial\theta}\right\|=a^2\sin\phi\sqrt{\sin^2\phi\cos^2\theta+\sin^2\phi\sin^2\theta+\cos^2\phi}$$

$$=a^2\sin\phi\sqrt{\sin^2\phi+\cos^2\phi}$$

$$=a^2\sin\phi.$$

Hence:

$$\text{Area}\,(S)=\iint_D a^2\sin\phi\,d\phi\,d\theta$$

$$=\int_0^{2\pi}\left(\int_0^{\pi}a^2\sin\phi\,d\phi\right)d\theta$$

$$=\int_0^{2\pi}\left(-a^2\cos\phi\Big|_{\phi=0}^{\phi=\pi}\right)d\theta$$

$$=\int_0^{2\pi}\left(a^2-(-a^2)\right)d\theta$$

$$=\int_0^{2\pi}2a^2\,d\theta$$

$$=2a^2\theta\Big|_0^{2\pi}$$

$$=4\pi a^2.$$

This formula gets used enough that we put it in a box:

> The surface area of a sphere of radius a is 4πa².

b. Here, $f(x, y, z) = z^2$, and substituting for z in terms of the parameters gives $f(\sigma(\phi, \theta)) = a^2\cos^2\phi$. Using the result of (5.4) for $\left\|\frac{\partial\sigma}{\partial\phi}\times\frac{\partial\sigma}{\partial\theta}\right\|$, we obtain:

$$\iint_S z^2 \, dS = \iint_D f(\sigma(\phi, \theta)) \left\| \frac{\partial \sigma}{\partial \phi} \times \frac{\partial \sigma}{\partial \theta} \right\| d\phi \, d\theta$$

$$= \iint_D (a^2 \cos^2 \phi) \cdot (a^2 \sin \phi) \, d\phi \, d\theta$$

$$= a^4 \iint_D \cos^2 \phi \sin \phi \, d\phi \, d\theta$$

$$= a^4 \int_0^{2\pi} \left(\int_0^{\pi} \cos^2 \phi \sin \phi \, d\phi \right) d\theta \qquad (\text{let } u = \cos \phi, \, du = -\sin \phi \, d\phi)$$

$$= a^4 \int_0^{2\pi} \left(-\frac{1}{3} \cos^3 \phi \Big|_{\phi=0}^{\phi=\pi} \right) d\theta$$

$$= a^4 \int_0^{2\pi} (\frac{1}{3} - (-\frac{1}{3})) \, d\theta$$

$$= a^4 \int_0^{2\pi} \frac{2}{3} \, d\theta$$

$$= \frac{2}{3} a^4 \theta \Big|_0^{2\pi}$$

$$= \frac{4}{3} \pi a^4.$$

This time, $f(x, y, z) = x^2 + y^2 + z^2$. We use a trick! On the sphere S, $x^2 + y^2 + z^2 = a^2$, so $\iint_S (x^2 + y^2 + z^2) \, dS = \iint_S a^2 \, dS = a^2 \iint_S 1 \, dS = a^2$ Area (S). Using the formula for the area from part (a) then gives:

$$\iint_S (x^2 + y^2 + z^2) \, dS = a^2 \cdot 4\pi a^2 = 4\pi a^4.$$

The trick enabled us to evaluate the integral with almost no calculation. This is a highly desirable situation. In retrospect, we can get even more mileage out of it by going back and using the result of part (c) to redo $\iint_S z^2 \, dS$ in part (b). For the sphere S is symmetric in x, y, and z, whence $\iint_S x^2 \, dS = \iint_S y^2 \, dS = \iint_S z^2 \, dS$. Therefore $\iint_S (x^2 + y^2 + z^2) \, dS = 3 \iint_S z^2 \, dS$. or:

$$\iint_S z^2 \, dS = \frac{1}{3} \iint_S (x^2 + y^2 + z^2) \, dS = \frac{1}{3} \cdot 4\pi a^4 = \frac{4}{3} \pi a^4.$$

where the next-to-last step used part (c). Happily, this is the same as the answer we obtained in part (b) originally by grinding it out with a parametrization.

5.6 TRIPLE INTEGRALS AND BEYOND

Integrals of real-valued functions of three or more variables are conceptually the same as double integrals, though the notation is more encumbered. We merely outline the definition. Given:

⊙ a bounded subset W of $\mathbb{R}^n$ (that is, there is an n-dimensional box R = $[a_1, b_1]$ × $[a_2, b_2]$ × $\cdots$ × $[a_n, b_n]$ such that W ⊂ R) and

⊙ a bounded real-valued function $f : W \to R$ (that is, there is a scalar M such that $|f(x)| \le M$ for all x in W), one subdivides W into small n-dimensional subboxes of dimensions Δx_1 by Δx_2 by $\cdots$ by Δx_n, chooses a sample point p in each subbox, and considers sums of the form $\sum f(\mathbf{p}) \Delta x_1 \Delta x_2 \cdots \Delta x_n$, where the summation is over all the subboxes. The integral is the limit of these sums as the dimensions of the subboxes go to zero. It is denoted by:

$$\iint \cdots \int_W f \, dV \quad \text{or} \quad \iint \cdots \int_W f \, dx_1 \, dx_2 \cdots dx_n.$$

For instance, if $W \subset R^3$ and $f = f(x, y, z)$ is a function of three variables, then $\iiint_W f \, dV$ is called a **triple integral**.

The higher-dimensional integral has the same sort of weighted sum intepretations as the double integral, such as:

◉ n-dimensional volume: $\operatorname{Vol}(W) = \iint \cdots \int_W 1 \, dV$,

◉ average value: $\overline{f} = \frac{1}{\operatorname{Vol}(W)} \iint \cdots \int_W f \, dV$.

For example, if $W \subset R^3$, the average x-coordinate on W is $\overline{x} = \frac{1}{\operatorname{Vol}(W)} \iiint_W x \, dV$, similarly for $\overline{y}$ and $\overline{z}$, and the point $(\overline{x}, \overline{y}, \overline{z})$ is called the centroid of W.

To calculate the integral, one usually evaluates a sequence of n one-dimensional partial integrals. This can be complicated, because it may be difficult to determine the nested limits of integration that describe the domain W . This is true even in three dimensions, where, in principle, it is still possible to visualize what W looks like. We illustrate this by examining in detail a single example of a triple integral.

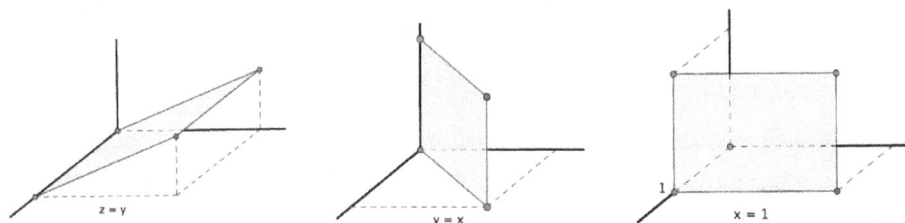

Figure 5.33: Three of the planes that bound W : z = y (left), y = x (middle), and x = 1 (right). The fourth is z = 0, the xy-plane.

Example 5.20. Find $\iiint_W (y + z) \, dV$ if W is the solid region in R^3 bounded by the planes z = y, y = x, x = 1, and z = 0. See Figure 5.33.

The planes in question slice through one another in such a way that the solid W that they bound is a tetrahedron. Each of its four triangular faces is contained in one of the planes. Looking at W towards the origin from the first octant, the z = y face is the top, z = 0 is the bottom, x = 1 is the front, and y = x is the back. This is shown on the left of Figure 5.34.

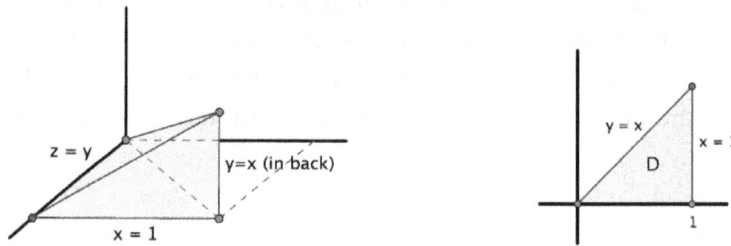

Figure 5.34: The region of integration W (left) and its projection D on the xy-plane (right)

To find the endpoints that describe W in an iterated integral, our approach is to focus on choosing which variable to integrate with respect to first. Once that variable is integrated out of the picture, what remains is a double integral, which by now is a much more familiar situation.

For instance, say we decide to integrate first with respect to z, treating x and y as constant. We need to know the values of x and y for which this process makes sense, that is, the pairs (x, y) for which there is at least one z such that (x, y, z) ∈ W . These pairs are precisely the projection of W on the xy-plane. In this example, this is the triangle D at the base of the tetrahedron, which is shown on the right of Figure 5.34. In the xy-plane, it is bounded by the x-axis and the lines x = 1 and y = x.

For each (x, y) in D, the values of z for which (x, y, z) ∈ W go from z = 0 to z = y. Thus:

$$\iiint_W (y+z)\,dV = \iint_D \left(\int_0^y (y+z)\,dz \right) dx\,dy.$$

The remaining limits of integration come from describing D as a double integral in the usual way, for instance, using the order of integration dy dx. After that, the integral can be evaluated one antidifferentiation at a time. Thus:

$$\iiint_W (y+z)\,dV = \int_0^1 \left(\int_0^x \left(\int_0^y (y+z)\,dz \right) dy \right) dx$$

$$= \int_0^1 \left(\int_0^x (yz + \tfrac{1}{2}z^2) \Big|_{z=0}^{z=y} dy \right) dx$$

$$= \int_0^1 \left(\int_0^x \tfrac{3}{2}y^2\,dy \right) dx$$

$$= \int_0^1 \left(\tfrac{1}{2}y^3 \Big|_{y=0}^{y=x} \right) dx$$

$$= \int_0^1 \tfrac{1}{2}x^3\,dx$$

$$= \tfrac{1}{8}x^4 \Big|_0^1$$

$$= \tfrac{1}{8}.$$

Example 5.21. Express the same integral $\iiint_W (y+z)\, dV$ from the previous example as an iterated integral in which:

       a. the first integration is with respect to x,
       b. the first integration is with respect to y.

To integrate first with respect to x, we treat y and z as constant. Proceeding as in the previous example, the relevant values of (y, z) are the points in the projection D of W on the yz-plane. As before, this is a triangle, though no longer one of the faces of the tetrahedron. The triangle is bounded in the yz-plane by the y-axis and the lines y = 1 and z = y. See Figure 5.35 on the left.

For each (y, z) in D, the point (x, y, z) is in W from x = y to x = 1. Hence:

$$\iiint_W (y+z)\, dV = \iint_D \left(\int_y^1 (y+z)\, dx \right) dy\, dz.$$

Describing the remaining double integral over D using the order of integration, say dz dy, gives:

$$\iiint_W (y+z)\, dV = \int_0^1 \left(\int_0^y \left(\int_y^1 (y+z)\, dx \right) dz \right) dy.$$

We won't carry out the iterated antidifferentiation, though we are optimistic that the final answer is again $\frac{1}{8}$.

This time, the integral will be a double integral of an integral with respect to y over the projection of W on the xz-plane . Again, the projection D is a triangle. Two sides of the triangle are contained in the x-axis and the line x = 1 in the xz-plane. The third side is the projection of the line where the planes z = y and y = x intersect. This line of intersection in R3 is described by z = y = x, and its projection on the xz-plane is z = x. This is the remaining side of D. See Figure 5.35 on the right.

For each (x, z) in D, y goes from y = z to y = x. Therefore:

$$\iiint_W (y+z)\, dV = \iint_D \left(\int_z^x (y+z)\, dy \right) dx\, dz.$$

Finally, integrating over D using the order dz dx results in:

$$\iiint_W (y+z)\, dV = \int_0^1 \left(\int_0^x \left(\int_z^x (y+z)\, dy \right) dz \right) dx.$$

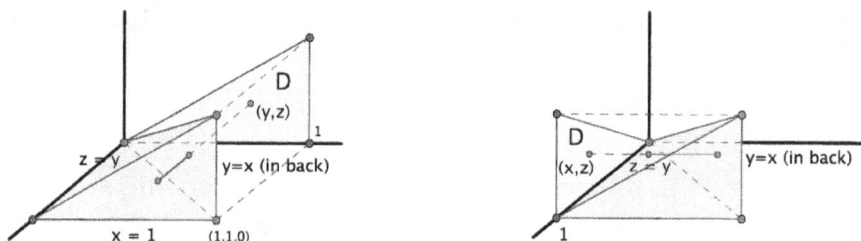

Figure 5.35: Integrating first with respect to x (left) or y (right)

5.7 EXERCISES FOR CHAPTER 5

Section 1 Volume and iterated integrals

In Exercises 1.1–1.4, evaluate the given iterated integral.

1.1. $\int_1^2 \left(\int_3^4 x^2 y \, dy \right) dx$

1.2. $\int_0^1 \left(\int_{x^2}^x (x + y) \, dy \right) dx$

1.3. $\int_0^{\pi/2} \left(\int_0^{\sin y} (x \cos y + 1) \, dx \right) dy$

1.4. $\int_1^2 \left(\int_{-y}^y \sin(3x + 2y) \, dx \right) dy$

1.5. Evaluate $\iint_D \frac{x^2}{y} \, dA$ if D is the rectangle described by $1 \le x \le 3$, $2 \le y \le 4$.

1.6. Evaluate $\iint_D (x + y) \, dA$ if D is the quarter-disk $x^2 + y^2 \le 1$, $x \ge 0$, $y \ge 0$.

1.7. Evaluate $\iint_D e^{x+y} \, dA$ if D is the triangular region with vertices $(0, 0)$, $(0, 2)$, $(1, 0)$.

1.8. Evaluate $\iint_D xy \, dA$ if D is the triangular region with vertices $(0, 0)$, $(1, 1)$, $(2, 0)$.

In Exercises 1.9–1.14, (a) sketch the domain of integration D in the xy-plane and (b) write an equivalent expression with the order of integration reversed.

1.9. $\int_1^2 \left(\int_3^4 f(x, y) \, dy \right) dx$

1.10. $\int_0^1 \left(\int_{e^x}^e f(x, y) \, dy \right) dx$

1.11. $\displaystyle\int_0^1 \left(\int_{y^2}^y f(x,y)\, dx \right) dy$

1.12. $\displaystyle\int_0^{\pi/2} \left(\int_0^{\sin y} f(x,y)\, dx \right) dy$

1.13. $\displaystyle\int_{-1}^1 \left(\int_0^{\sqrt{1-x^2}} f(x,y)\, dy \right) dx$

1.14. $\displaystyle\int_{-3}^6 \left(\int_{\frac{1}{3}y}^{-1+\sqrt{3+y}} f(x,y)\, dx \right) dy$

1.15. Consider the iterated integral $\displaystyle\int_{\frac{1}{4}}^1 \left(\int_{\frac{1}{x}}^4 y e^{xy}\, dy \right) dx.$

    a. Sketch the domain of integration D in the xy-plane.

    b. Write an equivalent expression with the order of integration reversed.

    c. Evaluate the integral using whatever method seems best.

1.16. Evaluate the integral $\displaystyle\int_0^6 \left(\int_{\frac{y}{2}}^3 \sin(x^2)\, dx \right) dy.$

1.17. Evaluate the integral $\displaystyle\iint_D \sqrt{1-x^2}\, dA,$ where D is the quarter-disk in the first quadrant described by $x^2 + y^2 \leq 1$, $x \geq 0$, and $y \geq 0$.

1.18. Find the volume of the wedge-shaped solid that lies above the xy-plane, below the plane $z = y$, and inside the cylinder $x^2 + y^2 = 1$.

1.19. Find the volume of the pyramid-shaped solid in the first octant bounded by the three coordinate planes and the planes $x + z = 1$ and $y + 2z = 2$.

1.20.   a. If c is a positive constant, find the volume of the tetrahedron in the first octant bounded by the plane $x + y + z = c$ and the three coordinate planes.

    b. Consider the tetrahedron W in the first octant bounded by the plane $x + y + z = 1$ and the three coordinate planes. Suppose that you want to divide W into three pieces of equal volume by slicing it with two planes parallel to $x + y + z = 1$, i.e., with planes of the form $x + y + z = c$. How should the slices be made?

1.21. Let D be the unit square $0 \leq x \leq 1$, $0 \leq y \leq 1$, and let $f: D \to R$ be the function given by $f(x,y) = \min \{x, y\}$. Find $\iint_D f(x,y)\, dx\, dy.$

Section 2 The double integral

Let $R = [0, 1] \times [0, 1]$, and let $f: R \to R$ be the function given by:

$$f(x,y) = \begin{cases} 1 & \text{if } (x,y) = (\tfrac{1}{2}, \tfrac{1}{2}), \\ 0 & \text{otherwise.} \end{cases}$$

    a. Let R be subdivided into a 3 by 3 grid of subrectangles of equal size (Figure 5.36). What are the possible values of Riemann sums based on this subdivision? (Different choices of sample points may give different values.)

b. Repeat for a 4 by 4 grid of subrectangles of equal size.

c. How about for an n by n grid of subrectangles of equal size?

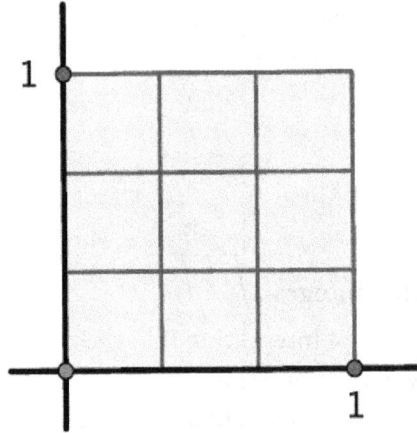

Figure 5.36: A 3 by 3 subdivision of R = [0, 1] × [0, 1]

a. Let δ be a positive real number. If R is subdivided into a grid of sub-rectangles of dimensions Δ_{xi} by Δ_{yj}, where $\Delta_{xi} < \delta$ and $\Delta_{yj} < \delta$ for all i, j, show that any Riemann sum based on the subdivision lies in the range $0 \leq \sum_{i,j} f(\mathbf{p}_{ij}) \Delta x_i \Delta y_j < 4\delta^2$.

a. Show that

$$\iint_R f(x,y)\,dA = 0, \text{ i.e., that } \lim_{\substack{\Delta x_i \to 0 \\ \Delta y_j \to 0}} \sum_{i,j} f(\mathbf{p}_{ij}) \Delta x_i \Delta y_j = 0.$$

Technically, this means you need to show that, given any $\epsilon > 0$, there exists a $\delta > 0$ such that, for any subdivision of R into subrectangles with $\Delta x_i < \delta$ and $\Delta y_j < \delta$ for all i, j, every Riemann sum based on the subdivision satisfies

$$\left| \sum_{i,j} f(\mathbf{p}_{ij}) \Delta x_i \Delta y_j - 0 \right| < \epsilon.$$

2.2. Let R = [0, 1] × [0, 1], and let $f : R \to R$ be the function given by:

$$f(x,y) = \begin{cases} 1 & \text{if } x = 0 \text{ or } y = 0, \\ 0 & \text{otherwise.} \end{cases}$$

a. Let R be subdivided into a 3 by 3 grid of subrectangles of equal size (Figure 5.36). What are the possible values of Riemann sums based on this subdivision?

b. Repeat for a 4 by 4 grid of subrectangles of equal size.

c. How about for an n by n grid of subrectangles of equal size?

d. Is f integrable on R? If so, what is the value of $\iint_R f(x,y)\,dA$?

2.3. Here is a result that will come in handy increasingly as we calculate more integrals. Let R = [a, b] × [c, d] be a rectangle, and let F : R → R be a real-valued function such that the variables x and y separate into two continuous factors, that is, F (x, y) = f (x)g(y), where f : [a, b] → R and g : [c, d] → R are continuous. Show that:

$$\iint_R f(x)g(y)\,dx\,dy = \left(\int_a^b f(x)\,dx\right)\left(\int_c^d g(y)\,dy\right).$$

2.4. Let f : U → R be a continuous function defined on an open set U in $\mathbb{R}^2$, and let a be a point of U .

If f (a) > 0, prove that there exists a rectangle R containing a such that:

$$\iint_R f\,dA > 0.$$

(Hint: Use Exercise 6.4 of Chapter 3.)

Let g : U → R be another continuous function on U . If f (a) > g(a), prove that there exists a rectangle R containing a such that:

$$\iint_R f\,dA > \iint_R g\,dA.$$

(Hint: Consider f - g.)

2.5. In this exercise, we use the double integral to give another proof of the equality of mixed partial derivatives. Let f (x, y) be a real-valued function defined on an open set U in $\mathbb{R}^2$ whose first and second-order partial derivatives are continuous on U .

Let R = [a, b] × [c, d] be a rectangle that is contained in U . Show that:

$$\iint_R \frac{\partial^2 f}{\partial x\,\partial y}(x,y)\,dA = \iint_R \frac{\partial^2 f}{\partial y\,\partial x}(x,y)\,dA.$$

(Hint: Use Fubini's theorem, with appropriate orders of integration, and the fundamental theorem of calculus to show that both sides are equal to f (b, d) - f (b, c) - f (a, d) + f (a, c).)

Show that $\frac{\partial^2 f}{\partial x\,\partial y}(\mathbf{a}) = \frac{\partial^2 f}{\partial y\,\partial x}(\mathbf{a})$ for all a in U . (Hint: Use Exercise 2.4 to show that something goes wrong if they are not equal.)

Section 3 Interpretations of the double integral

3.1. The population density of birds in a wildlife refuge decreases at a uniform rate with the distance from a river. If the river is modeled as the x-axis in the plane, then the density at the point (x, y) is given by f (x, y) = 5 - |y| hundred birds per square mile, where x and y are measured in miles. Find the total number of birds in the rectangle R = [-2, 2] × [0, 1].

3.2. A small cookie has the shape of the region in the first quadrant bounded by the curves y = x² and x = y², where x and y are measured in inches. Chocolate is poured unevenly on top of the cookie in such a way that the density of chocolate at the point (x, y) is given by $f(x, y) = 100(x + y)$ grams per square inch. Find the total mass of chocolate on the cookie.

3.3. The eye of a tornado is positioned directly over the origin in the plane. Suppose that the wind speed on the ground at the point (x, y) is given by v(x, y) = 30(x² + y²) miles per hour.

  a. Find the average wind speed on the square R = [0, 2] × [0, 2].
  b. Find all points (x, y) of R at which the wind speed equals the average.

3.4. Find the centroid of the half-disk of radius a in $\mathbb{R}^2$ described by x² + y² ≤ a², y ≥ 0.

3.5. Find the centroid of the triangular region in $\mathbb{R}^2$ with vertices (0, 0), (1, 2), and (1, 3).

3.6. Find the centroid of the triangular region in $\mathbb{R}^2$ with vertices (0, 0), (2, 2), and (1, 3).

3.7. Let D be the fudgsicle-shaped region of $\mathbb{R}^2$ that consists of the rectangle [-1, 1] × [0, h] of height h topped by a half-disk of radius 1, as shown in Figure 5.37. Find the value of h such that the point (0, h) is the centroid of D.

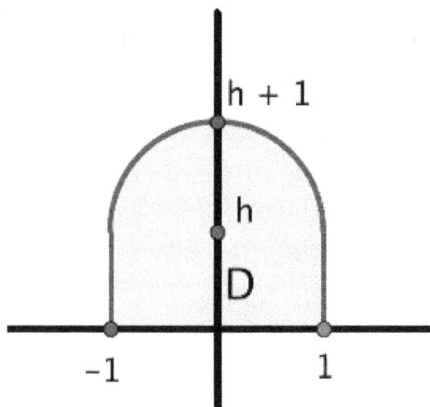

Figure 5.37: The fudgsicle

3.8. Let R be a rectangle in $\mathbb{R}^2$, and let $f: R \to R$ be an integrable function. If f is continuous, then it is true that there is a point of R at which f assumes its average value, that is, there exists a point c of R such that $f(c) = \bar{f}$. The proof is not hard, but it depends on subtle properties of the real numbers that are beyond what we cover here. The same conclusion need not hold, however, for discontinuous functions. Find an example of a rectangle R and an integrable function $f: R \to R$ such that there is no point c of R at which $f(c) = \bar{f}$.

Section 4 Parametrization of surfaces

4.1. Find a parametrization of the upper hemisphere of radius a described by $x^2 + y^2 + z^2 = a^2$, $z \geq 0$, using the following coordinates as parameters. Be sure to state what the domain of your parametrization is.

- a. the rectangular coordinates x, y
- b. the polar/cylindrical coordinates r, θ
- c. the spherical coordinates φ, θ

4.2. Let S be the surface described in cylindrical coordinates by z = θ, $0 \leq r \leq 1$, $0 \leq \theta \leq 4\pi$. It is called a **helicoid**. Converting to rectangular coordinates gives a parametrization:

$$\sigma: [0,1] \times [0,4\pi] \to \mathbb{R}^3, \qquad \sigma(r,\theta) = (r\cos\theta, r\sin\theta, \theta).$$

- a. Describe the curve that is traced out by σ if $r = \frac{1}{2}$ is held fixed and θ varies, that is, the curve parametrized by the path $\alpha(\theta) = \sigma(\frac{1}{2},\theta)$, $0 \leq \theta \leq 4\pi$.

- b. Similarly, describe the curve that is traced out by σ if $\theta = \frac{\pi}{4}$ is held fixed and r varies, that is, the curve parametrized by $\beta(r) = \sigma(r, \frac{\pi}{4})$, $0 \leq r \leq 1$.

- c. Describe S in a few words, and draw a sketch.

Section 5 Integrals with respect to surface area

5.1. Let S be the triangular surface in $\mathbb{R}^3$ whose vertices are (1, 0, 0), (0, 1, 0), and (0, 0, 1).

- a. Find a parametrization of S using x and y as parameters. What is the domain of your parametrization? (Hint: The triangle is contained in a plane.)
- b. Use your parametrization and formula (5.3) to find the area of S. Check that your answer agrees with what you would get using the formula: $\text{Area} = \frac{1}{2}$ (base)(height).
- c. Find $\iint_S 7\, dS$. (Hint: Use your answer to part (b).)

5.2. Let S be the cylindrical surface $x^2 + y^2 = 4$, $0 \leq z \leq 3$. Find the integral $\iint_S z^2\, dS$.

5.3. Consider the graph z = f (x, y) of a smooth real-valued function f defined on a bounded subset D of $\mathbb{R}^2$. Show that the surface area of the graph is given by the formula:

$$\text{Surface area} = \iint_D \sqrt{1 + \left(\frac{\partial f}{\partial x}\right)^2 + \left(\frac{\partial f}{\partial y}\right)^2}\, dx\, dy.$$

5.4. Let f : [a, b] → R be a smooth real-valued function of one variable, where a ≥ 0, and let S be the surface of revolution that is swept out when the curve z = f (x) in the xz-plane is rotated all the way around the z-axis. Show that the surface area of S is given by the formula:

$$\text{Surface area} \;=\; 2\pi \int_a^b x\sqrt{1 + f'(x)^2}\, dx.$$

(Hint: Parametrize S using cylindrical coordinates.)

5.5. Let W be the solid bounded on top by the plane z = y + 5, on the sides by the cylinder $x^2 + y^2 = 4$, and on the bottom by the plane z = 0.

  a. Sketch W .
  b. Let S be the surface that bounds W (top, sides, and bottom). Find the surface area of S.
  c. Find $\iint_S z\, dS$, where S is the surface in part (b).

5.6. Let S be the helicoid from Exercise 4.2, parametrized by:

$$\sigma \colon [0,1] \times [0,4\pi] \to \mathbb{R}^3, \qquad \sigma(r,\theta) = (r\cos\theta, r\sin\theta, \theta).$$

Find $\iint_S z\sqrt{x^2 + y^2}\, dS.$

5.7. Let W be the solid region of Example 5.2 that lies inside the cylinders $x^2 + z^2 = 1$ and $y^2 + z^2 = 1$ and above the xy-plane, and let S be the exposed part of the surface that bounds W , that is, the cylindrical surfaces but not the base.

  a. Find the surface area of S.
  b. Find $\iint_S z\, dS.$

5.8. The sphere $x^2 + y^2 + z^2 = a^2$ of radius a sits inscribed in the circular cylinder $x^2 + y^2 = a^2$. Given real numbers c and d, where $-a \le c < d \le a$, show that the portions of the sphere and the cylinder lying between the planes z = c and z = d have equal surface area. (Hint: Parametrize both surfaces using the cylindrical coordinates θ and z as parameters.)

Section 6 Triple integrals and beyond

In Exercises 6.1–6.2, evaluate the given iterated integral.

6.1. $\displaystyle \int_0^1 \left(\int_0^x \left(\int_y^x (x + y - z)\, dz \right) dy \right) dx$

6.2. $\displaystyle \int_0^1 \left(\int_x^{2x} \left(\int_0^{xy} x^3 y^2 z\, dz \right) dy \right) dx$

6.3. Find $\iiint_W xyz\, dV$ if W is the solid region in $\mathbb{R}^3$ below the surface $z = x^2 + y^2$ and above the square $0 \le x \le 1, 0 \le y \le 1$.

6.4. Find $\iiint_W (xz + y)\, dV$ if W is the solid region in $\mathbb{R}^3$ above the xy-plane bounded by the surface y = x2 and the planes z = y, y = 1, and z = 0.

6.5. Find $\iiint_W x\,dV$ if W is the region in the first octant bounded by the saddle surface z = x²-y² and the planes x = 2, y = 0, and z = 0.

6.6. Find $\iiint_W (1+x+y+z)^2\,dV$ if W is the solid tetrahedron in the first octant bounded by the plane x + y + z = 1 and the three coordinate planes.

6.7. Let a be a positive real number, and let W = [0, a] [0, a] [0, a] in R³. Find the average value of f (x, y, z) = x² + y² + z² on W .

For each of the triple integrals in Exercises 6.8–6.11, (a) sketch the domain of integration in R³, (b) write an equivalent expression in which the first integration is with respect to x, and (c) write an equivalent expression in which the first integration is with respect to y.

6.8. $\displaystyle\int_{-2}^{2}\left(\int_{0}^{\sqrt{2-\frac{x^2}{2}}}\left(\int_{0}^{\sqrt{1-\frac{x^2}{4}-\frac{y^2}{2}}} f(x,y,z)\,dz\right)dy\right)dx$

6.9. $\displaystyle\int_{-1}^{1}\left(\int_{0}^{\sqrt{1-x^2}}\left(\int_{0}^{1} f(x,y,z)\,dz\right)dy\right)dx$

6.10. $\displaystyle\int_{0}^{1}\left(\int_{x}^{1}\left(\int_{0}^{x} f(x,y,z)\,dz\right)dy\right)dx$

6.11. $\displaystyle\int_{0}^{1}\left(\int_{x}^{1}\left(\int_{0}^{y} f(x,y,z)\,dz\right)dy\right)dx$

PART-IV
VECTOR-VALUED FUNCTIONS

6 Chapter

DIFFERENTIABILITY AND THE CHAIN RULE

So far, we have studied functions of which at least one of the domain or codomain is a subset of R, in other words:

- ◉ vector-valued functions of one variable, also known as paths, $\alpha : I \to \mathbb{R}^n$, where $I \subset R$, or

- ◉ real-valued functions of n variables $f : U \to R$, where $U \subset \mathbb{R}^n$.

Now, we consider the general case of vector-valued functions of n variables, that is, functions of the form $f : U \to Rm$, where $U \subset \mathbb{R}^n$ and both n and m are allowed to be greater than 1.

Given such a vector-valued function, we can leverage what we know about real-valued functions because, if $x \in U$, then, as an element of Rm, f (x) has m coordinates:

$$f(\mathbf{x}) = (f_1(\mathbf{x}), f_2(\mathbf{x}), \ldots, f_m(\mathbf{x})).$$

where each $f_i(x)$ is a real number. In other words, each component is a real-valued function $f_i : U \to R$. Thus a vector-valued function is also a sequence of real-valued functions. We use the notation above for the components consistently from now on. For example, if $f : \mathbb{R}^2 \to \mathbb{R}^2$ is the polar coordinate transformation $f(r, \theta) = (r \cos \theta, r \sin \theta)$, then $f_1(r, \theta) = r \cos \theta$ and $f_2(r, \theta) = r \sin \theta$. Also, as with paths, we denote f in plainface type, reserving boldface for a special kind of vector-valued function that we study beginning in Chapter 8.

We treated the case of real-valued functions fairly rigorously, and we shall see that much of the theory carries over to vector-valued functions without incident. On the other hand, we discussed paths back in Chapter 2 more informally, so what we are about to do here in the vector-valued case can be taken as establishing the theory behind what we did then.

6.1 CONTINUITY REVISITED

The motivation and definition for continuity of vector-valued functions carry over almost verbatim from the real-valued case.

Definition. Let U be an open set in $\mathbb{R}^n$, and let $f: U \to Rm$ be a function. We say that f is **continuous at a point** a of U if, given any open ball $B(f(a), \epsilon)$ about f (a), there exists an open ball $B(a, \delta)$ about a such that:

$$f(B(\mathbf{a}, \delta)) \subset B(f(\mathbf{a}), \epsilon).$$

In other words, if $x \in B(a, \delta)$, then f (x) $\in$ B(f (a), ϵ). See Figure 6.1. Equivalently, given any $\epsilon > 0$, there exists a $\delta > 0$ such that:

$$\text{if } \|\mathbf{x} - \mathbf{a}\| < \delta, \text{ then } \|f(\mathbf{x}) - f(\mathbf{a})\| < \epsilon.$$

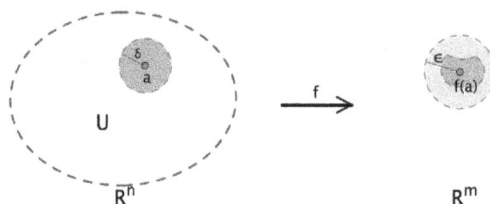

Figure 6.1: Continuity of a vector-valued function: given any ball B(f (a), ϵ) about f (a), there is a ball B(a, δ) about a such that f (B(a, δ)) $\subset$ B(f (a), ϵ).

We say that f is a **continuous function** if it is continuous at every point of its domain.

Proposition 6.1. If $f(x) = (f_1(x), f_2(x), \ldots, f_m(x))$, then f is continuous at a if and only if $f_1, f_2, \ldots, f_m$ are all continuous at a.

Proof. Intuitively, this is because $f(x) - f(a)$ can be made small if and only if all of its components fi(x)-fi(a) can be made small, but this needs to be made more precise. There are two implications that need to be proven here, that each condition implies the other. We give the details for one of the implications and leave the other for the exercises.

Namely, we show that, if $f_1, f_2, \ldots, f_m$ are continuous at a, then so is f. Let $\epsilon > 0$ be given. We want to ensure that $\|$f (x) - f (a)$\| < \epsilon$. Note that:

$$\|f(\mathbf{x}) - f(\mathbf{a})\| = \sqrt{(f_1(\mathbf{x}) - f_1(\mathbf{a}))^2 + (f_2(\mathbf{x}) - f_2(\mathbf{a}))^2 + \cdots + (f_m(\mathbf{x}) - f_m(\mathbf{a}))^2}.$$

If we could arrange that $|f_i(\mathbf{x}) - f_i(\mathbf{a})| < \frac{\epsilon}{\sqrt{m}}$ for i = 1, 2, ..., m, then $\|$f (x) - $f(a)\| < \sqrt{\frac{\epsilon^2}{m} + \frac{\epsilon^2}{m} + \cdots + \frac{\epsilon^2}{m}} = \sqrt{\epsilon^2} = \epsilon$ would follow.

This we can do. For taking $\sqrt{\epsilon}$ as the given input in the definition of continuity, the fact that f_i is continuous at a for i = 1, 2, ..., m means that there exists a $\delta i > 0$ such that $|f_i(\mathbf{x}) - f_i(\mathbf{a})| < \frac{\epsilon}{\sqrt{m}}$ whenever $\|$x - a$\| < \delta_i$. Let $\delta = \min\{\delta_1, \delta_2, \ldots, \delta_m\}$. If $\|$x - a$\| < \delta$, it remains true that $|f_i(\mathbf{x}) - f_i(\mathbf{a})| < \frac{\epsilon}{\sqrt{m}}$ for all i, and hence, as noted above, $\|$f (x) - f (a)$\| < \epsilon$. We have found a δ that satisfies the definition of continuity for f.

The proof of the converse is Exercise 1.1.

The proposition allows us to determine if f is continuous by looking at its real-valued component functions, which is something we have studied before. For instance, for the function $f(r, \theta) = (r \cos \theta, r \sin \theta)$, both component functions $f_1(r, \theta) = r \cos \theta$ and $f_2(r, \theta) = r \sin \theta$ are continuous as real-valued algebraic combinations and compositions of continuous pieces. Therefore f is continuous as well.

Similarly, the definition of limit for vector-valued functions carries over from the real-valued case essentially unchanged.

Definition. Let U be an open set in $\mathbb{R}^n$, and let a be a point of U . If $f: U - \{a\} \to Rm$ is a vector-valued function defined on U , except possibly at the point a, then we say that $\lim_{x \to a} f(x)$ exists if there is a vector L in Rm such that the function $\tilde{f}: U \to \mathbb{R}^m$ defined by

$$\tilde{f}(x) = \begin{cases} f(x) & \text{if } x \neq a, \\ L & \text{if } x = a \end{cases}$$

is continuous at a. When this happens, we write $\lim_{x \to a} f(x) = L$.

As before, it is immediate that $f: U \to Rm$ is continuous at a if and only if $\lim_{x \to a} f(x) = f(a)$. One consequence of this is that Proposition 6.1 then implies that a vector-valued limit can also be viewed as a sequence of real-valued limits.

Proposition 6.2. If $f(x) = (f_1(x), f_2(x), \ldots, f_m(x))$ and $L = (L_1, L_2, , L_m)$, then $\lim_{x \to a} f(x) = L$ if and only if $\lim_{x \to a} fi(x) = Li$ for all $i = 1, 2, , m$.

6.2 DIFFERENTIABILITY REVISITED

Recall that, in the real-valued case, $f : U \to R$ is differentiable at a if it has a first-order approximation $f(x) \approx \ell(x) = f(a) + \nabla f(a) \cdot (x - a)$ such that $\lim_{x \to a} \frac{f(x) - \ell(x)}{\|x - a\|} = 0$. For a vector-valued function $f: U \to Rm$, we find an approximation, again denoted by ℓ, by compiling the approximations of the component functions $f_1, f_2, \ldots, f_m$ into one vector:

$$\ell(x) = \begin{bmatrix} \ell_1(x) \\ \ell_2(x) \\ \vdots \\ \ell_m(a) \end{bmatrix} = \begin{bmatrix} f_1(a) + \nabla f_1(a) \cdot (x - a) \\ f_2(a) + \nabla f_2(a) \cdot (x - a) \\ \vdots \\ f_m(a) + \nabla f_m(a) \cdot (x - a) \end{bmatrix}$$

$$= \begin{bmatrix} f_1(a) \\ f_2(a) \\ \vdots \\ f_m(a) \end{bmatrix} + \begin{bmatrix} \nabla f_1(a) \cdot (x - a) \\ \nabla f_2(a) \cdot (x - a) \\ \vdots \\ \nabla f_m(a) \cdot (x - a) \end{bmatrix}$$

$$= f(a) + \begin{bmatrix} \nabla f_1(a) \\ \nabla f_2(a) \\ \vdots \\ \nabla f_m(a) \end{bmatrix} \cdot (x - a)$$

$$= f(a) + Df(a) \cdot (x - a).$$

where, in the last two lines,

$$Df(\mathbf{a}) = \begin{bmatrix} \nabla f_1(\mathbf{a}) \\ \nabla f_2(\mathbf{a}) \\ \vdots \\ \nabla f_m(\mathbf{a}) \end{bmatrix}$$

is the matrix whose rows are the gradients of the component functions and x - a is a column vector.

We consider the error $f(\mathbf{x}) - \ell(\mathbf{x}) = f(\mathbf{x}) - f(\mathbf{a}) - Df(\mathbf{a}) \cdot (\mathbf{x} - \mathbf{a})$ in using the approximation and proceed as in the real-valued case.

Definition. Let U be an open set in $\mathbb{R}^n$, and let $f: U \to \mathbb{R}m$ be a function. If $a \in U$, consider the matrix:

$$Df(\mathbf{a}) = \begin{bmatrix} \frac{\partial f_1}{\partial x_1}(\mathbf{a}) & \cdots & \frac{\partial f_1}{\partial x_n}(\mathbf{a}) \\ \vdots & \ddots & \vdots \\ \frac{\partial f_m}{\partial x_1}(\mathbf{a}) & \cdots & \frac{\partial f_m}{\partial x_n}(\mathbf{a}) \end{bmatrix}.$$

We say that f is differentiable at a if:

$$\lim_{\mathbf{x} \to \mathbf{a}} \frac{f(\mathbf{x}) - f(\mathbf{a}) - Df(\mathbf{a}) \cdot (\mathbf{x} - \mathbf{a})}{\|\mathbf{x} - \mathbf{a}\|} = 0. \tag{6.1}$$

When this happens, $Df(\mathbf{a})$ is called the **derivative**, or **Jacobian matrix**, of f at a.

To repeat a point made in the real-valued case, the derivative is not a single number, but rather a matrix or, even better, the linear part of a good affine approximation of f near a.

We begin by calculating a few simple examples of $Df(\mathbf{a})$. In general, it is an m by n matrix whose entries are various partial derivatives. The ith row is the gradient ∇f_i, and the jth column is the "velocity" $\frac{\partial}{\partial x_j}$ of f with respect to x_j.

Example 6.3. Let $f: \mathbb{R}^2 \to \mathbb{R}^2$ be $f(r, \theta) = (r \cos \theta, r \sin \theta)$. Then:

$$Df(r, \theta) = \begin{bmatrix} \cos \theta & -r \sin \theta \\ \sin \theta & r \cos \theta \end{bmatrix}.$$

The formula above gives what the matrix looks like at a general point a = (r, θ). At a specific point, such as $a = (2, \frac{\pi}{4})$, we have

$$Df(\mathbf{a}) = Df(2, \tfrac{\pi}{4}) = \begin{bmatrix} \frac{\sqrt{2}}{2} & -2 \cdot \frac{\sqrt{2}}{2} \\ \frac{\sqrt{2}}{2} & 2 \cdot \frac{\sqrt{2}}{2} \end{bmatrix} = \begin{bmatrix} \frac{\sqrt{2}}{2} & -\sqrt{2} \\ \frac{\sqrt{2}}{2} & \sqrt{2} \end{bmatrix}.$$

Example 6.4. Let $g: \mathbb{R}^3 \to \mathbb{R}^2$ be the projection of xyz-space onto the xy-plane, g(x, y, z) = (x, y). Then:

$$Dg(x, y, z) = \begin{bmatrix} 1 & 0 & 0 \\ 0 & 1 & 0 \end{bmatrix}.$$

In other words, $Dg(\mathbf{a}) = \begin{bmatrix} 1 & 0 & 0 \\ 0 & 1 & 0 \end{bmatrix}$ for all a in 3.

Example 6.5. Let h : $\mathbb{R}^2 \to \mathbb{R}^3$ be h(x, y) = $(x^2 + y^3, x^4y^5, e^{6x+7y})$. Then, if a = (x, y):

$$Dh(\mathbf{a}) = Dh(x, y) = \begin{bmatrix} 2x & 3y^2 \\ 4x^3y^5 & 5x^4y^4 \\ 6e^{6x+7y} & 7e^{6x+7y} \end{bmatrix}.$$

To determine whether a function is differentiable without going through the definition, we may use criteria similar to those in the real-valued case. For instance, the same reasoning as before shows that a differentiable $\frac{\partial f_i}{\partial x_j}$ tion must be continuous and that it is necessary for all the partial derivatives to exist in order for f to be differentiable.

In addition, the condition for differentiability in the definition (6.1) requires that a certain limit be the zero vector. This can happen if and only if the limit of each of the components is zero, too. This follows from Proposition 6.2. In other words, we have the following componentwise criterion for differentiability.

Proposition 6.6. If $f(x) = (f_1(x), f_2(x), \ldots, f_m(x))$, then f is differentiable at a if and only if $f_1, f_2, \ldots, f_m$ are all differentiable at a.

Example 6.7. Let I be an open interval in R, and let $\alpha : I \to \mathbb{R}^n$ be a path in $\mathbb{R}^n$, where $\alpha(t) = (x_1(t), x_2(t), \ldots, x_n(t))$. By the last proposition, α is differentiable if and only if each of its components $x_1, x_2, \ldots, x_n$ is differentiable. The latter condition is essentially how we defined differentiability of paths in Chapter 2, so what we did there is supported by the theory we now have in place. Moreover, by definition of the derivative matrix:

$$D\alpha(t) = \begin{bmatrix} x_1'(t) \\ x_2'(t) \\ \vdots \\ x_n'(t) \end{bmatrix}.$$

This agrees with $\alpha'(t)$, the velocity, under the usual identification of a vector in $\mathbb{R}^n$ with a column matrix.

According to Proposition 6.6, we may apply the C^1 test for real-valued functions to each of the components of a vector-valued function to obtain the following criterion.

Theorem 6.8 (The C^1 test). If all the partial derivatives $\frac{\partial f_i}{\partial x_j}$ (i.e., the entries of Df (x)) exist and are continuous on U, then f is differentiable at every point of U.

To illustrate this, in each of Examples 6.3–6.5 above, all entries of the D matrices are continuous, so those three functions f, g, h are differentiable at every point of their domains.

A function f is called smooth if each of its component functions $f_1, f_2, \ldots, f_m$ is smooth in the sense we have discussed previously (see Section 4.9), that is, they have continuous partial derivatives of all orders. As before, we generally state our results about differentiability for smooth functions without worrying about whether this is a stronger assumption than necessary.

6.3 THE CHAIN RULE: A CONCEPTUAL APPROACH

We now discuss the all-important rule for the derivative of a composition. We first take a fairly high-level approach to understand where the rule comes from.

Let U and V be open sets in $\mathbb{R}^n$ and Rm, respectively, and let $f: U \to V$ and $g: V \to R^p$ be functions. The composition $g \circ f: U \to R^p$ is then defined, and by definition, if a ∈ U , the composition is differentiable at a if there is a good first-order approximation of the form:

$$g(f(\mathbf{x})) \approx g(f(\mathbf{a})) + \boxed{D(g \circ f)(\mathbf{a})} \cdot (\mathbf{x} - \mathbf{a}) \tag{6.2}$$

when x is near a. The challenge is to figure out the expression that goes into the box for D(g°f)(a).

The idea is that the first-order approximation of the composition g°f should be the composition of the first-order approximations of g and f individually. Let b = f (a), and assume that f is differentiable at a and g is differentable at b. Then there are good first-order approximations:

- for $f: f(x) \approx \ell f(x)$
  $\approx f(a) + Df(a) \cdot (x - a)$ when x is near a, (6.3)
- for $g: g(y) \approx \ell g(y)$
  $\approx g(b) + Dg(b) \cdot (y - b)$ when y is near b
  $\approx g(f(a)) + Dg(f(a)) \cdot (y - f(a)).$ (6.4)

Note that, by continuity, when x is near a, f (x) is near f (a) = b, so substituting y = f (x) in (6.4) gives:

$$g(f(\mathbf{x})) \approx g(f(\mathbf{a})) + Dg(f(\mathbf{a})) \cdot (f(\mathbf{x}) - f(\mathbf{a})).$$

Then, using (6.3) to substitute f (x) - f (a) ≈ Df (a) · (x - a) yields:

$$g(f(\mathbf{x})) \approx g(f(\mathbf{a})) + \boxed{Dg(f(\mathbf{a})) \cdot Df(\mathbf{a})} \cdot (\mathbf{x} - \mathbf{a}).$$

If you like, you can check that the right side of this last approximation is also the composition $\ell g(\ell f(x))$ of the first-order approximations of g and f. In any case, the upshot is that we have identified something that fits naturally in the box in equation (6.2).

Theorem 6.9 (Chain rule). Let U and V be open sets in $\mathbb{R}^n$ and Rm, respectively, and let $f: U \to V$ and $g: V \to R^p$ be functions. Let a be a point of U. If f is differentiable at a and g is differentiable at f (a), then $g \circ f$ is differentiable at a and:

$$D(g \circ f)(a) = Dg(f(a)) \cdot Df(a),$$

where the product on the right is matrix multiplication.

In terms of the sizes of the various parties involved, the left side of the chain rule is a p by n matrix, while the right side is a product (p by m) × (m by n). This need not be memorized. In practice, it usually takes care of itself.

The chain rule says that the derivative of a composition is the product of the derivatives, or, to be even more sophisticated, the composition of the linear parts of the associated first-order approximations. Though the objects involved are more complicated, this is actually the same as what happens back in first-year calculus. There, if y is a function of u, say y = g(u), and u is a function of x, say u = f (x), then y becomes a function of x via composition: y = g(f (x)). The one-variable chain rule says:

$$\frac{dy}{dx} = \frac{dy}{du}\frac{du}{dx}.$$

Again, the derivative of a composition is the product of the derivatives of the composed steps.

The line of reasoning we used to obtain the chain rule is natural and reasonably straightforward, but a proper proof requires greater attention to the various approximations involved. The letter ϵ is likely to appear. The arguments are a little like those needed for a correct proof of the Little Chain Rule back in Exercise 5.6 of Chapter 4, though the details here are more complicated. We shall say more about the proof of the chain rule in the next section.

The following example is meant to illustrate how the pieces of the chain rule fit together.

Example 6.10. Let $f: \mathbb{R}^2 \to \mathbb{R}^2$ be given by f (r, θ) = (r cos θ, r sin θ) and g : $\mathbb{R}^2 \to R^3$ by g(x, y) = (x² + y², 2x + 3y, xy²). Find $D(g \circ f)(2, \frac{\pi}{2})$.

By the chain rule,

$$D(g \circ f)(2, \frac{\pi}{2}) = Dg\big(f(2, \frac{\pi}{2})\big) \cdot Df(2, \frac{\pi}{2}) = Dg(0, 2) \cdot Df(2, \frac{\pi}{2}),$$

since $f(2, \frac{\pi}{2}) = (2\cos\frac{\pi}{2}, 2\sin\frac{\pi}{2}) = (0, 2)$. Calculating the necessary derivatives is straightforward:

$$Dg(x, y) = \begin{bmatrix} 2x & 2y \\ 2 & 3 \\ y^2 & 2xy \end{bmatrix} \Rightarrow Dg(0, 2) = \begin{bmatrix} 0 & 4 \\ 2 & 3 \\ 4 & 0 \end{bmatrix},$$

$$Df(r, \theta) = \begin{bmatrix} \cos\theta & -r\sin\theta \\ \sin\theta & r\cos\theta \end{bmatrix} \Rightarrow Df(2, \frac{\pi}{2}) = \begin{bmatrix} 0 & -2 \\ 1 & 0 \end{bmatrix}.$$

Therefore:

$$D(g \circ f)(2, \frac{\pi}{2}) = \begin{bmatrix} 0 & 4 \\ 2 & 3 \\ 4 & 0 \end{bmatrix} \begin{bmatrix} 0 & -2 \\ 1 & 0 \end{bmatrix} = \begin{bmatrix} 4 & 0 \\ 3 & -4 \\ 0 & -8 \end{bmatrix}.$$

6.4 THE CHAIN RULE: A COMPUTATIONAL APPROACH

The chain rule states that the derivative of a composition is the product of the derivatives:

$$D(g \circ f)(\mathbf{a}) = Dg(f(\mathbf{a})) \cdot Df(\mathbf{a}).$$

This is remarkably concise and elegant. Nevertheless, under the hood, there is a lot going on. Each entry of a matrix product involves considerable information: the (i, j)th entry of the product is the dot product of the ith row and jth column:

$$\begin{bmatrix} \cdots & \boxed{(i,j)} & \cdots \end{bmatrix} = \begin{bmatrix} * & * & \cdots & * & * \end{bmatrix} \begin{bmatrix} * \\ * \\ \vdots \\ * \\ * \end{bmatrix}.$$

There is one such dot product for each entry of D(g ° f)(a). We shall write out these entries in a few cases in a moment and discover that we have seen this type of product before. You can think of the discovery either as an alternative way to derive the chain rule or as confirmation of the consistency of the theory.

We take a simple situation that we have studied previously and then tweak it a couple of times.

Example 6.11. Let $\alpha : R \to R^3$ be a smooth path in R^3, $\alpha(t) = (x(t), y(t), z(t))$, and let g: $R^3 \to R$ be a smooth real-valued function of three variables, w = g(x, y, z). Then w becomes a function of t by composition: w = g(α(t)) = g(x(t), y(t), z(t)). As a matter of notation, from here on, we use w to denote the value of the composition, reserving g to stand for the original function of x, y, z.

The composition is a real-valued function of one variable, so the Little Chain Rule applies. It says that the derivative of the compositon is "gradient dot velocity":

$$\frac{dw}{dt} = \nabla g(\alpha(t)) \cdot \alpha'(t) = \left(\frac{\partial g}{\partial x}, \frac{\partial g}{\partial y}, \frac{\partial g}{\partial z}\right) \cdot \left(\frac{dx}{dt}, \frac{dy}{dt}, \frac{dz}{dt}\right)$$

$$= \frac{\partial g}{\partial x}\frac{dx}{dt} + \frac{\partial g}{\partial y}\frac{dy}{dt} + \frac{\partial g}{\partial z}\frac{dz}{dt}, \tag{6.5}$$

where the partial derivatives $\frac{\partial g}{\partial x}, \frac{\partial g}{\partial y}, \frac{\partial g}{\partial z}$ are evaluated at the point α(t) = (x(t), y(t), z(t)). This calculation can be visualized using a "dependence diagram," as shown in Figure 6.2. It illustrates the dependence of the various variables involved: g is a function of x, y, z, and in turns x, y, z are functions of t. The Little Chain Rule (6.5) says that, to find the derivative of the top with respect to the bottom, multiply the derivatives along each possible path from top to bottom and add these products.

$$w = g$$

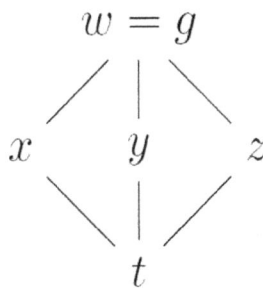

Figure 6.2: A dependence diagram for the Little Chain Rule

Tweak 1. Suppose that x, y, z are functions of two variables s and t. In other words, replace α : R → R³ with a smooth function f : ℝ² → R³, where f (s, t) = (x(s, t), y(s, t), z(s, t)). Then the composition g°f : ℝ² → R³ → R is a function of s and t: w = (g°f) (s, t) = g(x(s, t), y(s, t), z(s, t)). The corresponding dependence diagram is shown in Figure 6.3.

To compute $\frac{\partial w}{\partial s}$ (or $\frac{\partial w}{\partial t}$), only one variable actually varies. Because the composition is again real-valued, the Little Chain Rule applies with respect to that variable, just as in the previous case. Notationally, because w, x, y, and z are no longer functions of just one variable, we need to replace the ordinary derivative terms $\frac{d}{ds}$ (or $\frac{d}{dt}$) in equation (6.5) with partial derivatives $\frac{\partial}{\partial s}$ (or $\frac{\partial}{\partial t}$). For instance, for $\frac{\partial w}{\partial s}$ the Little Chain Rule says:

$$\frac{\partial w}{\partial s} = \frac{\partial g}{\partial x}\frac{\partial x}{\partial s} + \frac{\partial g}{\partial y}\frac{\partial y}{\partial s} + \frac{\partial g}{\partial z}\frac{\partial z}{\partial s}. \tag{6.6}$$

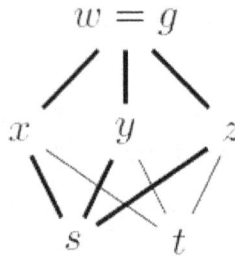

Figure 6.3: Dependence diagram for tweak 1

The relevant paths in the dependence diagram are shown in bold in the figure. Similarly:

$$\frac{\partial w}{\partial t} = \frac{\partial g}{\partial x}\frac{\partial x}{\partial t} + \frac{\partial g}{\partial y}\frac{\partial y}{\partial t} + \frac{\partial g}{\partial z}\frac{\partial z}{\partial t}. \tag{6.7}$$

In fact, equations (6.6) and (6.7) can be rewritten as a single matrix equation:

$$\begin{bmatrix} \frac{\partial w}{\partial s} & \frac{\partial w}{\partial t} \end{bmatrix} = \begin{bmatrix} \frac{\partial g}{\partial x} & \frac{\partial g}{\partial y} & \frac{\partial g}{\partial z} \end{bmatrix} \begin{bmatrix} \frac{\partial x}{\partial s} & \frac{\partial x}{\partial t} \\ \frac{\partial y}{\partial s} & \frac{\partial y}{\partial t} \\ \frac{\partial z}{\partial s} & \frac{\partial z}{\partial t} \end{bmatrix}. \tag{6.8}$$

The entries corresponding to equation (6.6) are shown in red. The point is that the matrix on the left of equation (6.8) is the derivative $D(g \circ f)$, while those on the right are Dg and Df, respectively.

Tweak 2. We alter g so that it is vector-valued, say $g : R^3 \to \mathbb{R}^2$, where $g(x, y, z) = (g_1(x, y, z), g_2(x, y, z))$, an)d lea ve $f: \mathbb{R}^2 \to R^3$ as in t)weak 1. The composition $g \circ f: \mathbb{R}^2 \to \mathbb{R}^2$ is vector-valued: $w = g(f(s, t)) = (g_1(f(s, t)), g_2(f(s, t))$. Each of the components is a real-valued function of s and t: $w_1 = g_1(f(s, t)) = g_1(x(s, t), y(s, t), z(s, t))$ and $w_2 = g_2(f(s, t)) = g_2(x(s, t), y(s, t), z(s, t))$. The dependence diagram is shown in Figure 6.4.

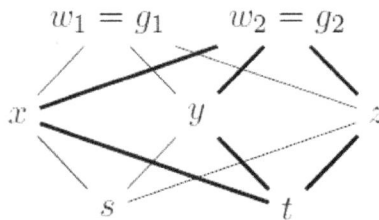

Figure 6.4: Dependence diagram for tweak 2

Both w_1 and w_2 have partial derivatives with respect to s and t, and each of them fits exactly into the context of tweak 1. For example:

$$\frac{\partial w_2}{\partial t} = \frac{\partial g_2}{\partial x}\frac{\partial x}{\partial t} + \frac{\partial g_2}{\partial y}\frac{\partial y}{\partial t} + \frac{\partial g_2}{\partial z}\frac{\partial z}{\partial t},$$

as highlighted in bold in the figure. Again, the collection of the four individual partial derivatives of the composition computed in this way can be organized into a single matrix equation:

$$\begin{bmatrix} \frac{\partial w_1}{\partial s} & \frac{\partial w_1}{\partial t} \\ \frac{\partial w_2}{\partial s} & \frac{\partial w_2}{\partial t} \end{bmatrix} = \begin{bmatrix} \frac{\partial g_1}{\partial x} & \frac{\partial g_1}{\partial y} & \frac{\partial g_1}{\partial z} \\ \frac{\partial g_2}{\partial x} & \frac{\partial g_2}{\partial y} & \frac{\partial g_2}{\partial z} \end{bmatrix} \begin{bmatrix} \frac{\partial x}{\partial s} & \frac{\partial x}{\partial t} \\ \frac{\partial y}{\partial s} & \frac{\partial y}{\partial t} \\ \frac{\partial z}{\partial s} & \frac{\partial z}{\partial t} \end{bmatrix}.$$

Once again, $D(g \circ f)$ appears on the left, while the product of Dg and Df is on the right.

The same pattern persists in general for any composition. The individual partial derivatives of the composition can be calculated using the Little Chain Rule and then combined as the matrix equation $D(g \circ f)(a) = Dg(f(a)) \cdot Df(a)$. In a way, the matrix version of the chain rule is a bookkeeping device that keeps track of many, many Little Chain Rule calculations!

Indeed, since we presented a rigorous proof of the Little Chain Rule in Exercise 5.6 of Chapter 4, we can think of this approach as supplying a proof of the general chain rule as well.

Example 6.12. The substitutions $x = r \cos \theta$ and $y = r \sin \theta$ convert a smooth real-valued function $f(x, y)$ of x and y into a function of r and θ: $w = f(r \cos \theta, r \sin \theta)$. For instance, if $f(x, y) = x^2 y^3$, then $w = (r \cos \theta)^2 (r \sin \theta)^3 = r^5 \cos^2 \theta \sin^3 \theta$.

We find general formulas for $\frac{\partial w}{\partial r}$ and $\frac{\partial w}{\partial \theta}$ in terms of $\frac{\partial f}{\partial x}$ and $\frac{\partial f}{\partial y}$.. As just discussed, we can compute these individual partials using the Little Chain Rule and a dependence diagram (Figure 6.5).

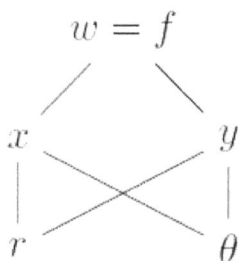

$$w = f$$

Figure 6.5: Finding partial derivatives with respect to polar coordinates

Hence:

$$\frac{\partial w}{\partial r} = \frac{\partial f}{\partial x}\frac{\partial x}{\partial r} + \frac{\partial f}{\partial y}\frac{\partial y}{\partial r} = \frac{\partial f}{\partial x}\cos\theta + \frac{\partial f}{\partial y}\sin\theta$$

and $$\frac{\partial w}{\partial \theta} = \frac{\partial f}{\partial x}\frac{\partial x}{\partial \theta} + \frac{\partial f}{\partial y}\frac{\partial y}{\partial \theta} = -\frac{\partial f}{\partial x}r\sin\theta + \frac{\partial f}{\partial y}r\cos\theta.$$

Example 6.13 (Implicit differentiation). Let $f : \mathbb{R}^2 \to \mathbb{R}$ be a smooth real-valued function of two variables, and let S be the level set corresponding to $f = c$. That is, S $= \{(x, y) \in \mathbb{R}^2 : f(x, y) = c\}$. Assume that the condition $f(x, y) = c$ defines y implicitly as a smooth function of x, $y = y(x)$, for some interval of x values. See Figure 6.6. Find the derivative $\frac{dy}{dx}$ in terms of f and its partial derivatives.

We start with the condition $f(x, y(x)) = c$ that defines $y(x)$. The expression on the left is the composition $x \to (x, y(x)) \to f(x, y(x))$. Call this last value w. The dependence diagram is in Figure 6.7.

The derivative $\frac{dw}{dx}$ can be computed using the chain rule:

$$\frac{dw}{dx} = \frac{\partial f}{\partial x}\frac{dx}{dx} + \frac{\partial f}{\partial y}\frac{dy}{dx} = \frac{\partial f}{\partial x} + \frac{\partial f}{\partial y}\frac{dy}{dx}.$$

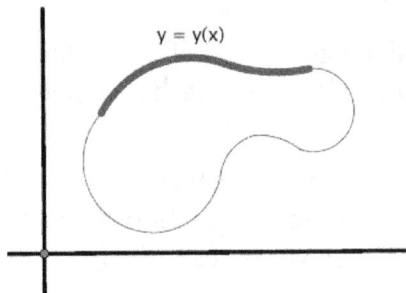

Figure 6.6: The level set f (x, y) = c defines y implicitly as a function of x on an interval of x values.

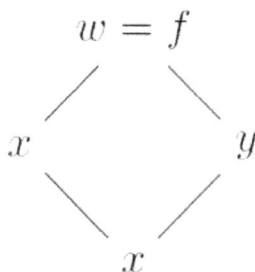

Figure 6.7: A dependence diagram for implicit differentiation

On the other hand, w has a constant value of c, so $\frac{dw}{dx} = 0$. Hence:

$$\frac{\partial f}{\partial x} + \frac{\partial f}{\partial y}\frac{dy}{dx} = 0. \tag{6.9}$$

For example, consider the ellipse $\frac{x^2}{9} + \frac{y^2}{4} = 1$. It's the level set of $f(x,y) = \frac{x^2}{9} + \frac{y^2}{4}$ corresponding to c = 1. Hence, by the previous example, along the ellipse, we have:

$$\frac{dy}{dx} = -\frac{\partial f/\partial x}{\partial f/\partial y} = -\frac{\frac{2}{9}x}{\frac{1}{2}y} = -\frac{4x}{9y}.$$

Of course, implicit differentiation problems of this type appear in first-year calculus, and the multivariable chain rule is never mentioned. There, to find the derivative along the ellipse, one simply differentiates both sides of the equation of the ellipse and uses the one-variable chain rule.

In the other words, one computes $\frac{d}{dx}\left(\frac{x^2}{9} + \frac{y^2}{4}\right) = \frac{d}{dx}(1)$, or:

$$\frac{2}{9}x + \frac{1}{2}y\frac{dy}{dx} = 0.$$

This is exactly the same as the result given by equation (6.9), from which the general formula for $\frac{dy}{dx}$ followed immediately. The multivariable chain rule gives us a way of articulating what the one-variable calculation is doing.

6.5 EXERCISES FOR CHAPTER 6

Section 1 Continuity revisited

1.1.   a. If $x = (x_1, x_2, \ldots, x_n)$ and $y = (y_1, y_2, \ldots, y_n)$ are elements of $\mathbb{R}^n$, show that $|xi - yi| \le \|x - y\|$ for i = 1, 2, ..., n.

   b. Let U be an open set in $\mathbb{R}^n$, and let $f : U \to Rm$ be a function, where $f(x) = (f_1(x), f_2(x), \ldots, f_m(x))$. Show that, if f is continuous at a point a of U , then so are $f_1, f_2, \ldots, f_m$.

1.2. Let U be an open set in $\mathbb{R}^n$, a a point of U, and $f: U - \{a\} \to Rm$ a function defined on U, expect possibly at a. Write out in terms $\in$'s δ's what it means to say that $\lim_{x \to a} f(x) = L$.

Section 2 Differentiability revisited

2.1. Let $f: \mathbb{R}^2 \to \mathbb{R}^2$ be given by $f(x, y) = (x^2 - y^2, 2xy)$.

   a. Find $Df(x, y,)$.
   b. Find $Df(1, 2)$.

2.2. Let $f: R^3 \to \mathbb{R}^2$ be given by $f(x, y, z) = (xy^2z^3 + 2, x\cos(yz))$.

   a. Find $Df(x, y, z)$.
   b. Find $Df(-\frac{\pi}{2}, 1, \frac{\pi}{2})$.

2.3. Find $Df(x, y, z)$ if $f(x, y, z) = x + y^2 + 3z^3$.

2.4. Find $Df(x, y, z)$ if $f(x, y, z) = (x, y^2, 3z^3)$.

2.5. Find $Df(x, y)$ if $f(x, y) = (x, y, 0)$.

2.6. Find $Df(x)$ if $f(x) = (x^2, x \sin x, 2x + 3)$.

2.7. Let $f: \mathbb{R}^2 \to R^3$ be given by $f(x, y) = (e^{x+y}, e^{-x} \cos y, e^{-x} \sin y)$.

    a. Find $Df(x, y)$.

    b. Show that f is differentiable at every point of $\mathbb{R}^2$.

2.8. Let $f: R^3 \to R^3$ be the spherical coordinate transformation from $\rho\phi\theta$-space to xyz-space: $f(\rho, \phi, \theta) = (\rho \sin \phi \cos \theta, \rho \sin \phi \sin \theta, \rho \cos \phi)$.

    a. Find $Df(\rho, \phi, \theta)$.

    b. Show that f is differentiable at every point of R^3.

    c. Show that $\det Df(\rho, \phi, \theta) = \rho^2 \sin \phi$. (This result will be used in the next chapter.)

2.9. Let $T: R^3 \to R^3$ be a linear transformation, and let A be its matrix with respect to the standard bases. Hence $T(x) = Ax$, where A is a 3 by 3 matrix,

$$A = \begin{bmatrix} a & b & c \\ d & e & f \\ g & h & i \end{bmatrix}.$$

    a. Find $DT(x, y, z)$.

    b. Use the definition of differentiability to show that T is differentiable at every point a of R^3.

2.10. Given points $x = (x_1, x_2, x_3)$ and $y = (y_1, y_2, y_3)$ in R^3, let us think of (x, y) as the point $(x_1, x_2, x_3, y_1, y_2, y_3)$ in R^6. Let $f: R^6 \to R$ and $g: R^6 \to R^3$ be the dot and cross products in R^3, respectively, that is:

$$f(x, y) = x \cdot y \quad \text{and} \quad g(x, y) = x \times y.$$

    a. Find $Df(x, y)$ and $Dg(x, y)$.

    b. Show that f and g are differentiable at every point (x, y) of R^6.

Section 3 The chain rule: a conceptual approach

3.1. Let $f: \mathbb{R}^2 \to R^3$ and $g: R^3 \to \mathbb{R}^2$ be given by:

$$f(s, t) = (s - t + 2, e^{2s+3t}, s \cos t) \quad \text{and} \quad g(x, y, z) = (xyz, x^2 + z^3).$$

    a. Find $Df(s, t)$ and $Dg(x, y, z)$.

    b. Find $D(g \circ f)(0, 0)$.

    c. Find $D(f \circ g)(0, 0, 0)$.

3.2. Let $f: \mathbb{R}^2 \to R^3$ and $g: R^3 \to R$ be the functions:

$$f(s, t) = (2s - t^2, st - 1, 2s^2 + st - t^2) \quad \text{and} \quad g(x, y, z) = (x + 1)^2 e^{yz}.$$

If h(s, t) = g(f (s, t)), find Dh(1, 2).

3.3. Let U and V be open sets in $\mathbb{R}^2$, and let $f : U \to V$ and $g : V \to U$ be inverse functions. That is:

- ⊙ g(f (x, y)) = (x, y) for all (x, y) in U and
- ⊙ f (g(x, y)) = (x, y) for all (x, y) in V .

Let a be a point of U , and let b = f (a). If f is differentiable at a and g is differentiable at b, what can you say about the following matrix products?

      a. Dg(b) · Df (a)

      b. Df (a) · Dg(b)

3.4. Exercise 5.5 of Chapter 4 described a generalization of the mean value theorem for real-valued functions of n variables. Does the same generalization apply to vector-valued functions? More precisely, let a be a point of $\mathbb{R}^n$, and let B = B(a, r) be an open ball centered at a. Is it necessarily true that, if $f : B \to Rm$ is a differentiable function on B and b is any point of B, then there exists a point c on the line segment connecting a and b such that:

$$f (b) - f (a) = Df (c) \cdot (b - a)?$$

If true, give a proof. Otherwise, find a counterexample.

Section 4 The chain rule: a computational approach

4.1. Let $f (x, y, z)$ be a smooth real-valued function of x, y, and z. The substitutions x = s + 2t, y = 3s +4t, and z = 5s +6t convert f into a function of s and t: w = f (s +2t, 3s +4t, 5s +6t).

Find expressions for $\frac{\partial w}{\partial s}$ and $\frac{\partial w}{\partial t}$ in terms of $\frac{\partial f}{\partial x}$, $\frac{\partial f}{\partial y}$, and $\frac{\partial f}{\partial z}$.

4.2. Let w = f (x, y) be a smooth real-valued function, where:

$$x = s^2 - t^2 + u^2 + 2s - 2t \text{ and } y = stu.$$

Find expressions for $\frac{\partial w}{\partial s}$, $\frac{\partial w}{\partial t}$, and $\frac{\partial w}{\partial u}$ in terms of $\frac{\partial f}{\partial x}$ and $\frac{\partial f}{\partial y}$.

4.3. The spherical substitutions x = ρ sin φ cos θ, y = ρ sin φ sin θ, and z = ρ cos φ convert a smooth real-valued function f (x, y, z) into a function of ρ, φ, and θ:

$$w = f (\rho \sin \phi \cos \theta, \rho \sin \phi \sin \theta, \rho \cos \phi).$$

    a. Find formulas for $\frac{\partial w}{\partial \rho}$, $\frac{\partial w}{\partial \phi}$, and $\frac{\partial w}{\partial \theta}$ in terms of $\frac{\partial f}{\partial x}$, $\frac{\partial f}{\partial y}$, and $\frac{\partial f}{\partial z}$.

    b. If f (x, y, z) = $x^2 + y^2 + z^2$, use your answer to part (a) to find $\frac{\partial w}{\partial \phi}$. and verify that it is the same as the result obtained if you first write w in terms of ρ, φ, and θ directly, say by substituting for x, y, and z, and then differentiate that expression with respect to φ.

4.4. The substitutions $x = s^2 - t^2$ and $y = 2st$ convert a smooth re al-valued function f (x, y) into a function of s and t: $w = f(s^2 - t^2, 2st)$. Find a formula for $\|\nabla w\| = \|(\frac{\partial w}{\partial s}, \frac{\partial w}{\partial t})\|$ in terms of $\frac{\partial f}{\partial x}$ and $\frac{\partial f}{\partial y}$.

4.5. Let $f(x, y)$ be a smooth real-valued function of x and y. The substitutions

$$x = e^{2s+t} - 1 \text{ and } y = s^2 - 3st + 2t$$

convert f into a function w of s and t. If $\frac{\partial w}{\partial s} = 5$ and $\frac{\partial w}{\partial t} = 4$ when s and t are both 0, find the values of $\frac{\partial f}{\partial x}$ and $\frac{\partial f}{\partial y}$ when x and y are both 0. (Note that when s and t are both 0, so are x and y.)

4.6. The substitutions $x = r \cos \theta$ and $y = r \sin \theta$ convert a smooth real-valued function $f(x, y)$ into a function of r and θ: $w = f(r \cos \theta, r \sin \theta)$.

Show that $r\frac{\partial w}{\partial r} = x\frac{\partial f}{\partial x} + y\frac{\partial f}{\partial y}$.

Find a similar expression for $\frac{\partial w}{\partial \theta}$ in terms of x, y, $\frac{\partial f}{\partial x}$, and $\frac{\partial f}{\partial y}$.

4.7. (a) Suppose that $w = f(u, v)$ is a smooth real-valued function, where $u = x/z$ and $v = y/z$. Show that:

$$x\frac{\partial w}{\partial x} + y\frac{\partial w}{\partial y} + z\frac{\partial w}{\partial z} = 0. \qquad (6.10)$$

(b) Without calculating any partial derivatives, deduce that $w = \frac{x^2+xy+y^2}{z^2}$ satisfies equation (6.10).

4.8. Let $f: U \to R$ be a smooth real-valued function of three variables defined on an open set U in R^3, and let c be a constant. Assume that the condition $f(x, y, z) = c$ defines z implicitly as a smooth function of x and y, $z = z(x, y)$, on some open set of points (x, y) in $\mathbb{R}^2$. Show that, on this open set:

$$\frac{\partial z}{\partial x} = -\frac{\partial f/\partial x}{\partial f/\partial z} \quad \text{and} \quad \frac{\partial z}{\partial y} = -\frac{\partial f/\partial y}{\partial f/\partial z}.$$

4.9. The condition $xy + xz + yz + e^{x+2y+3z} = 1$ defines z implicitly as a smooth function of x and y on some open set of points (x, y) containing $(0, 0)$ in $\mathbb{R}^2$. Find $\frac{\partial z}{\partial x}$ and $\frac{\partial z}{\partial y}$ when $x = 0$ and $y = 0$.

4.10. a. Let $f: R^3 \to \mathbb{R}^2$ be a smooth function. By definition, the level set S of f corresponding to the value $c = (c_1, c_2)$ is the set of all (x, y, z) such that $f(x, y, z) = c$ or, equivalently, where $f_1(x, y, z) = c_1$ and $f_2(x, y, z) = c_2$. Typically, the last two equations define a curve in R^3, the curve of intersection of the surfaces defined by $f_1 = c_1$ and $f_2 = c_2$.

Assume that the condition $f(x, y, z) = c$ defines x and y implicitly as smooth functions of z on some interval of z values in R, that is, $x = x(z)$ and $y = y(z)$ are functions that satisfy $f(x(z), y(z), z) = c$. Then the corresponding portion of the level set is parametrized by $\alpha(z) = (x(z), y(z), z)$.

Show that $x'(z)$ and $y'(z)$ satisfy the matrix equation:

$$\begin{bmatrix} \dfrac{\partial f_1}{\partial x} & \dfrac{\partial f_1}{\partial y} \\ \dfrac{\partial f_2}{\partial x} & \dfrac{\partial f_2}{\partial y} \end{bmatrix} \begin{bmatrix} x'(z) \\ y'(z) \end{bmatrix} = - \begin{bmatrix} \dfrac{\partial f_1}{\partial z} \\ \dfrac{\partial f_2}{\partial z} \end{bmatrix},$$

where the partial derivatives are evaluated at the point $(x(z), y(z), z)$. (Hint: Let w $= (w_1, w_2) = f(x(z), y(z), z)$.)

b. Let $f(x, y, z) = (x - \cos z, y - \sin z)$, and let C be the level set of f corresponding to $c = (0, 0)$. Find the functions $x = x(z)$ and $y = y(z)$ and the parametrization α of C described in part (a), and describe C geometrically. Then, use part (a) to help find $D\alpha(z)$, and verify that your answer makes sense.

4.11. Let $f(x, y)$ be a smooth real-valued function of x and y defined on an open $\frac{\partial f}{\partial y}$ U in $\mathbb{R}^2$. The partial derivative $\frac{\partial f}{\partial x}$ is also a real-valued function of x and y (as is $\frac{\partial f}{\partial y}$). If $\alpha(t) = (x(t), y(t))$ is a smooth path in U , then substituting for x and y in terms of t converts $\frac{\partial f}{\partial x}$,into a function of t, say $w = \frac{\partial f}{\partial x}(x(t), y(t))$.

  a. Show that:

$$\frac{dw}{dt} = \frac{\partial^2 f}{\partial x^2} \frac{dx}{dt} + \frac{\partial^2 f}{\partial y \, \partial x} \frac{dy}{dt}.$$

  where the partial derivatives are evaluated at $\alpha(t)$.

  b. Now, consider the composition $g(t) = (f \circ \alpha)(t)$. Find a formula for the second derivative $\frac{d^2 g}{dt^2}$ in terms of the partial derivatives of f as a function of x and y and the derivatives of x and y as functions of t. (Hint: Differentiate the Little Chain Rule. Note that this involves differentiating some products.)

  c. Let a be a point of U . If $\alpha : I \to U$, $\alpha(t) = a + tv = (a_1 + tv_1, a_2 + tv_2)$, is a parametrization of a line segment containing a and if $g(t) = (f \circ \alpha)(t)$, use your formula from part

  to show that:

$$\frac{d^2 g}{dt^2} = v_1^2 \frac{\partial^2 f}{\partial x^2} + 2v_1 v_2 \frac{\partial^2 f}{\partial x \, \partial y} + v_2^2 \frac{\partial^2 f}{\partial y^2}.$$

  where the partial derivatives are evaluated at $\alpha(t)$. (Compare this with equation (4.22) in Chapter 4 regarding the second-order approximation of f at a.)

4.12. Let $f(x, y)$ be a smooth real-valued function defined on an open set in $\mathbb{R}^2$. The polar coordinate substitutions $x = r \cos \theta$ and $y = r \sin \theta$ convert f into a function of r and θ: w = $f(r \cos θ, r \sin θ)$.

a. Show that:

$$\frac{\partial^2 w}{\partial \theta^2} = -r \cos\theta \frac{\partial f}{\partial x} - r \sin\theta \frac{\partial f}{\partial y} + r^2 \sin^2\theta \frac{\partial^2 f}{\partial x^2} - 2r^2 \sin\theta \cos\theta \frac{\partial^2 f}{\partial x \, \partial y}$$
$$+ r^2 \cos^2\theta \frac{\partial^2 f}{\partial y^2}.$$

(Hint: Start with the results of Example 6.12.)

b. Find an analogous expression for the mixed partial derivative $\frac{\partial^2 w}{\partial r \, \partial \theta}$.

c. Find an analogous expression for $\frac{\partial^2 w}{\partial r^2}$.

d. If $f(x, y) = x^2 + y^2 + xy$, use your answer to part (b) to find $\frac{\partial^2 w}{\partial r \, \partial \theta}$. Then, verify that it is the same as the result obtained by writing w in terms of r and θ directly by substituting for x and y and then differentiating that expression with respect to r and θ.

7
Chapter

CHANGE OF VARIABLES

We now turns to the counterpart for integrals of the chain rule. It is called the change of variables theorem. For functions of one variable, the corresponding notion is the method of substitution. The one-variable and multivariable versions are expressed in equations that are similar formally, but the approaches one might take to understand where they come from are quite different. It's exciting when a change in perspective still leads to a result of recognizable form. We make a few remarks comparing the two versions later, after the theorem has been stated.

In many respects, the change of variables theorem and the chain rule hold together the foundation of the more advanced theory of multivariable calculus. We begin, however, with a concrete example where changing variables just seems like the right thing to do.

Example 7.1. Let D be the disk $x^2 + y^2 \le 4$ in the xy-plane. Find $\iint_D \sqrt{x^2 + y^2}\, dx\, dy$. For instance, if one wanted to know the average distance to the origin on D, one would evaluate this integral and divide by Area (D) = 4π.

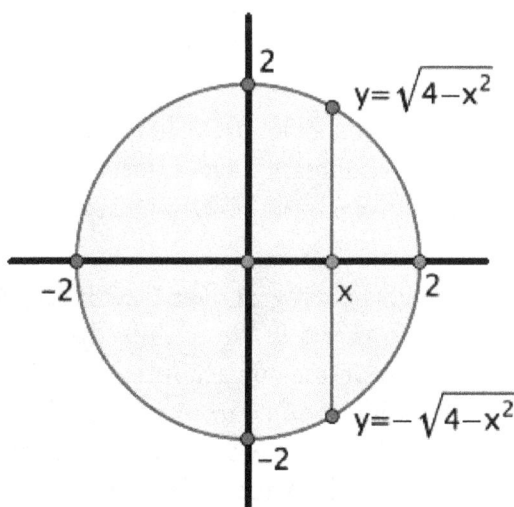

Figure 7.1: The disk $x^2 + y^2 \le 4$

Setting up the integral with the order of integration dy dx in the usual way (Figure 7.1) gives:

$$\iint_D \sqrt{x^2 + y^2}\, dx\, dy = \int_{-2}^{2} \left(\int_{-\sqrt{4-x^2}}^{\sqrt{4-x^2}} \sqrt{x^2 + y^2}\, dy \right) dx.$$

While not impossible, this looks potentially messy. On the other hand, expressing the problem in terms of polar coordinates seems promising for two reasons.

- ◉ The function to be integrated is simpler: $\sqrt{x^2 + y^2} = r$.

- ◉ The region of integration is simpler: in polar coordinates, D is described by $0 \le r \le 2$, $0 \le \theta \le 2\pi$, in other words, by a rectangle in the rθ-plane.

The obstacle is what to do about the dA = dx dy part of the integral. For this, we need to see how small pieces of area in the xy-plane are related to those in the rθ-plane. We returns to this example later.

7.1 CHANGE OF VARIABLES FOR DOUBLE INTEGRALS

Let $\iint_D f(x, y)\, dx\, dy$ be a given double integral. We want to describe how to convert the integral to an equivalent one with a different domain of integration and a correspondingly different integrand. The setup is to assume that there is a smooth function T : D* → D, where D* is a subset of $\mathbb{R}^2$. We think of T as transforming D* onto D, so we assume that T (D*) = D and that T is one-to-one, except possibly along the boundary of D*. This last part means that whenever a and b are distinct points of D*, not on the boundary, then the values T (a) and T (b) are distinct, too: a≠ b ⇒ T (a)≠ T (b). (For the purposes of integration, anomalous behavior confined to the boundary can be allowed typically without messing things up. See the remarks on page 130.)

The new integral will be an integral over D*. It helps keep things straight if the coordinates in the domain and codomain have different names, so we think of T as a transformation from the uv-plane to the xy-plane and write T (u, v) = (x(u, v), y(u, v)).

Thus we want to convert the given integral over D with respect to x and y to an integral over D* with respect to u and v. We outline the main idea. It is similar to what we did in Section 5.5 to define surface integrals with respect to surface area using a parametrization, though of course there is a big difference between motivating a definition and justifying a theorem. Another difference is that, in the interim, we have learned about the derivative of a vector-valued function.

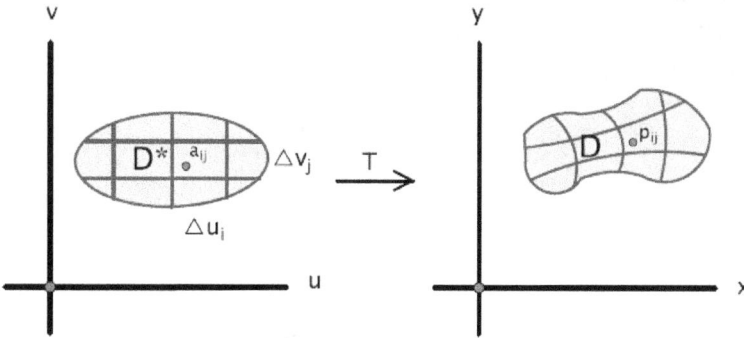

Figure 7.2: Change of variables: x and y as functions of u and v via transformation T

To begin, subdivide D* into small subrectangles of dimensions Δu_i by Δv_j, as in Figure 7.2. The transformation T sends a typical subrectangle of area $\Delta u_i \Delta v_j$ in the uv-plane to a small curvy quadrilateral in the xy-plane. Let ΔA_{ij} denote the area of the curvy quadrilateral. If we choose a sample point p_{ij} in each curvy quadrilateral, we can form a Riemann sum-like approximation of the integral over D:

$$\iint_D f(x, y)\, dx\, dy \approx \sum_{i,j} f(\mathbf{p}_{ij})\, \Delta A_{ij}. \qquad (7.1)$$

To turns this into a Riemann sum over D∗, we need to relate the areas of the subrectangles and the curvy quadrilaterals.

To do so, we use the first-order approximation of T . Choose a point a_{ij} in the (i, j) th subrectangle of D* such that $T(a_{ij}) = p_{ij}$. Then the first-order approximation says that, for points u in D* near a_{ij}:

$$\begin{aligned} T(\mathbf{u}) &\approx T(\mathbf{a}_{ij}) + DT(\mathbf{a}_{ij}) \cdot (\mathbf{u} - \mathbf{a}_{ij}) \\ &\approx \mathbf{p}_{ij} + DT(\mathbf{a}_{ij}) \cdot (\mathbf{u} - \mathbf{a}_{ij}) \\ &\approx (\mathbf{p}_{ij} - DT(\mathbf{a}_{ij}) \cdot \mathbf{a}_{ij}) + DT(\mathbf{a}_{ij}) \cdot \mathbf{u}. \end{aligned} \qquad (7.2)$$

It may be difficult to get a foothold on the behavior of T itself, but describing how the approximation transforms the (i, j)th subrectangle is quite tractable.

For this, we bring in some linear algebra. Note that the derivative DT (a_{ij}) is a 2 by 2 matrix. As shown in Chapter 1, any 2 by 2 matrix $A = \left[\begin{smallmatrix} a & b \\ c & d \end{smallmatrix}\right]$ determines a linear transformation $L : \mathbb{R}^2 \to \mathbb{R}^2$ given by matrix multiplication: L(x) = Ax. This transformation sends the unit square determined by e1 and e2 to the parallelogram determined by $L(\mathbf{e}_1) = \left[\begin{smallmatrix} a \\ c \end{smallmatrix}\right]$ and $L(\mathbf{e}_2) = \left[\begin{smallmatrix} b \\ d \end{smallmatrix}\right]$, which are the columns of A (Figure 7.3). We saw Proposition 1.13 of Chapter 1 that the area of this parallelogram is det L(e₁) L(e₂) , where L(e₁) L(e₂) is the matrix whose rows are L(e₁) and L(e₂). This is precisely the matrix transpose At. Thus the area of the parallelogram is | det(At)| = | det A|, where we have used the general fact that an n by n matrix and its transpose have the same determinant(Proposition 1.14, Chapter 1). In particular, L has changed the area by a factor of $|\det A|$.

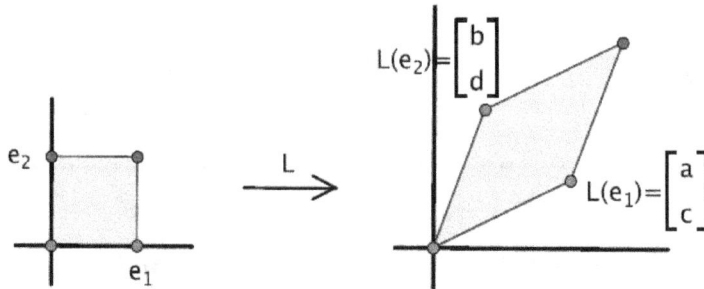

Figure 7.3: A linear transformation L sends the square determined by e1 and e2 to the parallelogram determined by $L(e_1)$ and $L(e_2)$.

This principle extends easily to a couple of slightly more general cases.

- ⊙ Given nonzero scalars ℓ and w, let R be the $|\ell|$ by $|w|$ rectangle determined by $\ell e1$ and $we2$. Then L sends R to the parallelogram P determined by $L(\ell e1) = \ell L(e_1)$ and $L(we_2) = wL(e_2)$. Compared to the original case, the areas of both rectangle and parallelogram have changed by a common factor of $|\ell w|$, hence they still differ by a factor of $|\det A|$.

- ⊙ Suppose that we translate the rectangle R in the previous case by a constant vector a. This results in another $|\ell|$ by $|w|$ rectangle R' consisting of all points of the form a+x, where x ∈ R. By linearity, $L(a+x) = L(a)+L(x)$, so L transforms R to the translation of the parallelogram $L(R) = P$ by $L(a)$. Let's call the translated parallelogram P'. Since translations don't affect area, the areas of R' and its image P' again differ by a factor of $|\det A|$.

This shows the following.

Lemma 7.2. A linear transformation $L : \mathbb{R}^2 \to \mathbb{R}^2$ represented with respect to the standard bases by a 2 by 2 matrix A sends rectangles whose sides are parallel to the coordinate axes to parallelograms and alters the area of the rectangles by a factor of $|\det A|$.

Returnsing to the first-order approximation (7.2) of T , we conclude that multiplication by DT (a_{ij}) transforms the subrectangle of area $\Delta u_i \, \Delta v_j$ that contains a_{ij} to a parallelogram of area $|\det DT (a_{ij})| \Delta u_i \, \Delta v_j$. Hence, after additional translation, the first-order approximation (7.2) sends the subrectangle containing a_{ij} to a parallelogram that:

- ⊙ approximates the curvy quadrilateral of area ΔA_{ij} that contains p_{ij} and

- ⊙ has area $|\det DT (a_{ij})| \Delta u_i \, \Delta v_j$.

Thus the Riemann sum approximation (7.1) becomes:

$$\iint_D f(x,y)\,dx\,dy \approx \sum_{i,j} f(\mathbf{p}_{ij})\,\triangle A_{ij} \approx \sum_{i,j} f(T(\mathbf{a}_{ij}))\,|\det DT(\mathbf{a}_{ij})|\,\triangle u_i\,\triangle v_j.$$

Letting $\triangle u_i$ and $\triangle v_j$ go to zero, this becomes an integral over D*, giving the following major result.

Theorem 7.3 (Change of variables theorem for double integrals). Let D and D* be bounded subsets of $\mathbb{R}^2$, and let T : D* D be a smooth function such that T (D*) = D and that is one-to-one, except possibly on the boundary of D*. If f is integrable on D, then:

$$\iint_D f(x,y)\,dx\,dy = \iint_{D^*} f(T(u,v))\,|\det DT(u,v)|\,du\,dv.$$

In other words, x and y are replaced in terms of u and v in the function f using (x, y) = T (u, v) and dx dy is replaced by

$$|\det DT(u,v)|\,du\,dv = \left|\det \begin{bmatrix} \frac{\partial x}{\partial u} & \frac{\partial x}{\partial v} \\ \frac{\partial y}{\partial u} & \frac{\partial y}{\partial v} \end{bmatrix}\right|\,du\,dv.$$

The determinant of the derivative, det DT (u, v), is sometimes referred to as the **Jacobian determinant.**

By comparison, in first-year calculus, if f (x) is a real-valued function of one variable and x is expressed in terms of another variable u, say as x = T (u), then the method of substitution says that:

$$\int_{T(a)}^{T(b)} f(x)\,dx = \int_a^b f(T(u))\,T'(u)\,du.$$

In this case, it seems simplest to understand the result in terms of antiderivatives using the chain rule and the fundamental theorem of calculus rather than through Riemann sums, though we won't present the details of the argument. In particular, substituting for x in terms of u includes the substitution $dx = T'(u)\,du = \frac{dx}{du}\,du$. The substitution dx dy = | det DT (u, v)| du dv in the previous paragraph is the analogue in the double integral case. We say more about the principle of substitution in the next section.

Returnsing to double integrals, in the case of polar coordinates, the substitutions for x and y are given by (x, y) = T (r, θ) = (r cos θ, r sin θ). Then
$$DT(r,\theta) = \begin{bmatrix} \cos\theta & -r\sin\theta \\ \sin\theta & r\cos\theta \end{bmatrix}. \text{ and:}$$

$$|\det DT(r,\theta)| = |r\cos^2\theta + r\sin^2\theta| = |r| = r.$$

By the change of variables theorem, this means that "dx dy = r dr dθ" and that polar coordinates transform integrals as follows.

Corollary 7.4. In polar coordinates:

$$\iint_D f(x,y)\,dx\,dy = \iint_{D^*} f(r\cos\theta, r\sin\theta)\,r\,dr\,d\theta,$$

where D^* is the region in the $r\theta$-plane that describes D in polar coordinates.

We now complete Example 7.1, which asked to evaluate $\iint_D \sqrt{x^2+y^2}\,dx\,dy$, where D is the disk $x^2 + y^2 \le 4$ in the xy-plane. As noted earlier, D is described in polar coordinates by the rectangle D* given by $0 \le r \le 2$, $0 \le \theta \le 2\pi$ in the $r\theta$-plane. The polar coordinate transformation T is shown in Figure 7.4.

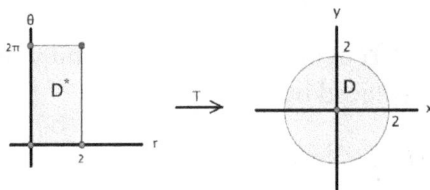

Figure 7.4: A change of variables in polar coordinates

We verify that T satisfies the hypotheses of the theorem, though we shall grow less meticulous about writing out the details on this point after a couple more examples. All the points on the left side of D*, where r = 0, are mapped to the origin, that is, T (0, θ) = (0, 0) for $0 \le \theta \le 2\pi$.

Also, along the top and bottom of D , θ = 0 and θ = 2π correspond to the same thing in the xy-plane: T (r, 0) = T (r, 2π) for $0 \le r \le 2$. Apart from this, distinct points in D* are mapped to distinct values in D. In other words, T is one-to-one away from the boundary of D^* , so the change of variables theorem applies.

Since $\sqrt{x^2+y^2} = r$, this gives:

$$\iint_D \sqrt{x^2+y^2}\,dx\,dy = \iint_{D^*} r \cdot r\,dr\,d\theta$$

$$= \int_0^{2\pi}\left(\int_0^2 r^2\,dr\right)d\theta$$

$$= \int_0^{2\pi}\left(\frac{1}{3}r^3\Big|_{r=0}^{r=2}\right)d\theta$$

$$= \int_0^{2\pi}\frac{8}{3}\,d\theta$$

$$= \frac{8}{3}\theta\Big|_0^{2\pi}$$

$$= \frac{16}{3}\pi.$$

7.2 A WORD ABOUT SUBSTITUTION

The change of variables theorem concerns how to convert an integral with respect to x and y into an integral with respect to two other variables, say u and v. Another name for this is substitution: we require expressions to substitute for x, y, and dA = dx dy in terms of u and v. Substituting for x and y means writing them as functions of u and v: x = x(u, v) and y = y(u, v). This is equivalent to having a function T (u, v) = (x(u, v), y(u, v)) from the uv-plane to the xy-plane. The logistics can be confusing and perhaps counterintuitive: the transformation of the integral goes from being with respect to x and y to being with respect to u and v, but, in order to accomplish this, the geometric transformation goes the other way, from (u, v) to (x, y). This is an inherent aspect of substitution.

Sometimes, the original integrand is denoted by an expression like $\eta = f(x, y)\, dx\, dy$, and then the transformed integrand $f(T(u, v)) \,|\det DT(u, v)|\, du\, dv$ is denoted by $T*(\eta)$. Another appropriate notation for it might be η^*. With this notation, the change of variables theorem becomes:

$$\iint_{T(D^*)} \eta = \iint_{D^*} T^*(\eta). \tag{7.3}$$

Actually, what we just said is not entirely accurate and the situation is a little more complicated,7 but it serves to illustrate the general principle: in a substitution, the geometry is "pushed forward," and the integral is "pulled back."

7.3. EXAMPLES: LINEAR CHANGES OF VARIABLES, SYMMETRY

We illustrate the change of variables theorem with three further examples.

Example 7.5. Let D be the region in $\mathbb{R}^2$ given $\frac{x^2}{9} + \frac{y^2}{4} \leq 1, y \geq 0.$ Evaluate $\iint_D (x^2 + y^2)\, dx\, dy.$ Here, the integrand $f(x, y) = x^2 + y^2$ is fairly simple, but the region of integration poses some problems. D is the region inside the upper half of an ellipse. The endpoints of the iterated integral with respect to x and y are somewhat hard to work with, and polar coordinates don't help either— the relationship between r and θ needed to describe D is a little complicated. On the other hand, we can think of D as a transformed semicircular disk, which is easy to describe in polar coordinates. Thus we attempt to map the half-disk D* of radius 1 given in the uv-plane by $u^2 + v^2 \leq 1, v \geq 0$, over to D and hope that this does not complicate the function to be integrated too much.

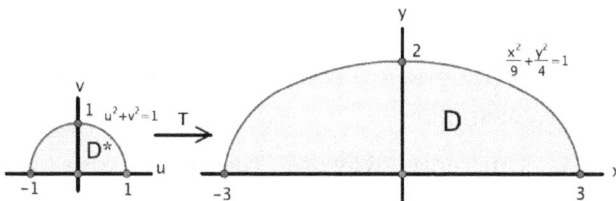

Figure 7.5: Changing an integral over a half-ellipse to an integral over a half-disk

The x and y-intercepts of the ellipse are ±3 and ±2, respectively, so to achieve the transformassstion, as illustrated in Figure 7.5, we stretch horizontally by a factor of 3 and vertically by a factor of 2. In other words, define:

$$T\,(u,\,v) = (3u,\,2v).$$

In effect, we are making the substitutions x = 3u and y = 2v. Then T transforms D* to D, that is, if (u, v) ∈ D*, then T (u, v) ∈ D. This is because if (u, v) satisfies $u^2 + v^2$ ≤ 1 and v ≥ 0, then (x, y) = T (u, v) satisfies $\frac{x^2}{9} + \frac{y^2}{4} = \frac{(3u)^2}{9} + \frac{(2v)^2}{4} = u^2 + v^2 \leq 1$ and y = 2v ≥ 0.

Moreover, T is one-to-one on the whole uv-plane. This seems reasonable geometrically, or, in terms of equations, if $(u_1, v_1) \neq (u_2, v_2)$, then $(3u_1, 2v_1) \neq (3u_2, 2v_2)$. Perhaps it is more readable to phrase this in the contrapositive: if $(3u_1, 2v_1) = (3u_2, 2v_2)$, then $(u_1, v_1) = (u_2, v_2)$. This is clear. For instance, $3u_1 = 3u_2 \Rightarrow u_1 = u_2$.

Finally, $DT(u,v) = \begin{bmatrix} 3 & 0 \\ 0 & 2 \end{bmatrix}$, so | det DT (u, v)| = 6, and f (T (u, v)) = $(3u)^2 + (2v)^2 = 9u^2 + 4v^2$.

Thus, by the change of variables theorem:

$$\iint_D (x^2 + y^2)\,dx\,dy = \iint_{D^*} (9u^2 + 4v^2) \cdot 6\,du\,dv.$$

We convert this to polar coordinates in the uv-plane using u = r cos θ, v = r sin θ, and du dv = r dr dθ. The half-disk D* is described in polar coordinates by 0 ≤ r ≤ 1, 0 ≤ θ ≤ π, in other words, by the rectangle [0, 1] × [0, π] in the rθ-plane. Thus:

$$\iint_D (x^2 + y^2)\,dx\,dy = \iint_{\substack{D^* \text{ in} \\ \text{polar}}} (9r^2 \cos^2 \theta + 4r^2 \sin^2 \theta) \cdot 6 \cdot r\,dr\,d\theta$$

$$= 6 \int_0^\pi \left(\int_0^1 r^3 (9 \cos^2 \theta + 4 \sin^2 \theta)\,dr \right) d\theta. \qquad (7.4)$$

At this point, we pause to put on the record a couple of facts from the exercises that we shall use freely from now on (see Exercise 2.3 in Chapter 5 and Exercise 1.1 in this chapter).

⊙ For a function F whose variables separate into two independent continuous factors, F (x, y) = f (x)g(y), the integral over a rectangle [a, b] × [c, d] is given by:

$$\int_c^d \left(\int_a^b f(x)g(y)\,dx \right) dy = \left(\int_a^b f(x)\,dx \right) \left(\int_c^d g(y)\,dy \right).$$

Note that integrals over rectangles are characterized by the property that all the limits of integration are constant.

$\int_0^\pi \cos^2 \theta\,d\theta = \int_0^\pi \sin^2 \theta\,d\theta = \frac{\pi}{2}$. (An easy way to remember this is to note that $\int_0^\pi \cos^2 \theta\,d\theta = \int_0^\pi \sin^2 \theta\,d\theta$ from the graphs of the sine and cosine functions and that

$\int_0^\pi (\cos^2 \theta + \sin^2 \theta) \, d\theta = \int_0^\pi 1 \, d\theta = \theta \big|_0^\pi = \pi.)$

Applying these facts to the present integral (7.4) gives:

$$\iint_D (x^2 + y^2) \, dx \, dy = 6 \left(\int_0^1 r^3 \, dr \right) \left(\int_0^\pi (9 \cos^2 \theta + 4 \sin^2 \theta) \, d\theta \right)$$

$$= 6 \cdot \left(\frac{1}{4} r^4 \big|_0^1 \right) \cdot \left(9 \cdot \frac{\pi}{2} + 4 \cdot \frac{\pi}{2} \right)$$

$$= 6 \cdot \frac{1}{4} \cdot \frac{13}{2} \pi$$

$$= \frac{39}{4} \pi.$$

Example 7.6. If D is the parallelogram with vertices (0, 0), (3, 2), (2, 3), (-1, 1), find $\iint_D xy \, dx \, dy$.

The difficulty here is that setting up an iterated integral in terms of x and y involves splitting into three integrals in order to describe D. To simplify the domain, we use the fact previously noted that a linear transformation $T : \mathbb{R}^2 \to \mathbb{R}^2$ transforms the unit square into the parallelogram determined by T (e_1) and T (e_2). Here, parallelogram D is determined by the vectors (3, 2) and (-1, 1), so we want T (e_1) = (3, 2) and T (e_2) = (-1, 1). U]sing Proposition 1.7 from Chapter 1, we put these vectors into the columns of a matrix $A = \begin{bmatrix} 3 & -1 \\ 2 & 1 \end{bmatrix}$ and define:

$$T(u, v) = \begin{bmatrix} 3 & -1 \\ 2 & 1 \end{bmatrix} \begin{bmatrix} u \\ v \end{bmatrix} = \begin{bmatrix} 3u - v \\ 2u + v \end{bmatrix} = (3u - v, 2u + v).$$

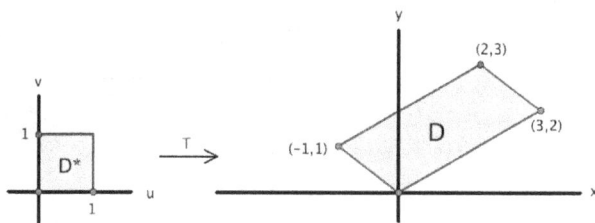

Figure 7.6: A linear change of variables

Then T maps the unit square D* = [0, 1] × [0, 1] in the uv-plane onto D. See Figure 7.6. We leave for the exercises the verification that T is one-to-one on $\mathbb{R}^2$ (Exercise 3.3), so we can use the change of variables theorem to pull back the original integral to an integral over D*.

The substitutions given by T are x = 3u – v and y = 2u + v. Also, $DT(u,v) = \begin{bmatrix} 3 & -1 \\ 2 & 1 \end{bmatrix}$, and | det DT (u, v)| = |3 + 2| = 5. Hence:

$$
\begin{aligned}
\iint_D xy\,dx\,dy &= \iint_{D^*} (3u - v)(2u + v) \cdot 5\,du\,dv \\
&= 5 \int_0^1 \left(\int_0^1 (6u^2 + uv - v^2)\,dv \right) du \\
&= 5 \int_0^1 \left(6u^2 v + \frac{1}{2}uv^2 - \frac{1}{3}v^3 \right)\Big|_{v=0}^{v=1} du \\
&= 5 \int_0^1 (6u^2 + \frac{1}{2}u - \frac{1}{3})\,du \\
&= 5 \left(2u^3 + \frac{1}{4}u^2 - \frac{1}{3}u \Big|_0^1 \right) \\
&= 5 \left(2 + \frac{1}{4} - \frac{1}{3} \right) \\
&= 5 \cdot \frac{24 + 3 - 4}{12} \\
&= \frac{115}{12}.
\end{aligned}
$$

Example 7.7. A subset D of the xy-plane is said to be **symmetric in the y-axis** if, whenever (x, y) is in D, so is (-x, y) (Figure 7.7). Let D be such a subset.

a. Show that $\iint_D x\,dx\,dy = 0$.
b. Let L = {(x, y) ∈ D : x ≤ 0} and R = { (x, y) ∈ D : x ≥ 0} be the left and right halves of D, respectively. Show that $\iint_L y\,dx\,dy = \iint_R y\,dx\,dy$.

The rough intuition here is that, in the Riemann sums $\sum x \triangle x \triangle y$ for $\iint_D x\,dx\,dy$, the symmetry of D means that each contribution for x > 0 is balanced by an opposite contribution for x < 0, and vice versa. For a more polished argument, let T : $\mathbb{R}^2 \to \mathbb{R}^2$ be the reflection in the y-axis, T (x, y) = (x, y). Since D is symmetric in the y-axis, T transforms D into itself, i.e., T (D) = D. Hence, in the notation of the change of variables theorem, D* = D. Since distinct points have distinct reflections, T is one-to-one. Also $DT(x,y) = \begin{bmatrix} -1 & 0 \\ 0 & 1 \end{bmatrix}$, which gives | det DT (x, y)| = |-1| = 1.

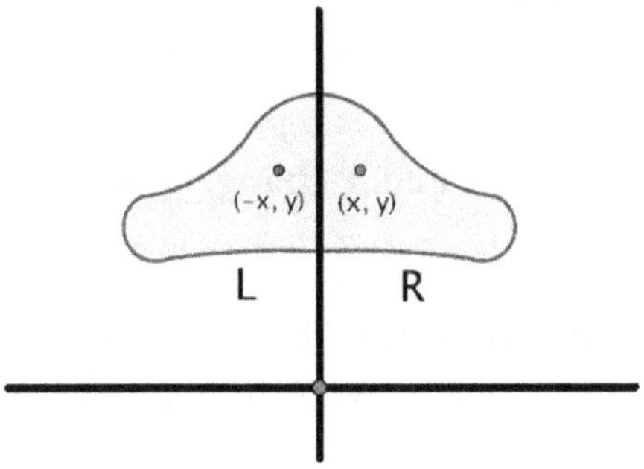

Figure 7.7: Symmetry in the y-axis

We apply the change of variables theorem with $f(x, y) = x$ so that $f(T(x, y)) = f(-x, y) = -x$.

Then:

$$\iint_D x\,dx\,dy = \iint_{D^*=D} f(T(x,y))\,|\det DT(x,y)|\,dx\,dy$$
$$= \iint_D -x \cdot 1\,dx\,dy$$
$$= -\iint_D x\,dx\,dy.$$

As a result, $2\iint_D x\,dx\,dy = 0$, so $\iint_D x\,dx\,dy = 0$.

Here, the intuition is that the contributions to the Riemann sums $\sum y\,\triangle x\,\triangle y$ for $\iint_D y\,dx\,dy$ are the same for (x, y) and $(-x, y)$, hence the sums converge to the same value on L and on R. More formally, again let $T(x, y) = (-x, y)$. Then T $(L) = R$, i.e., $R^* = L$, and $|\det DT(x, y)| = 1$ as before. Now, $f(x, y) = y$, so $f(T(x, y)) = f(-x, y) = y$. Consequently:

$$\iint_R y\,dx\,dy = \iint_{R^*=L} f(T(x,y))\,|\det DT(x,y)|\,dx\,dy$$
$$= \iint_L y \cdot 1\,dx\,dy$$
$$= \iint_L y\,dx\,dy.$$

This is exactly what we wanted to show.

This example can be generalized. Namely, let D be symmetric in the y-axis, and let $f: D \to R$ be an integrable real-valued function.

- If $f(-x, y) = -f(x, y)$ for all (x, y) in D, then:

$$\iint_D f(x,y)\,dx\,dy = 0.$$

- If $f(-x, y) = f(x, y)$ for all (x, y) in D, then:

$$\iint_L f(x,y)\,dx\,dy = \iint_R f(x,y)\,dx\,dy.$$

The justifications use essentially the same reasoning as in the example.

7.4 CHANGE OF VARIABLES FOR N-FOLD INTEGRALS

The change of variables theorem for double integrals has a natural analogue for functions of n variables. Let T be a smooth transformation that sends a subset W* of $\mathbb{R}^n$ onto a subset W of $\mathbb{R}^n$ and that is one-to-one, except possibly on the boundary of W*. See Figure 7.8. We think of T as a function from $(u_1, u_2, \ldots, u_n)$-space to $(x_1, x_2, \ldots, x_n)$-space, so:

$$T(u_1, u_2, \ldots, u_n) = \big(x_1(u_1, u_2, \ldots, u_n), x_2(u_1, u_2, \ldots, u_n), \ldots, x_n(u_1, u_2, \ldots, u_n)\big).$$

Then the change of variables formula reads:

$$\iint \cdots \int_W f(\mathbf{x})\, dx_1\, dx_2 \cdots dx_n = \iint \cdots \int_{W*} f(T(\mathbf{u}))\, \big|\det DT(\mathbf{u})\big|\, du_1\, du_2 \cdots du_n.$$

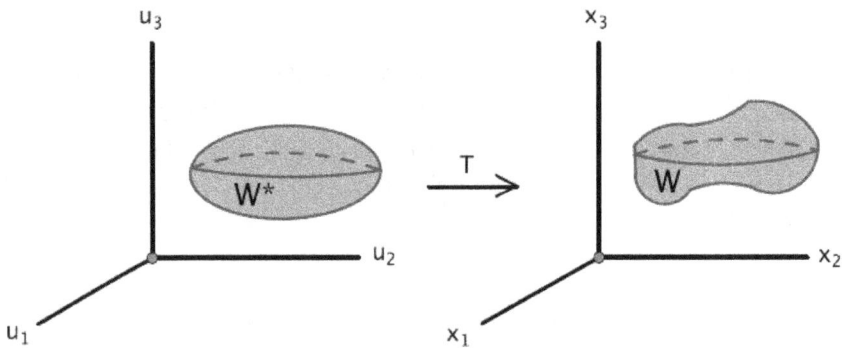

Figure 7.8: A three-dimensional change of variables

More informally, one uses T to substitute for $x_1, x_2, \ldots, x_n$ in terms of $u_1, u_2, \ldots, u_n$ and sets $dx_1\, dx_2\, dx_n = |\det DT(u)|\, du_1\, du_2 \ldots \ldots du_n$. This pulls back the integral over W to an integral over W*.

The intuition behind this generalization is the same as for double integrals: A linear transformation $L : \mathbb{R}^n \to \mathbb{R}^n$ represented by a matrix A alters n-dimensional volume by a factor of $|\det A|$. So, to obtain the integral, one chops up W* into small pieces and, on each one, approximates T by its first-order approximation, which alters volume by a factor of $|\det DT(u)|$.

For triple integrals, there are two standard sets of alternate variables.

- ⊙ **Cylindrical coordinates.** For this, the change of variables is T $(r, \theta, z) = (x, y, z)$, where, as we saw in Section 5.4.2, the substitutions are given by:

$$\begin{cases} x = r\cos\theta \\ y = r\sin\theta \\ z = z. \end{cases}$$

Therefore $DT(r, \theta, z) = \begin{bmatrix} \cos\theta & -r\sin\theta & 0 \\ \sin\theta & r\cos\theta & 0 \\ 0 & 0 & 1 \end{bmatrix}$.

After expanding along the third column, this gives

$$|\det DT(r, \theta, z)| = \left| \det \begin{bmatrix} \cos\theta & -r\sin\theta \\ \sin\theta & r\cos\theta \end{bmatrix} \right| = r,$$

i.e:

$$\boxed{dx\,dy\,dz = r\,dr\,d\theta\,dz.}$$

⊙ **Spherical coordinates.** This time, T (ρ, ϕ, θ) = (x, y, z) with conversions:

$$\begin{cases} x = \rho\sin\phi\cos\theta \\ y = \rho\sin\phi\sin\theta \\ z = \rho\cos\phi. \end{cases}$$

See Section 5.4.3. Then $DT(\rho, \phi, \theta) = \begin{bmatrix} \sin\phi\cos\theta & \rho\cos\phi\cos\theta & -\rho\sin\phi\sin\theta \\ \sin\phi\sin\theta & \rho\cos\phi\sin\theta & \rho\sin\phi\cos\theta \\ \cos\phi & -\rho\sin\phi & 0 \end{bmatrix}$.

We leave as an exercise the calculation that | det DT (ρ, ϕ, θ)| = ρ^2 sin ϕ (Exercise 2.8 in Chapter 6), i.e.:

$$\boxed{dx\,dy\,dz = \rho^2 \sin\phi\,d\rho\,d\phi\,d\theta.}$$

Example 7.8. Find the volume of a closed ball of radius a in R³, that is, the volume of:

$$W = \{(x, y, z) \in \mathbb{R}^3 : x^2 + y^2 + z^2 \le a^2\}.$$

See Figure 7.9.

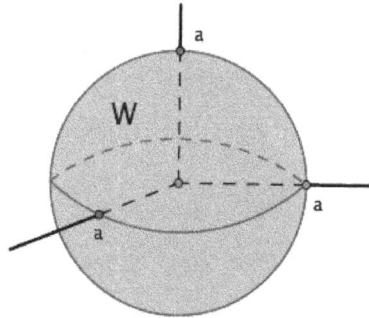

Figure 7.9: A ball of radius a in R³

We begin with $\mathrm{Vol}\,(W) = \iiint_W 1\,dx\,dy\,dz$. As we are about to see, W is easy to describe in spherical coordinates, so we convert to an integral in spherical coordinates. The endpoints depend on the order of integration, and here we choose $d\rho\,d\phi\,d\theta$. Thus the first integration is with respect to ρ. For fixed ϕ and θ, ρ varies along a radial line segment from $\rho = 0$ to $\rho = a$. Then, for fixed θ, ϕ rotates all the way down from the positive z-axis, $\phi = 0$, to the negative z-axis, $\phi = \pi$. Finally, θ varies from $\theta = 0$ to $\theta = 2\pi$. In other words, W is described in $\rho\phi\theta$-space by the three-dimensional box $W^* = [0, a] \times [0, \pi] \times [0, 2\pi]$. Hence:

$$
\begin{aligned}
\mathrm{Vol}\,(W) &= \iiint_W 1\,dx\,dy\,dz \\
&= \iiint_{W^*} 1 \cdot \rho^2 \sin\phi\,d\rho\,d\phi\,d\theta \\
&= \int_0^{2\pi} \left(\int_0^{\pi} \left(\int_0^a \rho^2 \sin\phi\,d\rho \right) d\phi \right) d\theta.
\end{aligned}
$$

The variables of the integrand separate into independent factors and all endpoints are constant, so, as in the two-dimensional case, we can separate the integrals as well:

$$
\begin{aligned}
\mathrm{Vol}\,(W) &= \left(\int_0^a \rho^2\,d\rho \right)\left(\int_0^{\pi} \sin\phi\,d\phi \right)\left(\int_0^{2\pi} 1\,d\theta \right) \\
&= \left(\frac{1}{3}\rho^3 \Big|_0^a \right)\left(-\cos\phi \Big|_0^{\pi} \right)\left(\theta \Big|_0^{2\pi} \right) \\
&= \frac{1}{3}a^3 \cdot \left(1 - (-1) \right) \cdot 2\pi \\
&= \frac{4}{3}\pi a^3.
\end{aligned}
$$

That is:

> The volume of a three-dimensional ball of radius a is $\frac{4}{3}\pi a^3$.

This is often referred to as the volume of a sphere, though that is a misnomer. A sphere is a surface, not a solid.

For those who would like more practice setting up these types of integrals, this example could have been solved using cylindrical coordinates as well, where dx dy dz = r dr dθ dz. We use the order of √integration dz dr dθ. For fixed r and θ, z varies from $z = -\sqrt{a^2 - x^2 - y^2} = -\sqrt{a^2 - r^2}$ to $z = +\sqrt{a^2 - r^2}$. Moreover, the possible values of r and θ come from the two-dimensional disk of radius a in the xy-plane, which in polar coordinates is described by $0 \le r \le a$, $0 \le \theta \le 2\pi$. As a result:

$$\mathrm{Vol}\,(W) = \int_0^{2\pi} \left(\int_0^a \left(\int_{-\sqrt{a^2-r^2}}^{\sqrt{a^2-r^2}} 1 \cdot r\, dz \right) dr \right) d\theta$$

$$= \int_0^{2\pi} \left(\int_0^a rz \Big|_{z=-\sqrt{a^2-r^2}}^{z=\sqrt{a^2-r^2}} dr \right) d\theta$$

$$= \int_0^{2\pi} \left(\int_0^a 2r\sqrt{a^2 - r^2}\, dr \right) d\theta.$$

Since we already calculated the answer, we won't complete the evaluation.

Example 7.9. Let W be the region in R³ inside both the circular cylinder x² + y² = 4 and the sphere x² + y² + z² = 9 and above the xy-plane. Find $\iiint_W (x^2 + y^2) z\, dx\, dy\, dz$.

The region of integration W is a silo-shaped solid bounded laterally by a circular cylinder and capped with a sphere, as drawn in Figure 7.10. Because the base is a disk in the xy-plane, perhaps the simplest way to describe W is with polar coordinates together with the z-coordinate, that is, with cylindrical coordinates. Substituting x = r cos θ, y = r sin θ, and dx dy dz = r dr dθ dz gives:

$$\iiint_W (x^2 + y^2) z\, dx\, dy\, dz = \iiint_{\substack{W\ \mathrm{in} \\ \mathrm{cylindrical}}} r^2 z \cdot r\, dr\, d\theta\, dz = \iiint_{\substack{W\ \mathrm{in} \\ \mathrm{cylindrical}}} r^3 z\, dr\, d\theta\, dz.$$

We describe W in cylindrical coordinates using the order of integration dz dr dθ. For fixed r and θ, z goes from z = 0 to $z = \sqrt{9 - x^2 - y^2} = \sqrt{9 - r^2}$. The domain of values of (r, θ) comes from the base of W , which is a disk of radius 2. Thus 0 ≤ r ≤ 2 and 0 ≤ θ ≤ 2π. Therefore:

$$\iiint_W (x^2 + y^2) z\, dx\, dy\, dz = \int_0^{2\pi} \left(\int_0^2 \left(\int_0^{\sqrt{9-r^2}} r^3 z\, dz \right) dr \right) d\theta$$

$$= \int_0^{2\pi} \left(\int_0^2 \frac{1}{2} r^3 z^2 \Big|_{z=0}^{z=\sqrt{9-r^2}} dr \right) d\theta$$

$$= \int_0^{2\pi} \left(\int_0^2 \frac{1}{2} r^3 (9 - r^2)\, dr \right) d\theta$$

$$= \frac{1}{2} \left(\int_0^2 r^3 (9 - r^2)\, dr \right) \left(\int_0^{2\pi} 1\, d\theta \right)$$

$$= \frac{1}{2} \left(\frac{9}{4} r^4 - \frac{1}{6} r^6 \Big|_0^2 \right) \left(\theta \Big|_0^{2\pi} \right)$$

$$= \frac{1}{2} \cdot \left(36 - \frac{32}{3} \right) \cdot 2\pi$$

$$= \frac{1}{2} \cdot \frac{76}{3} \cdot 2\pi$$

$$= \frac{76}{3} \pi.$$

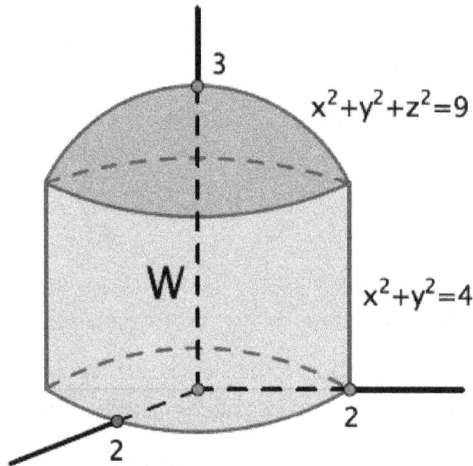

Figure 7.10: The region bounded by a circular cylinder of radius 2, a sphere of radius 3, and the xy-plane

Example 7.10. Find the centroid $(\overline{x}, \overline{y}, \overline{z})$ of the solid W bounded above by the sphere x²+y²+z² = 8 and below by the cone $z = \sqrt{x^2 + y^2}$, shown in Figure 7.11. First, by symmetry, $\overline{x} = 0$ and $\overline{y} = 0$. More precisely, consider:

$$\overline{x} = \frac{1}{\text{Vol}(W)} \iiint_W x \, dx \, dy \, dz.$$

Note that W is symmetric in the yz-plane in the sense that (-x, y, z) is in W whenever (x, y, z) is in W . In addition, the function being integrated, namely, f (x, y, z) = x, satisfies f (-x, y, z) = -x = -f (x, y, z). The same sort of change of variab le s argument that we applied in Example 7.7 to study symmetry for double integrals shows that $\iiint_W x \, dx \, dy \, dz = 0$, and hence $\overline{x} = 0$ The argument that y = 0 is similar using symmetry in the xz-plane.

To find $\overline{z} = \frac{1}{\text{Vol}(W)} \iiint_W z \, dx \, dy \, dz$, we actually must do some calculation. We begin with $\text{Vol}(W) = \iiint_W 1 \, dx \, dy \, dz$. It is simplest to describe W using spherical coordinates. Using the order of integration dρ dφ dθ, for fixed φ and θ, ρ goes along a radial segment from ρ = 0 to $\rho = 2\sqrt{2}$, the radius of the spherical cap. Then, for fixed θ, the angle φ varies from φ = 0 to $\phi = \frac{\pi}{4}$. Lastly, θ goes all the way around from θ = 0 to θ = 2π. Hence:

$$\mathrm{Vol}\,(W) = \iiint_W 1\,dx\,dy\,dz = \iiint_{\substack{W\,\mathrm{in}\\ \mathrm{spherical}}} 1 \cdot \rho^2 \sin\phi\,d\rho\,d\phi\,d\theta$$

$$= \int_0^{2\pi} \left(\int_0^{\frac{\pi}{4}} \left(\int_0^{2\sqrt{2}} \rho^2 \sin\phi\,d\rho \right) d\phi \right) d\theta$$

$$= \left(\int_0^{2\sqrt{2}} \rho^2\,d\rho \right) \left(\int_0^{\frac{\pi}{4}} \sin\phi\,d\phi \right) \left(\int_0^{2\pi} 1\,d\theta \right)$$

$$= \left(\frac{1}{3}\rho^3 \Big|_0^{2\sqrt{2}} \right) \left(-\cos\phi \Big|_0^{\frac{\pi}{4}} \right) \left(\theta \Big|_0^{2\pi} \right)$$

$$= \frac{16\sqrt{2}}{3} \cdot \left(-\frac{\sqrt{2}}{2} - (-1) \right) \cdot 2\pi$$

$$= \frac{32}{3}(\sqrt{2} - 1)\pi.$$

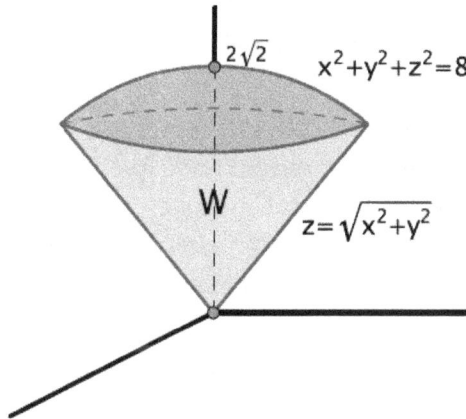

Figure 7.11: An ice cream cone-shaped solid, capped by a sphere of radius $2\sqrt{2}$

Similarly, substituting the conversion z = ρ cos φ:

$$\iiint_W z\,dx\,dy\,dz = \iiint_{\substack{W\,\mathrm{in}\\ \mathrm{spherical}}} \rho\cos\phi \cdot \rho^2 \sin\phi\,d\rho\,d\phi\,d\theta$$

$$= \int_0^{2\pi} \left(\int_0^{\frac{\pi}{4}} \left(\int_0^{2\sqrt{2}} \rho^3 \cos\phi \sin\phi\,d\rho \right) d\phi \right) d\theta$$

$$= \left(\int_0^{2\sqrt{2}} \rho^3\,d\rho \right) \left(\int_0^{\frac{\pi}{4}} \cos\phi \sin\phi\,d\phi \right) \left(\int_0^{2\pi} 1\,d\theta \right)$$

$$= \left(\frac{1}{4}\rho^4 \Big|_0^{2\sqrt{2}} \right) \left(\frac{1}{2}\sin^2\phi \Big|_0^{\frac{\pi}{4}} \right) \left(\theta \Big|_0^{2\pi} \right)$$

$$= 16 \cdot \frac{1}{4} \cdot 2\pi$$

$$= 8\pi.$$

As a result, $\overline{z} = \dfrac{1}{\frac{32}{3}(\sqrt{2}-1)\pi} \cdot 8\pi = \dfrac{3}{4(\sqrt{2}-1)}$, and the centroid of W is the point:

$$(\overline{x}, \overline{y}, \overline{z}) = \left(0, 0, \frac{3}{4(\sqrt{2}-1)}\right).$$

As a reality check, note that this is approximately the point (0, 0, 1.8), whereas the radius of the sphere is $2\sqrt{2} \approx 2.8$. Hence the centroid is roughly two-thirds of the way to the spherical top of the region. Given the top-heavy shape of W , this is plausible.

We close with a quadruple integral.

Example 7.11. Find the volume of the four-dimensional closed unit ball:

$$W = \{(x_1, x_2, x_3, x_4) \in \mathbb{R}^4 : x_1^2 + x_2^2 + x_3^2 + x_4^2 \le 1\}.$$

As usual, we begin with Vol $(W) = \iiiint_W 1\,dx_1\,dx_2\,dx_3\,dx_4$. This can be analyzed with techniques similar to those we used for triple integrals. For instance, say we integrate first with respect to x_4, treating x_1, x_2, and x_3 as fixed. The projection of W on the $x_1x_2x_3$-subspace is the three- dimensional ball B given by $x_1^2 + x_2^2 + x_3^3 \le 1$. For each (x_1, x_2, x_3) in B, t he values of x_4 for which $(x_1, x_2, x_3, x_4) \in$ W range from $x_4 = -\sqrt{1 - x_1^2 - x_2^2 - x_3^2}$ to $x_4 = +\sqrt{1 - x_1^2 - x_2^2 - x_3^2}$.

This is indicated in Figure 7.12, which is supposed to be a drawing in four-dimensional space $\mathbb{R}^4$. Of course, this should be taken with a grain of salt, but we're doing the best we can.

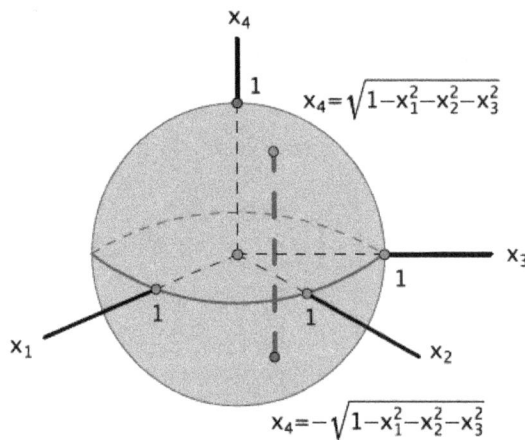

Figure 7.12: The volume of a four-dimensional ball as a triple integral of a single integral, $\iiint(\int \;\dots\; dx_4)\,dx_1\,dx_2\,dx_3$

Hence:

$$\text{Vol}(W) = \iiint_B \left(\int_{-\sqrt{1-x_1^2-x_2^2-x_3^2}}^{\sqrt{1-x_1^2-x_2^2-x_3^2}} 1 \, dx_4 \right) dx_1 \, dx_2 \, dx_3 \qquad (7.5)$$

$$= \iiint_B 2\sqrt{1 - x_1^2 - x_2^2 - x_3^2} \, dx_1 \, dx_2 \, dx_3.$$

This triple integral can be evaluated, say using spherical coordinates, though it gets a little messy. Here is an alternative. One can think of equation (7.5) above as expressing the volume of W as the triple integral of a single integral. We look at it instead as a double integral of a double integral, along the lines of Vol

$$(W) = \iint_{x_3,x_4} \left(\iint_{x_1,x_2} 1 \, dx_1 \, dx_2 \right) dx_3 \, dx_4.$$

More precisely, the projection of W on the x_3x_4-plane is the set of all pairs (x_3, x_4) for which there is at least one (x_1, x_2) such that (x_1, x_2, x_3, x_4) is in W . This is satisfied by all points in the unit disk $x_3^2 + x_4^2 \le 1.$. In fact, given such a (x_3, x_4), the corresponding points (x_1, x_2) are those such that $x_1^2 + x_2^2 \le 1 - x_3^2 - x_4^2$. We write this way of describing W as:

$$\text{Vol}(W) = \iint_{x_3^2+x_4^2 \le 1} \left(\iint_{x_1^2+x_2^2 \le 1-x_3^2-x_4^2} 1 \, dx_1 \, dx_2 \right) dx_3 \, dx_4.$$

See Figure 7.13 (more nonsense).

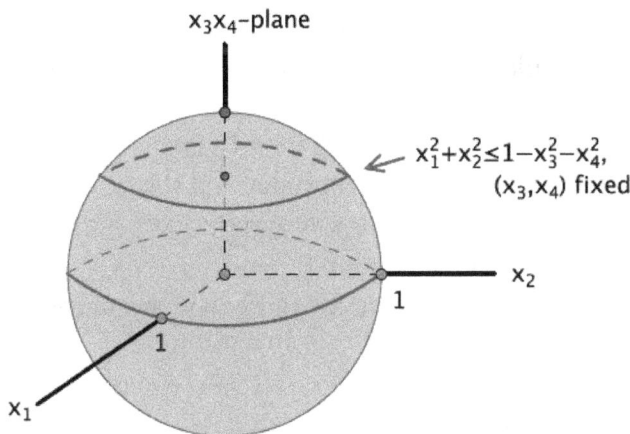

Figure 7.13: The volume of a four-dimensional ball as a double integral of a double integral, $\iint (\iint \ldots dx_1 \, dx_2) \, dx_3 \, dx_4$

The inner integral represents the area of a two-dimensional disk of radius $a = \sqrt{1 - x_3^2 - x_4^2}$ in the x_1x_2-plane. Hence $\iint_{x_1^2+x_2^2 \le 1-x_3^2-x_4^2} 1 \, dx_1 \, dx_2 = \pi a^2 = \pi(1 - x_3^2 - x_4^2).$

whence:

$$\text{Vol}\,(W) = \iint_{x_3^2+x_4^2 \le 1} \pi(1 - x_3^2 - x_4^2)\,dx_3\,dx_4.$$

This is an integral over the unit disk in the $x_3 x_4$-plane, so it can be evaluated using polar coordinates with $dx_3\,dx_4 = r\,dr\,d\theta$:

$$\begin{aligned}
\text{Vol}\,(W) &= \int_0^{2\pi}\left(\int_0^1 \pi(1 - r^2)\cdot r\,dr\right)d\theta \\
&= \pi\left(\int_0^1 r(1 - r^2)\,dr\right)\left(\int_0^{2\pi} 1\,d\theta\right) \\
&= \pi\left(\frac{1}{2}r^2 - \frac{1}{4}r^4\Big|_0^1\right)\left(\theta\Big|_0^{2\pi}\right) \\
&= \pi\cdot\frac{1}{4}\cdot 2\pi \\
&= \frac{\pi^2}{2}.
\end{aligned}$$

We leave for the exercises the problems of finding the volume of a five-dimensional ball and, more broadly, a strategy for approaching the volume of an n-dimensional ball in general. See Exercises 4.9 and 4.10.

7.5 EXERCISES FOR CHAPTER 7

Section 1 Change of variables for double integrals

This exercise involves two standard one-variable integrals that appear regularly enough that it seems like a good idea to get out into the open how they can be evaluated quickly. Namely, we compute the integrals $\int_0^{n\pi} \cos^2 x\,dx$ and $\int_0^{n\pi} \sin^2 x\,dx$. where n is a positive integer. They can be found using trigonometric identities, but there is another way that is easier to reproduce on the spot.

a. First, sketch the graphs $y = \cos^2 x$ and $y = \sin^2 x$ for x in the interval $[0, n\pi]$, and use them to illustrate that $\int_0^{n\pi} \cos^2 x\,dx = \int_0^{n\pi} \sin^2 x\,dx$.

b. This is an optional exercise for those who are uneasy about drawing the conclusion in part (a) based only on a picture.

    i.   Show that $\int_0^{n\pi}\sin^2 x\,dx = \int_{\frac{\pi}{2}}^{n\pi+\frac{\pi}{2}}\cos^2 x\,dx$. (Hint: Use the substitution $u = x + \frac{\pi}{2}$ and the identify $\sin(\theta - \frac{\pi}{2}) = -\cos\theta$.)

    ii. Show that $\int_0^{\frac{\pi}{2}}\cos^2 x\,dx = \int_{n\pi}^{n\pi+\frac{\pi}{2}}\cos^2 x\,dx$. (Hint: $\cos(\theta - \pi) = -\cos\theta$.)
Deduce that $\int_0^{n\pi}\cos^2 x\,dx = \int_0^{n\pi}\sin^2 x\,dx$.

c. Integrate the identity $\cos^2 x + \sin^2 x = 1$, and use part (a) (or (b)) to show that:

$$\int_0^{n\pi}\cos^2 x\,dx = \frac{n\pi}{2} \qquad \text{and} \qquad \int_0^{n\pi}\sin^2 x\,dx = \frac{n\pi}{2}.$$

1.2. Find $\iint_D y^2 \, dx \, dy$, where D is the upper half-disk described by x2 + y2 ≤ 1 and y ≥ 0.

1.3. Find $\iint_D xy \, dx \, dy$, where D is the region in the first quadrant lying inside the circle $x^2 + y^2 = 4$ and below the line y = x.

1.4. Find $\iint_D \cos(x^2 + y^2) \, dx \, dy$, where D is the region described by 4 ≤ x2 + y2 ≤ 16.

1.5. Let D be the region in the xy-plane satisfying x2 + y2 ≤ 2, y ≥ x, and x ≥ 0. See Figure 7.14.

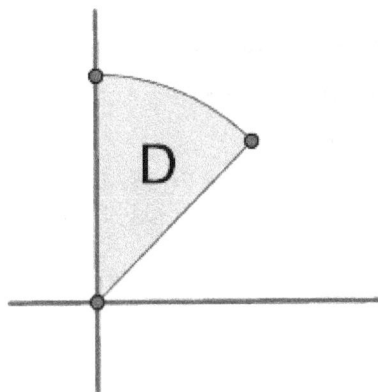

Figure 7.14: The region $x^2 + y^2$ ≤ 2, y ≥ x, and x ≥ 0

Consider the integral $\iint_D \sqrt{2 - x^2} \, dx \, dy$.

    a. Write an expression for the integral using the order of integration dy dx.

    b. Write an expression for the integral using the order of integration dx dy.

    c. Write an expression for the integral in polar coordinates in whatever order of integration you prefer.

    d. Evaluate the integral using whichever approach seems best.

1.6. Let W be the region in R^3 lying above the xy-plane, inside the cylinder $x^2 + y^2 = 1$, and below the plane x + y + z = 2. Find the volume of W .

1.7. Let W be the region in R^3 lying above the xy-plane, inside the cylinder $x^2 + y^2 = 1$, and below the plane x + y + z = 1. Find the volume of W .

1.8. In this exercise, we evaluate the improper one-variable integral $\int_{-\infty}^{\infty} e^{-x^2} \, dx$ by following the unlikely strategy of relating it to an improper double integral – th∞at turns out to be more tractable. Let a be a positive real number.

    a. Let R_a be the rectangle $R_a = [-a, a] \times [-a, a]$. Show that:

$$\iint_{R_a} e^{-x^2 - y^2} \, dx \, dy = \left(\int_{-a}^{a} e^{-x^2} \, dx \right)^2 .$$

    b. Let D_a be the disk $x^2 + y^2$ ≤ a^2. Use polar coordinates to evaluate

$$\iint_{D_a} e^{-x^2-y^2} \, dx \, dy.$$

c. Note that, as a goes to ∞, both Ra and Da fill out all of $\mathbb{R}^2$. It is true that both $\lim_{a\to\infty} \iint_{R_a} e^{-x^2-y^2} \, dx \, dy$ and $\lim_{a\to\infty} \iint_{D_a} e^{-x^2-y^2} \, dx \, dy$ exist and that they are equal. Their common value the improper integral $\iint_{\mathbb{R}^2} e^{-x^2-y^2} \, dx \, dy$. Us this information along with your answers (a) and (b) to show that:

$$\int_{-\infty}^{\infty} e^{-x^2} \, dx = \sqrt{\pi}.$$

Section 3 Examples: linear changes of variables, symmetry

3.1. Let D be the set of points (x, y) in $\mathbb{R}^2$ such that $(x - 3)^2 + (y - 2)^2 \le 4$.

  a. Sketch and describe D.
  b. Let T : $\mathbb{R}^2 \to \mathbb{R}^2$ be the translation T (u, v) = (u, v) + (3, 2) = (u + 3, v + 2). Describe the region D* in the uv-plane such that T (D*) = D.

  c. Use the change of variables in part (b) to convert the integral $\iint_D (x+y) \, dx \, dy$ over D to an integral over D*. Then, evaluate the integral using whatever techniques seem best.

3.2. Let a and b be positive real numbers. Find the area of the region $\frac{x^2}{a^2} + \frac{y^2}{b^2} \le 1$ inside an ellipse by using the change of variables T (u, v) = (au, bv). (Recall from Chapter 5 that Area $(D) = \iint_D 1 \, dA$.)

3.3. Let T : $\mathbb{R}^2 \to \mathbb{R}^2$ be the linear change of variables T (u, v) = (3u − v, 2u + v) of Example 7.6.

  a. Show that the only point (u, v) that satisfies T (u, v) = 0 is (u, v) = 0.
  b. Show that T is one-to-one on $\mathbb{R}^2$. (Hint: Start by assuming that T (a) = T (b), and use the linearity of T and part (a).)

3.4. Find $\iint_D (2x + 4y) \, dx \, dy$ if:

  a. D is the parallelogram with vertices (0, 0), (3, 1), (5, 5), and (2, 4),
  b. D is the triangular region with vertices (3, 1), (5, 5), and (2, 4). (Hint: Take advantage of the work you've already done in part (a). You should be able to use quite a bit of it.)

3.5. Find $\iint_D (x+y) \, e^{x^2-y^2} \, dx \, dy$, where D is the rectangle with vertices (0, 0), (1, −1), (3, 1), and (2, 2).

3.6. Find $\iint_D \sqrt{12 + x^2 + 3y^2} \, dx \, dy$, where D is the region in $\mathbb{R}^2$ described by $x^2 + 3y^2 \le 12$ and $0 \le y \le \frac{1}{\sqrt{3}} x$.

3.7. This problem concerns the integral:

$$\iint_D \frac{y}{x} \, e^{xy} \, dx \, dy. \tag{7.6}$$

where D is the region in the first quadrant of the xy-plane that is bounded by the curves y = x, y = 4x, xy = 1, and xy = 2. See Figure 7.15, left. The main idea is to simplify the integrand by making the substitutions u = xy and $v = \frac{y}{x}$.

    a. With these substitutions, solve for x and y as functions of u and v: (x, y) = T (u, v).

    b. Describe the region D∗ in the uv-plane such that T (D') = D.

    c. Evaluate the integral (7.6).

Figure 7.15: The region bounded by y = x, y = 4x, xy = 1, and xy = 2 (left) and the region x² - 4xy + 8y² ≤ 4 (right)

3.8. Let D be the region in $\mathbb{R}^2$ described by x² - 4xy + 8y² ≤ 4. See Figure 7.15, right. Find $\iint_D y^2 \, dx \, dy.$ (Hint: x² - 4xy + 8y² = (x - 2y)² + 4y².)

3.9. Find $\iint_D \frac{(x-y)^2}{(x+y)^2} \, dx \, dy.$ where D is the square with vertices (2, 0), (4, 2), (2, 4), and (0, 2).

3.10. Let T : $\mathbb{R}^2 \to \mathbb{R}^2$ be a one-to -one] linear transformation from the uv-plane to the xy-lane represented by a matrix $A = \begin{bmatrix} a & b \\ c & d \end{bmatrix}$, i.e., $(x, y) = T(u, v) = A \cdot \begin{bmatrix} u \\ v \end{bmatrix}$. Let D∗ and D be bounded regions in $\mathbb{R}^2$ of positive area such that T maps D∗ onto D, as shown in Figure 7.16.

    a. Let k = | det A|. Use the change of variables theorem to show that:

$$\text{Area (D)} = k \cdot \text{Area (D')}).$$

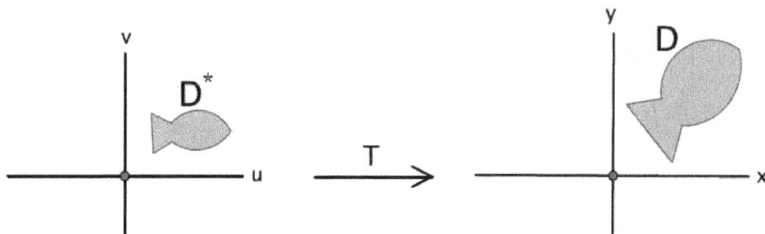

Figure 7.16: A linear change of variables

Let $(\bar{u}, \bar{v})$ and $(\bar{x}, \bar{y})$ denote the centroids of D* and D, respectively. Show that:

$$(\bar{x}, \bar{y}) = T(\bar{u}, \bar{v}).$$

In other words, linear transformations map centroids to centroids.

3.11. A region D in the xy-plane is called **symmetric in the line y = x** if, whenever a point (x, y) is in D, so is its reflection (y, x). If D is such a region, prove that its centroid lies on the line y = x.

3.12. Let D be a subset of $\mathbb{R}^2$ that is symmetric about the origin, that is, whenever (x, y) is in D, so is (-x, -y).

   a. Let $f: D \to R$ be an integrable function on D. Find a condition on f that ensures that $\iint_D f(x, y)\, dx\, dy = 0.$ Justify your answer.
   b. Find an example of such a function other than the constant function $f = 0$.

Section 4 Change of variables for n-fold integrals

4.1. Find $\iiint_W (x^2 + y^2 + z^2)\, dx\, dy\, dz$ if W is the region in R^3 that lies above the cone z = $\sqrt{x^2 + y^2}$ and below the plane z = 2.

4.2. Find $\iiint_W x^2 z\, dx\, dy\, dz$ if W is the region in R3 described by $x^2 + y^2 + z^2 \le 4$, y ≥ 0, and z ≥ 0.

4.3. Find $\iiint_W \frac{z}{(3 + x^2 + y^2)^2}\, dx\, dy\, dz$ if W is the region in R^3 given by $x^2 + y^2 + z^2$ 1 and z ≥ 0. (Hint: The proper coordinate system and order of integration can make a difference.)

4.4. A class of enthusiastic multivariable calculus students celebrates the change of variables theorem by drilling a cylindrical hole of radius 1 straight through the center of the earth. Assuming that the earth is a three-dimensional closed ball of radius 2, find the volume of the portion of the earth that remains.

4.5. Find the centroid of the half-ball of radius a in R3 that lies inside the sphere $x^2 + y^2 + z^2 = a^2$ and above the xy-plane.

4.6. Let W be the wedge-shaped solid in R3 that lies under the plane z = y, inside the cylinder $x^2 + y^2 = 1$, and above the xy-plane. Find the centroid of W .

4.7. Let W be the region in R^3 that lies inside the three cylinders $x^2 + z^2 = 1$, $y^2 + z^2 = 1$, $x^2 + y^2 = 1$, and above the xy-plane. Find the volume of W . (Hints: The integral

$$\int \frac{1}{\cos^2 \theta}\, d\theta = \int \sec^2\theta\, d\theta = \tan\theta + C$$

might be helpful. Also, $\sin^3 \theta = \sin^2 \theta \cdot \sin \theta = (1 - \cos^2 \theta) \sin \theta$.)

4.8. Within the solid ball $x^2 + y^2 + z^2 \le a^2$ of radius a in R^3, find the average distance to the origin.

4.9. Find the volume of the five-dimensional closed unit ball:

$$W = \{(x_1, x_2, x_3, x_4, x_5) \in \mathbb{R}^5 : x_1^2 + x_2^2 + x_3^2 + x_4^2 + x_5^2 \le 1\}.$$

(Hint: Organize the calculation as a double integral of a triple integral: $\iint (\iiint \ldots).$)

4.10. If a > 0, let Wn(a) denote the closed ball $x_1^2 + x_2^2 + \cdots + x_n^2 \le a^2$ in $\mathbb{R}^n$ of radius a centered at the origin, and let Vol (W$_n$(a)) denote its n-dimensional volume. For instance, we showed in Example 7.11 that Vol $(W_4(1)) = \frac{\pi^2}{2}$.

    a. Use the change of variables theorem to show that $\mathrm{Vol}\,(W_n(a)) = a^n\,\mathrm{Vol}\,(W_n(1))$.

    b. By thinking of Vol (W$_n$(1)) as a double integral of an (n - 2)-fold integral, find a relationship between Vol (W$_n$(1)) and Vol (W$_{n-2}$(1)).

    c. Verify that your answer to part (b) predicts a correct relationship between Vol (W$_4$(1)) and Vol (W$_2$(1)).

    d. Find Vol (W$_6$(1)), the volume of the six-dimensional closed unit ball.

PART-V
INTEGRALS OF VECTOR FIELDS

8 Chapter

VECTOR FIELDS

Now that we know what it means to differentiate vector-valued functions of more than one variable, we turns to integrating them. The functions that we integrate, however, are of a special type. The integrals we consider are rather specialized, too. The functions are called vector fields, and this short chapter is devoted to introducing them. The integration begins in the next chapter.

8.1 EXAMPLES OF VECTOR FIELDS

The domain of a vector field F is a subset U of $\mathbb{R}^n$. The values of F are in $\mathbb{R}^n$ as well, but we don't think so much of F as transforming U into its image. Rather, for a point x of U , we visualize the value F(x) as an arrow translated to start at x.

For example, F might represent a force exerted on an object, and the force might depend on where the object is. Then F(x) could represent the force exerted if the object is at the point x. Or some medium might be flowing through U , and F(x) could represent the velocity of the medium at x.

By taking a sample of points throughout U and drawing the corresponding arrows translated to start at those points, one can see how some vector quantity acts on U. For instance, with the aid of the arrows, one might be able to visualize the path an object would follow under the influence of a force or a flowing medium.

With this prelude, the actual definition is quite simple.

Definition. Let U be a subset of $\mathbb{R}^n$. A function $F : U \to \mathbb{R}^n$ is called a vector field on U .

A distinctive feature of vector fields is that both the elements x of the domain and their values F(x) are in $\mathbb{R}^n$, as opposed to a function in general from $\mathbb{R}^n$ to Rm.

Example 8.1. Let $F : \mathbb{R}^2 \to \mathbb{R}^2$ be given by F(x, y) = (1, 0).

This vector field assigns the vector (1, 0) = i to every point of the plane. To visualize the vector field, we draw the vector (1, 0), or rather a translated copy of it, at a representative collection of points. That is, at each such point, we draw an arrow of length 1 pointing to the right. This is a constant vector field. It is sketched on the left of Figure 8.1.

As the figure shows, the sample vectors often end up running into each other, making the picture a little muddled. As a result, we usually depict a vector field by drawing a constant positive scalar multiple of the vectors F(x), where the scalar factor is chosen to improve the readability of the picture. For the vector field in the last example, this is done on the right of the figure. The lengths of the vectors are scaled down, but the arrows still point in the correct direction.

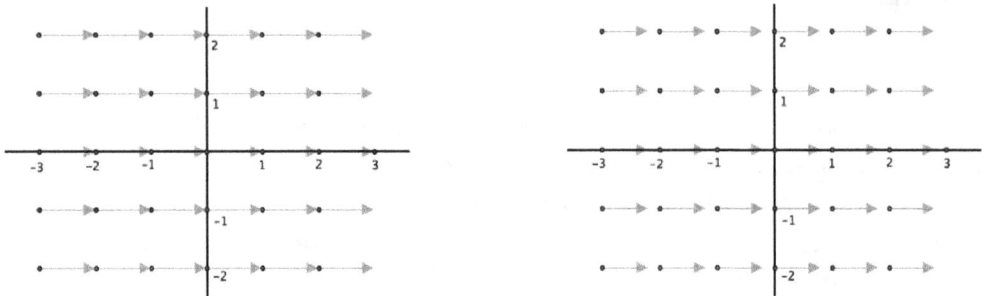

Figure 8.1: The constant vector field F(x, y) = (1, 0) = i literally (left) and scaled down (right)

Example 8.2. Let $F : \mathbb{R}^2 \to \mathbb{R}^2$ be given by F(x, y) = (-y, x).

For example, F(1, 0) = (-0, 1) = (0, 1) and F(0, 1) = (-1, 0), as in Figure 8.2.

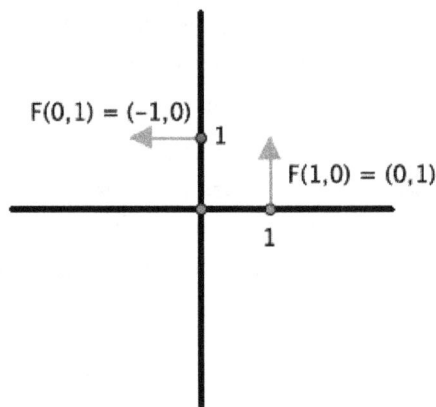

Figure 8.2: The vector field F(x, y) = (-y, x) at the points (1, 0) and (0, 1)

We can get some qualitative information of how F acts by noting, for example, that in the first quadrant, where both x and y are positive, F(x, y) = (-y, x) has a negative first component and positive second component, i.e., F(+, +) = (-, +), so the arrow points to the left and up.

Similarly, in the second quadrant F(-, +) = (-, -) (to the left and down), in the third quadrant F(-, -) = (+, -) (to the right and down), and in the fourth F(+, -) = (+, +) (to the right and up).

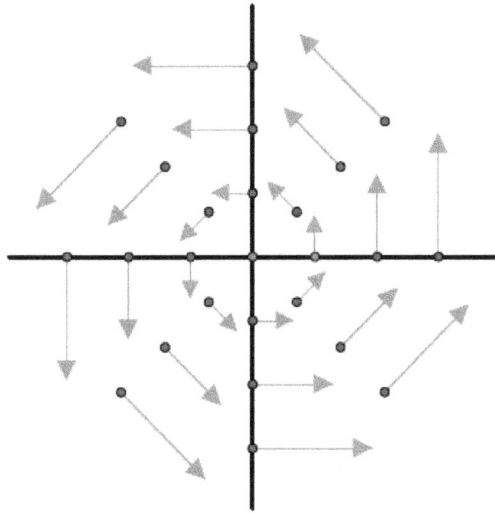

Figure 8.3: The vector field F(x, y) = (-y, x)

In fact, we can identify both the length and direction of F(x, y) more precisely:

- $\|\mathbf{F}(x,y)\| = \sqrt{(-y)^2 + x^2} = \|(x,y)\|,$

- F(x, y) · (x, y) = (-y, x) · (x, y) = -yx + xy = 0.

In other words, the length of F(x, y) equals the distance of (x, y) to the origin, and its direction is orthogonal to (x, y) in the counterclockwise direction. After drawing in a sample of arrows, one can imagine F as describing a circular counterclockwise flow, or vortex, around the origin, growing in magnitude as one moves further out, as in Figure 8.3.

Example 8.3. Let $W : \mathbb{R}^2 - \{(0, 0)\} \to \mathbb{R}^2$ be given by $\mathbf{W}(x,y) = \left(-\frac{y}{x^2+y^2}, \frac{x}{x^2+y^2}\right),$ where (x, y)≠(0,0)

This is a nonconstant positive scalar multiple of the previous example. The scalar factor is $\frac{1}{x^2+y^2}$. Thus the vectors point in the same direction as before, but now:

$$\|\mathbf{W}(x,y)\| = \frac{1}{x^2 + y^2}\|(x,y)\| = \frac{\|(x,y)\|}{\|(x,y)\|^2} = \frac{1}{\|(x,y)\|}.$$

Hence the vector field also circulates about the origin, but its length is inversely proportional to the distance to the origin. See Figure 8.4. We shall see later that this vector field has some notable features.

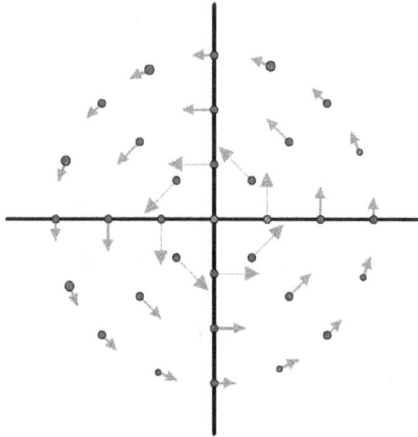

Figure 8.4: The vector field $\mathbf{W}(x, y) = \left(-\frac{y}{x^2+y^2}, \frac{x}{x^2+y^2}\right)$

Example 8.4 (The inverse square field). An important vector field G from classical physics is characterized by the following properties.

- ⊙ G is defined at every point of $\mathbb{R}^3$ except the origin.

- ⊙ The length of G is inversely proportional to the square of the distance to the origin. In other words, $\|\mathbf{G}(x, y, z)\| = \frac{c}{\|(x,y,z)\|^2}$ for some positive constant c.

- ⊙ The direction of G(x, y, z) is from (x, y, z) to the origin. That is, G(x, y, z) is a negative scalar multiple of (x, y, z).

See Figure 8.5. For instance, in physics, the gravitational force due to a mass at the origin and the electrostatic force due to a charged particle at the origin are both modeled by such a field.

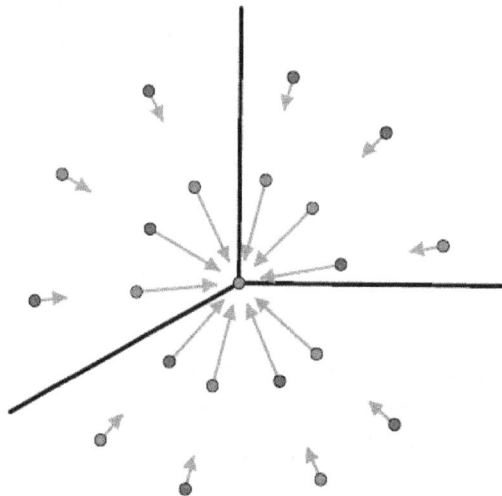

Figure 8.5: An inverse square vector field

To find a formula for G, we write $G(x,y,z) = \frac{c}{\|(x,y,z)\|^2}u$, where u is the unit vector in the direction from (x, y, z) to (0, 0, 0), that is, $u = -\frac{(x,y,z)}{\|(x,y,z)\|}$. Hence

$$G(x,y,z) = -\frac{c}{\|(x,y,z)\|^2} \cdot \frac{(x,y,z)}{\|(x,y,z)\|} = -\frac{c}{\|(x,y,z)\|^3}(x,y,z).$$

This is called an inverse square field. To summarize, it is the vector field $G : \mathbb{R}^3 -\{(0, 0, 0)\} \to \mathbb{R}^3$ given variously by:

$$G(x,y,z) = -\frac{c}{\|(x,y,z)\|^3}(x,y,z)$$

$$= -\frac{c}{(x^2+y^2+z^2)^{3/2}}(x,y,z)$$

$$= \left(-\frac{cx}{(x^2+y^2+z^2)^{3/2}}, -\frac{cy}{(x^2+y^2+z^2)^{3/2}}, -\frac{cz}{(x^2+y^2+z^2)^{3/2}}\right).$$

for some positive constant c.

In the future, we take the constant of proportionality to be c = 1 and refer to the resulting vector field G as "the" inverse square field.

These examples shall often serve as test cases for the concepts we are about to develop.

8.2 EXERCISES FOR CHAPTER 8

In Exercises 1.1–1.6, sketch the given vector field F. To get started, it may be helpful to get an idea of the general direction of the arrows F(x, y) at points on the x and y-axes and in each of the four quadrants.

1.1. F(x, y) = (x, y)

1.2. F(x, y) = (2x, y)

1.3. F(x, y) = (y, -x)

1.4. F(x, y) = (-2y, x)

1.5. F(x, y) = (y, x)

1.6. F(x, y) = (-x, y)

1.7   a. Find a smooth vector field F on $\mathbb{R}^2$ such that, at each point (x, y), F(x, y) is a unit vector normal to the parabola of the form $y = x^2 + c$ that passes through that point.

   b. Find a smooth vector field F on $\mathbb{R}^2$ such that, at each point (x, y), F(x, y) is a unit vector tangent to the parabola of the form $y = x^2 + c$ that passes through that point.

1.8.   a. Find a smooth vector field F on $\mathbb{R}^2$ such that $\|F(x, y)\| = \|(x, y)\|$ and, at each point (x, y) other than the origin, F(x, y) is a vector normal to the curve of the form $xy = c$ that passes through that point.

b. Find a smooth vector field F on $\mathbb{R}^2$ such that $\|F(x, y)\| = \|(x, y)\|$ and, at each point (x, y) other than the origin, F(x, y) is a vector tangent to the curve of the form xy = c that passes through that point.

Let F be a vector field on an open set U in $\mathbb{R}^n$. If we think of F as the velocity of a medium flowing through U , then an object dropped into U will follow a path along which the velocity at each point is the value of the vector field at the point. More precisely, a smooth path α : I U defined on an interval I is called an **integral path** of F if α'(t) = F(α(t)) for all t in I. The curve traced out by α is called an **integral curve**.

For example, let F(x, y) = (-y, x) be the circulating vector field from Example 8.2. If a > 0, let α : $\mathbb{R} \to \mathbb{R}^2$ be the parametrization of a circle of radius a given by:

$$\alpha(t) = (a \cos t, a \sin t). \tag{8.1}$$

Then α'(t) = (-a sin t, a cos t), and F(α(t)) = F(a cos t, a sin t) = (-a sin t, a cos t). Since the two are equal, α is an integral path of F. The integral curves are circles centered at the origin. This reinforces our sense that F describes counterclockwise circular flow around the origin.

1.9. Let F(x, y) = (-2y, x). (This vector field also appears in Exercise 1.4.)

a. If a is a real number, show that $\alpha(t) = (\sqrt{2}a \cos(\sqrt{2}t), a \sin(\sqrt{2}t))$ is an integral path of F.

b. Show that the integral curves satisfy the equation $\frac{x^2}{2} + y^2 = a^2$, and sketch a representative sample of them. Indicate the direction in which the curves are traversed by the integral paths.

1.10. Let F(x, y) = (y, x). (This vector field also appears in Exercise 1.5.)

a. If a is a real number, show that α(t) = a($e^t + e^{-t}$), a($e^t - e^{-t}$)) and α(t) = a($e^t - e^{-t}$), a($e^t + e^{-t}$) are integral paths of F.

b. Describe the integral curves, and sketch a representative sample of them. Indicate the direction in which the curves are traversed by the integral paths. (Hint: To find a relationship between x and y on the integral paths, compute x(t)² and y(t)².)

1.11. Find integral paths for the vector field $W(x, y) = (-\frac{y}{x^2+y^2}, \frac{x}{x^2+y^2})$, where (x, y)≠ (0, 0), from Example 8.3 by modifying the integral paths for F(x, y) = (-y, x) given in equation (8.1).

Finding explicit formulas for the integral paths of a vector field may be difficult in practice. To simplify the discussion, we restrict ourselves to two-dimensional vector fields F(x, y) = (F_1(x, y), F_2(x, y)) defined on open sets in $\mathbb{R}^2$. If α(t) = (x(t), y(t)) is a path, then the condition α'(t) = F(α(t)) that it must satisfy to be an integral path can be written in terms of coordinates as $(\frac{dx}{dt}, \frac{dy}{dt}) = (F_1(x(t), y(t)), F_2(x(t), y(t)))$. In other words, the components x(t) and y(t) of α(t) must satisfy the differential equations:

$$\begin{cases} \dfrac{dx}{dt} = F_1(x, y) \\[2mm] \dfrac{dy}{dt} = F_2(x, y). \end{cases} \tag{8.2}$$

Let (x_0, y_0) be a point of $\mathbb{R}^2$. For the vector fields in Exercises 1.12–1.13, (a) solve the differential equations (8.2) to find the integral path $\alpha(t) = (x(t), y(t))$ of F that satisfies the condition $\alpha(0) = (x_0, y_0)$ and (b) describe the corresponding integral curves of F geometrically and sketch a representative sample of them, including the direction in which they are traversed by the integral paths. (Note that these vector fields also appear in Exercises 1.1 and 1.6, respectively.)

1.12. $F(x, y) = (x, y)$

1.13. $F(x, y) = (-x, y)$

9

Chapter

LINE INTEGRALS

We begin integrating vector fields in the case that the domain of integration is a curve. The integrals are called line integrals. By comparison, we learned in Section 2.3 about the integral over a curve C of a real-valued function f. By definition, $\int_C f\,ds = \int_a^b f(\alpha(t))\,\|\alpha'(t)\|\,dt$. where $\alpha : [a, b] \to \mathbb{R}^n$ is a smooth parametrization of C. This is called the integral with respect to arclength.

The integral of a vector field is a new concept—for example, we don't simply integrate each of the real-valued component functions of the vector field independ,ently using the integral with respect to arclength. In the grand scheme of things, line integrals are a richer subject than integrals with respect to arclength. In part, this is because of the ways in which the line integral can be interpreted and applied, many of which originate in physics (for instance, see Exercise 3.19 at the end of the chapter). But in addition, on a more abstract level, line integrals provide us with a first peek into a line of unifying results that relate various types of integrals to one another. By the end of the book, we shall be in a position to see that these connections run quite deep.

We open the chapter by defining the line integral and discussing what it represents. There can be more than one option for how to compute a given line integral, and we identify the circumstances under which the different approaches apply. These computational techniques, however, follow from theoretical principles that are interesting in their own right. For instance, for vector fields in the plane, some line integrals can be expressed as double integrals. This result is known as Green's theorem. Throughout the chapter, the discussion reveals an active interplay between geometric features of the curve, the vector field, and, on occasion, the geometry of the domain on which the vector field is defined as well.

9.1 DEFINITIONS AND EXAMPLES

Let $F : U \to \mathbb{R}^n$ be a vector field on an open set U in $\mathbb{R}^n$, and let C be a curve in U . Roughly speaking, the integral of F over C is meant to represent the "effectiveness" of F in contributing to motion along C. By this, we mean the degree of alignment of F with the direction of motion. For instance, suppose that F is a force field and that $\alpha(t)$ is the position of a moving object at time t. Then the line integral of F over C is called the work done by F. If F points in the same direction as the direction of motion, then F helps move the object along. If it points in the opposite direction, then it impedes the motion. If it is orthogonal, it neither helps nor hinders.

In general, we measure the contribution of F to the motion at each point of C by finding the scalar component F_{tan} of F in the direction tangent to the curve. See Figure 9.1. Then, we integrate with respect to arclength, obtaining $\int_C F_{tan}\, ds$.

Note that $F_{tan} = \|F\| \cos \theta$, where θ is the angle between F and the tangent direction. Hence, If $T = \frac{\alpha'}{\|\alpha'\|}$ is the unit tangent vector, then:

$$F_{tan} = \|\mathbf{F}\| \cos \theta = \|\mathbf{F}\| \, \|\mathbf{T}\| \cos \theta = \mathbf{F} \cdot \mathbf{T}.$$

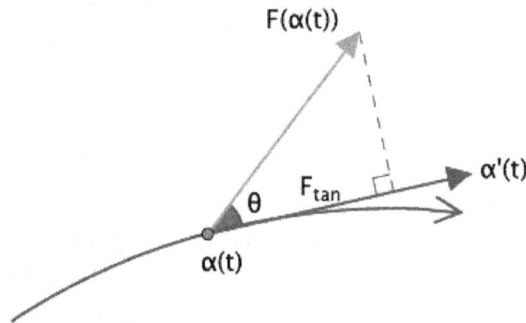

Figure 9.1: The component F_{tan} of a vector field in the tangent direction

By the definition of the integral with respect to arclength with $f = F_{tan}$, we obtain:

$$\int_C F_{tan}\, ds = \int_C \mathbf{F} \cdot \mathbf{T}\, ds = \int_a^b \left(\mathbf{F}(\alpha(t)) \cdot \mathbf{T}(t) \right) \|\alpha'(t)\|\, dt$$

$$= \int_a^b \left(\mathbf{F}(\alpha(t)) \cdot \frac{\alpha'(t)}{\|\alpha'(t)\|} \right) \|\alpha'(t)\|\, dt$$

$$= \int_a^b \mathbf{F}(\alpha(t)) \cdot \alpha'(t)\, dt.$$

This motivates the following definition.

Definition. Let U be an open set in $\mathbb{R}^n$, and let $\mathbf{F} : U \to \mathbb{R}^n$ be a continuous vector field on U . If $\alpha : [a, b] \to U$ is a smooth path in U , then the **line integral**, denoted by $\int_\alpha \mathbf{F} \cdot ds$, is defined to be:

$$\int_\alpha \mathbf{F} \cdot ds = \int_a^b \mathbf{F}(\alpha(t)) \cdot \alpha'(t)\, dt.$$

Written out in coordinates, say as $F = (F_1, F_2, \ldots, F_n)$ and $\alpha = (x_1, x_2, \ldots, x_n)$, the definition becomes:

$$\int_\alpha \mathbf{F} \cdot ds = \int_a^b (F_1 \circ \alpha, F_2 \circ \alpha, \dots, F_n \circ \alpha) \cdot \left(\frac{dx_1}{dt}, \frac{dx_2}{dt}, \dots, \frac{dx_n}{dt}\right) dt$$

$$= \int_a^b \left((F_1 \circ \alpha)\frac{dx_1}{dt} + (F_2 \circ \alpha)\frac{dx_2}{dt} + \cdots + (F_n \circ \alpha)\frac{dx_n}{dt}\right) dt. \qquad (9.1)$$

This suggests an alternative notation for the line integral:

$$\int_\alpha \mathbf{F} \cdot ds = \int_\alpha F_1\, dx_1 + F_2\, dx_2 + \cdots + F_n\, dx_n. \qquad (9.2)$$

The integrand $F_1\, dx_1 + F_2\, dx_2 + \cdots + F_n$ dxn is called a **differential form**, or, more precisely, a **differential 1-form**. This notation is very convenient, because it is easy to remember how to get from (9.2) to (9.1) in order to evaluate the integral. One simply uses the parametrization to substitute for $x_1, x_2, \dots, x_n$ in terms of the parameter t wherever they appear in the integrand. This includes the substitution $dx_i = \frac{dx_i}{dt}\, dt$ for i = 1, 2, ..., n. We use differential form notation for line integrals whenever possible.

One final word about notation: we have chosen to write the line integral as an integral over α rather than over the underlying curve C to acknowledge that the definition makes use of the specific parametrization α. This is an important point, and we elaborate on it momentarily.

Just as in first-year calculus a function can be integrated over many intervals, so can a vector field be integrated over many paths. The relation between the vector field and the path is reflected in the value of the integral.

Example 9.1. Let F be the vector field on $\mathbb{R}^2$ given by F(x, y) = (-y, x). This is the vector field from Chapter 8 that circulates around the origin, shown again in Figure 9.2.

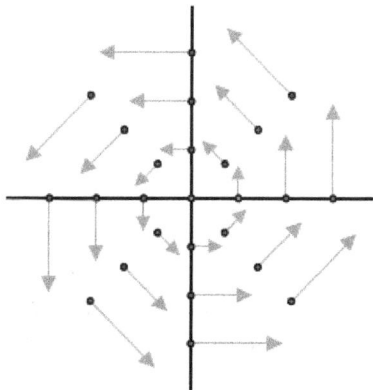

Figure 9.2: The circulating vector field F(x, y) = (-y, x) Integrate F over the following paths.

a. $\alpha_1(t) = (\cos t, \sin t), 0 \le t \le 2\pi$

b. $\alpha_2(t) = (t, t), 0 \le t \le 3$

c. $\alpha_3(t) = (\cos t, \sin t), 0 \le t \le \pi/2$ (d)

d. $\alpha_4(t) = (1 - t, t), 0 \le t \le 1$

e. $\alpha_5(t) = (\cos 2t, \sin 2t), 0 \le t \le \pi/2$

In differential form notation, the integrals $\alpha \int_\alpha F \cdot ds$ we seek are written $\int_\alpha -y\, dx + x\, dy$.

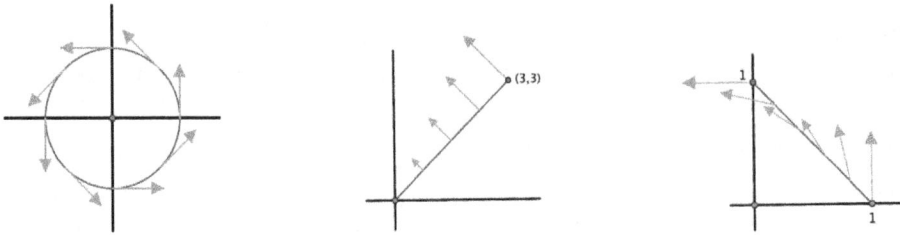

Figure 9.3: The vector field F(x, y) = (-y, x) along the unit circle (left), the line y = x (middle), and the line x + y = 1 (right)

a. The path α_1 traces out the unit circle counterclockwise. The direction of the vector field and the direction of motion are aligned at every point. See Figure 9.3 at left. As a result, $F \cdot T$ is always positive, and we expect the integral to be positive as well. In fact, for this parametrization, x = cos t and y = sin t, so substitution gives:

$$\int_{\alpha_1} -y\, dx + x\, dy = \int_0^{2\pi} \left(-y(t)\frac{dx}{dt} + x(t)\frac{dy}{dt}\right) dt$$

$$= \int_0^{2\pi} \left(-\sin t \cdot (-\sin t) + \cos t \cdot \cos t\right) dt$$

$$= \int_0^{2\pi} 1\, dt$$

$$= t\Big|_0^{2\pi}$$

$$= 2\pi.$$

b. The path α2 traverses a line segment radiating out from (0, 0) to (3, 3) along the line y = x. At every point of this path, the vector field is orthogonal to the direction of motion (Figure 9.3, middle). In other words, $F \cdot T = 0$ at every point, so we would expect the integral to be zero, too.

To confirm this:

$$\int_{\alpha_2} -y\,dx + x\,dy = \int_0^3 (-t \cdot 1 + t \cdot 1)\,dt = \int_0^3 0\,dt = 0.$$

c. This again moves along the unit circle, but only over the arc from (1, 0) to (0, 1). Using the work from part (a):

$$\int_{\alpha_3} -y\,dx + x\,dy = \int_0^{\frac{\pi}{2}} 1\,dt = t\Big|_0^{\frac{\pi}{2}} = \frac{\pi}{2}.$$

d. The path α4 also goes from (1, 0) to (0, 1), but here the parametrization satisfies x + y = (1 - t) + t = 1 so the curve is a segment contained in the line x + y = 1. The vector field is neither perfectly aligned with the direction of motion nor perpendicular to it, but it does have a positive component in the direction of motion at every point (Figure 9.3, right). Thus we expect the line integral to be positive. Indeed:

$$\int_{\alpha_4} -y\,dx + x\,dy = \int_0^1 (-t \cdot (-1) + (1 - t) \cdot 1))\,dt = \int_0^1 1\,dt = 1.$$

e. α_5 is another path from (1, 0) to (0, 1), and, again, x + y = cos² t + sin² t = 1. Thus the path traces out the same line segment as in part (d), though in a different way. The value of the line integral is:

$$\int_{\alpha_5} -y\,dx + x\,dy = \int_0^{\frac{\pi}{2}} \left(-\sin^2 t \cdot (2\cos t \cdot (-\sin t)) + \cos^2 t \cdot 2\sin t \cos t \right) dt$$

$$= \int_0^{\frac{\pi}{2}} (2\sin^3 t \cos t + 2\cos^3 t \sin t)\,dt$$

$$= \int_0^{\frac{\pi}{2}} 2\sin t \cos t (\sin^2 t + \cos^2 t)\,dt$$

$$= \int_0^{\frac{\pi}{2}} 2\sin t \cos t\,dt$$

$$= \sin^2 t\Big|_0^{\frac{\pi}{2}}$$

$$= 1.$$

Note that each of the paths α_3, α_4, and α_5 above goes from (1, 0) to (0, 1). The values of the line integral are not all equal, however. Thus changing the path between the endpoints can change the integral. On the other hand, α4 and α5 not only go between the same two points but also trace out the same curve, just parametrized in different ways. The integrals over these two paths are equal. Is this a coincidence?

In general, suppose that $\alpha : [a, b] \to U$ and $\beta : [c, d] \to U$ are smooth parametrizations of the same curve C. To help distinguish between them, we use t to denote the parameter of α and u for the parameter of β. It simplifies the argument somewhat to assume that α is one-to-one, except possibly at the endpoints, where $\alpha(a) = \alpha(b)$ is allowed, and likewise for β. This is not strictly necessary, though it has been true of all the examples of parametrizations we have encountered thus far.

We want to compare the integrals α $\int_\alpha F \cdot ds = \int_a^b F(\alpha(t)) \cdot \alpha'(t)\, dt$ and $\int_\beta F \cdot ds = \int_c^d F(\beta(u)) \cdot \beta'(u)\, du$ obtained using the respective parametrizations. Consider first the case that α and β trace out C in the same direction, that is, both start at a point $p = \alpha(a) = \beta(c)$ and end at a point $q = \alpha(b) = \beta(d)$, as in Figure 9.4.

Let x be a point of C other than the endpoints. Then $x = \alpha(t)$ for some value of t and $x = \beta(u)$ for some value of u. We write this as $t = g(u)$, that is, for a given value of u, g(u) is the value of t such that $\alpha(t) = \beta(u)$. We also set $g(c) = a$ and $g(d) = b$. Thus:

$$\beta(u) = \alpha(g(u)) \text{ for all } u \text{ in } [c, d]. \tag{9.3}$$

For instance, in the case of the line segments α_4 and α_5 above, $\alpha5(u) = (\cos^2 u, \sin^2 u) = 1 \sin^2 u, \sin^2 u) = \alpha4(\sin^2 u)$, so $g(u) = \sin^2 u$.

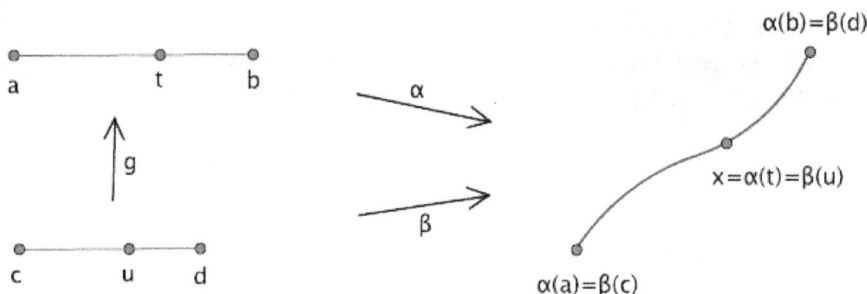

Figure 9.4: Two parametrizations α and β that trace out the same curve in the same direction: g(c) = a and g(d) = b

By equation (9.3) and the chain rule:

$$\beta'(u) = \alpha'(g(u))\, g'(u). \tag{9.4}$$

Hence:

$$\int_\beta F \cdot ds = \int_c^d F(\beta(u)) \cdot \beta'(u)\, du \quad \text{(by definition)}$$

$$= \int_c^d F(\alpha(g(u))) \cdot \alpha'(g(u))\, g'(u)\, du \quad (\text{let } t = g(u),\, dt = g'(u)\, du)$$

$$= \int_{g(c)}^{g(d)} F(\alpha(t)) \cdot \alpha'(t)\, dt = \int_a^b F(\alpha(t)) \cdot \alpha'(t)\, dt = \int_\alpha F \cdot ds.$$

In other words, both parametrization give the same value of the line integral.

If the parametrizations traverse C in opposite directions so that, say p = α(a) = β(d) and q = α(b) = β(c), then the only difference is that the endpoints are reversed. That is, g(d) = a and g(c) = b, so:

$$
\begin{aligned}
\int_\beta \mathbf{F} \cdot d\mathbf{s} &= \int_{g(c)}^{g(d)} \mathbf{F}(\alpha(t)) \cdot \alpha'(t)\, dt \qquad \text{(just as before, but \dots)} \\
&= \int_b^a \mathbf{F}(\alpha(t)) \cdot \alpha'(t)\, dt \\
&= -\int_a^b \mathbf{F}(\alpha(t)) \cdot \alpha'(t)\, dt \qquad\qquad\qquad (9.5) \\
&= -\int_\alpha \mathbf{F} \cdot d\mathbf{s}.
\end{aligned}
$$

The integrals differ by a sign.

We summarize this as follows.

Proposition 9.2. If α and β are smooth parametrizations of the same curve C, then:

$$
\int_\beta \mathbf{F} \cdot d\mathbf{s} = \pm \int_\alpha \mathbf{F} \cdot d\mathbf{s}.
$$

where the choice of sign depends on whether they traverse C in the same or opposite direction.

This allows us to define a line integral over a curve, as opposed to over a parametrization.

Definition. An oriented curve is a curve C with a specified direction of traversal. Given such a C, define:

$$
\int_C \mathbf{F} \cdot d\mathbf{s} = \int_\alpha \mathbf{F} \cdot d\mathbf{s}.
$$

where α is any smooth parametrization that traverses C in the given direction. By the preceding proposition, any two such parametrizations give the same answer.

The point is that the line integral depends only on the underlying curve and the direction in which it is traversed. This is critical in that it allows us to think of the line integral as an integral over an oriented geometric set, independent of the parametrization that is used to trace it out. This is an issue that we overlooked previously when we discussed integrals over curves with respect to arclength and integrals over surfaces with respect to surface area. In both of those cases, the integrals were defined using parametrizations, but we did not check that the answers were independent of the particular parametrization used to trace out the curve or surface.

In the case of the integral with respect to arclength, the story is that $\int_\alpha f\,ds = \int_\beta f\,ds$ for any smooth parametrizations α and β of C. The new twist is that the integral is not only independent of the parametrization but also of the direction in which it traverses C. The proof proceeds as before, but a difference arises because the definition $\int_\beta f\,ds = \int_c^d f(\beta(u))\,\|\beta'(u)\|\,du$ involves the norm $\|\beta'(u))\|$, not $\beta'(u)$ itself. This introduces a factor of $|g'(u)|$ after applying the chain rule as in equation (9.4) and taking norms. If α and β traverse C in opposite directions, then $g(u) < 0$, so $|g'(u)| = -g'(u)$. This minus sign is counteracted by the minus sign that appears in this case when reversing the limits of integration in (9.5), so the two minus signs cancel. We leave the details for the exercises (Exercise 1.12). The role of parametrizations in surface integrals is discussed in Chapter 10.

Having cleared the air on that lingering matter, we now return to studying line integrals.

Example 9.3. Evaluate $\int_C x\,dx + 2y\,dy + 3z\,dz$ if C is the oriented curve consisting of three quarter- circular arcs on the unit sphere in $\mathbb{R}^3$ from $(1, 0, 0)$ to $(0, 1, 0)$ to $(0, 0, 1)$ and then back to $(1, 0, 0)$.

See Figure 9.5.

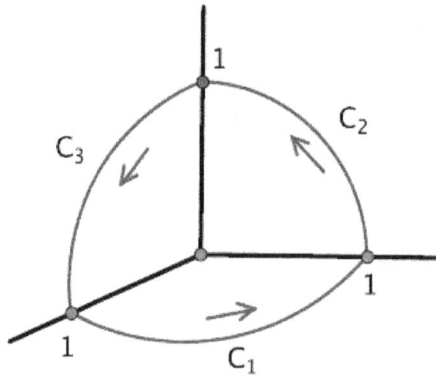

Figure 9.5: A curve consisting of three arcs C_1, C_2, and C_3

The vector field being integrated here is given by $F(x, y, z) = (x, 2y, 3z)$. Since C consists of three arcs C_1, C_2, C_3, we can imagine traversing each of them in succession and breaking up the integral as $\int_C F \cdot ds = \int_{C_1} F \cdot ds + \int_{C_2} F \cdot ds + \int_{C_3} F \cdot ds$. Moreover, we may parametrize each arc however we like as long as it is traversed in the proper direction. In this way, we can avoid having to parametrize all of C over a single parameter interval.

For C_1, we use the usual parametrization of the unit circle in the xy-plane from $(1, 0, 0)$ to $(0, 1, 0)$: $\alpha_1(t) = (\cos t, \sin t, 0)$, $0 \le t \le \frac{\pi}{2}$.

$$\int_{C_1} x\,dx + 2y\,dy + 3z\,dz = \int_0^{\frac{\pi}{2}} \left(\cos t \cdot (-\sin t) + 2\sin t \cdot \cos t + 3 \cdot 0 \cdot 0\right) dt$$

$$= \int_0^{\frac{\pi}{2}} \sin t \cos t\,dt = \frac{1}{2}\sin^2 t \Big|_0^{\frac{\pi}{2}} = \frac{1}{2}.$$

For C_2, we use a similar parametrization, only now in the yz-plane from (0, 1, 0) to (0, 0, 1): $\alpha_2(t) = (0, \cos t, \sin t)$, $0 \le t \le \frac{\pi}{2}$.

$$\int_{C_2} x\,dx + 2y\,dy + 3z\,dz = \int_0^{\frac{\pi}{2}} \left(0 \cdot 0 + 2\cos t \cdot (-\sin t) + 3\sin t \cdot \cos t\right) dt$$

$$= \int_0^{\frac{\pi}{2}} \sin t \cos t\,dt = \frac{1}{2}\sin^2 t \Big|_0^{\frac{\pi}{2}} = \frac{1}{2}.$$

Lastly, for C_3, we travel in the xz-plane from (0, 0, 1) to (1, 0, 0): $\alpha_3(t) = (\sin t, 0, \cos t)$, $0 \le t \le \frac{\pi}{2}$.

$$\int_{C_3} x\,dx + 2y\,dy + 3z\,dz = \int_0^{\frac{\pi}{2}} \left(\sin t \cdot \cos t + 2 \cdot 0 \cdot 0 + 3\cos t \cdot (-\sin t)\right) dt$$

$$= -2\int_0^{\frac{\pi}{2}} \sin t \cos t\,dt = -\sin^2 t \Big|_0^{\frac{\pi}{2}} = -1.$$

Adding these three calculations gives $\int_C x\,dx + 2y\,dy + 3z\,dz = \frac{1}{2} + \frac{1}{2} - 1 = 0$.

This example illustrates a general circumstance that we should make explicit. In addition to integrating over smooth oriented curves, we can integrate over curves C that are concatenations of smooth oriented curves, that is, curves of the form C $= C_1 \cup C_2 \cup \cdots \cup C_k$, where $C_1, C_2, \ldots, C_k$ are traversed in succession and each of them has a smooth parametrization. Such a curve C is called **piecewise smooth**. Under these conditions, we define:

$$\int_C \mathbf{F} \cdot d\mathbf{s} = \int_{C_1} \mathbf{F} \cdot d\mathbf{s} + \int_{C_2} \mathbf{F} \cdot d\mathbf{s} + \cdots + \int_{C_k} \mathbf{F} \cdot d\mathbf{s}.$$

This allows us to deal with curves that have corners, as in the example, without worrying about whether the entire curve can be parametrized in a smooth way.

9.2 ANOTHER WORD ABOUT SUBSTITUTION

Before moving on, we comment on the approach we took to defining the line integral. A curve is one-dimensional in the sense that it can be traced out using one parameter. Thus we might think of an integral over a curve as essentially a one-variable integral. The one-variable integrals that we are used to are those from first-year calculus, integrals over intervals in the real line.

The integral of a vector field F in $\mathbb{R}^n$ over a curve C is a new type of integral, however. F itself is a function of n variables $x_1, x_2, \ldots, x_n$, but, along C, each of these variables can be expressed in terms of the parameter t. This is exactly the information provided by the parametrization

$\alpha(t) = (x_1(t), x_2(t), \ldots, x_n(t))$, a ≤ t ≤ b. So the idea is that substituting for $x_1, x_2, \ldots, x_n$ in terms of t converts an integral over C to an integral over the interval I = [a, b], where things are familiar.

This fits into a typical pattern. Whereas the parametrization α transforms the interval I into the curve C, substitution has the effect of pulling back an integral over C to an integral over I. For instance, after substitution, the component function $F_i(x_1, x_2, \ldots, x_n)$ of $F(x_1, x_2, \ldots, x_n)$ is converted to the function $F_i(x_1(t), x_2(t), \ldots, x_n(t)) = F_i(\alpha(t))$. We say that α pulls back F_i to $F_i \circ \alpha$ and write:

$$\alpha^*(F_i) = F_i \circ \alpha.$$

Similarly, the expression dx_i is pulled back to $\frac{dx_i}{dt}\, dt$, and we write:

$$\alpha^*(dx_i) = \frac{dx_i}{dt}\, dt.$$

Actually, the xi's on each side of this equation mean slightly different things (coordinate label vs. function of t) and it would be better to write $\alpha^*(dx_i) = \frac{d\alpha_i}{dt}\, dt$, but hopefully our cavalier attitude will not lead us astray.

For a line integral, by pulling back the individual terms, the entire integrand $F_1\, dx_1 + F_2\, dx_2 + \cdots + F_n\, dx_n$ in terms of $x_1, x_2, \ldots, x_n$ is pulled back to the expression:

$$\alpha^*(F_1\, dx_1 + F_2\, dx_2 + \cdots + F_n\, dx_n) = \left((F_1 \circ \alpha)\frac{dx_1}{dt} + (F_2 \circ \alpha)\frac{dx_2}{dt} + \cdots + (F_n \circ \alpha)\frac{dx_n}{dt}\right) dt$$

in terms of t. If we denote $F_1\, dx_1 + F_2\, dx_2 + \cdots + F_n$ dxn by ω, then the definition of the line integral reads:

$$\int_{\alpha(I)} \omega = \int_I \alpha^*(\omega). \qquad (9.6)$$

This looks a lot like the change of variables theorem. For instance, compare with equation (7.3) of Chapter 7. We verified that the value of the integral is independent of the parametrization α so that it is well-defined as an integral over the oriented curve C = $\alpha(I)$. The chain rule played an essential, though easily missed, role in the verification. See equation (9.4).

This approach recurs more generally. In order to define a new type of integral, we use substitution to pull the integral back to a familiar situation. This is done in a way that allows the change of variables theorem and the chain rule to conspire to make the process legitimate.

9.3 CONSERVATIVE FIELDS

If C is a smooth oriented curve in an open set U in $\mathbb{R}^n$ from a point p to a point q as in Figure 9.6, the value of the line integral $\int_C \mathbf{F} \cdot d\mathbf{s}$ does not depend on how C is parametrized as long as it is traversed in the correct direction. We have seen, however, that the value does depend on the actual curve C: choosing a different curve from p to q can change the integral. Are there situations where even changing the curve doesn't matter?

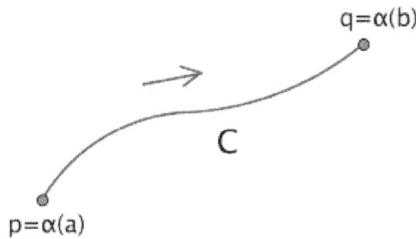

Figure 9.6: An oriented curve C from p to q

By definition $\int_C \mathbf{F} \cdot d\mathbf{s} = \int_a^b \mathbf{F}(\alpha(t)) \cdot \alpha'(t)\, dt$, where $\alpha : [a, b] \to U$ is a smooth parametrization that traces out C with the proper orientation. Now, suppose—and this is a big assumption—that F is the gradient of a smooth real-valued function f, that is, $\mathbf{F} = \nabla f$, or, in terms of components, $(F_1, F_2, \ldots, F_n) = (\frac{\partial f}{\partial x_1}, \frac{\partial f}{\partial x_2}, \ldots, \frac{\partial f}{\partial x_n})$. Then $\mathbf{F}(\alpha(t)) \cdot \alpha'(t) = \nabla f(\alpha(t)) \cdot \alpha'(t) = (f \circ \alpha)'(t)$, where the last equality is the Little Chain Rule. In other words, the integrand in the definition of the line integral is a derivative. So:

$$\int_C \mathbf{F} \cdot d\mathbf{s} = \int_a^b (f \circ \alpha)'(t)\, dt = (f \circ \alpha)(t)\big|_a^b = f(\alpha(b)) - f(\alpha(a)) = f(\mathbf{q}) - f(\mathbf{p}).$$

This depends only on the endpoints p and q and not at all on the curve that connects them.

In order to apply the Little Chain Rule, this argument requires that C have a smooth parametrization. If C is piecewise smooth instead, say $C = C_1 \cup C_2 \cup \cdots \cup C_k$, where each C_i is smooth and C_1 goes from p to p_1, C_2 from p_1 to p_2, and so on until C_k goes from p_{k-1} to q, then applying the previous result to the pieces gives a telescoping sum in which the intermediate endpoints appear twice with opposite signs:

$$\int_C \mathbf{F} \cdot ds = \int_{C_1} \mathbf{F} \cdot ds + \int_{C_2} \mathbf{F} \cdot ds + \cdots + \int_{C_k} \mathbf{F} \cdot ds$$
$$= \big(f(\mathbf{p}_1) - f(\mathbf{p})\big) + \big(f(\mathbf{p}_2) - f(\mathbf{p}_1)\big) + \cdots + \big(f(\mathbf{q}) - f(\mathbf{p}_{k-1})\big)$$
$$= -f(\mathbf{p}) + f(\mathbf{q})$$
$$= f(\mathbf{q}) - f(\mathbf{p}).$$

In other words, in the end, the same result holds, and again only p and q matter.

Definition. Let U be an open set in $\mathbb{R}^n$. A vector field $\mathbf{F} : U \to \mathbb{R}^n$ is called a **conservative**, or a **gradient**, **field** on U if there exists a smooth real-valued function $f : U \to R$ such that $\mathbf{F} = \nabla f$. Such an f is called a **potential function** for F.

We have proven the following.

Theorem 9.4 (Conservative vector field theorem). *If* F *is a conservative vector field on* U *with potential function* f, *then for any piecewise smooth oriented curve* C *in* U:

$$\int_C \mathbf{F} \cdot ds = f(\mathbf{q}) - f(\mathbf{p}),$$

where p and q are the starting and ending points, respectively, of C.

A couple of other ways to write this are:

$$\int_C \nabla f \cdot ds = f(\mathbf{q}) - f(\mathbf{p})$$

or
$$\int_C \frac{\partial f}{\partial x_1} dx_1 + \frac{\partial f}{\partial x_2} dx_2 + \cdots + \frac{\partial f}{\partial x_n} dx_n = f(\mathbf{q}) - f(\mathbf{p}).$$

We shall give some examples to illustrate the theorem shortly, but first we present an immediate general consequence that also helps explain in part why the word conservative is used. (See Exercise 3.19 for another reason.)

Definition. A curve C is called **closed** if it starts and ends at the same point. See Figure 9.7 for some examples.

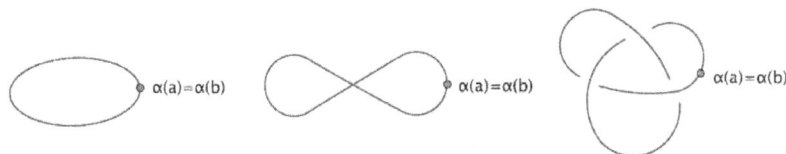

Figure 9.7: Examples of closed curves: an ellipse (left) and a figure eight (middle) in $\mathbb{R}^2$ and a knot in $\mathbb{R}^3$ (right)

Corollary 9.5. If F is conservative on U, then for any piecewise smooth oriented closed curve C in U:

$$\int_C \mathbf{F} \cdot d\mathbf{s} = 0.$$

Example 9.6. Let F be the vector field on $\mathbb{R}^3$ given by F(x, y, z) = (x, 2y, 3z). Is F a conservative field?

We look to see if there is a real-valued function f such that ∇f=F, that is, $\frac{\partial f}{\partial x} = x$, $\frac{\partial f}{\partial y} = 2y$, and $\frac{\partial f}{\partial z} = 3z$. By inspection, it's easy to see that $f(x, y, z) = \frac{1}{2}x^2 + y^2 + \frac{3}{2}z^2$ works. Thus F is conservative with f as potential function.

It follows $\int_C x\,dx + 2y\,dy + 3z\,dz = 0$ for any piecewise smooth oriented closed curve C in $\mathbb{R}^3$. This explains the answer of 0 that we obtained in Example 9.3 when we integrated F around the closed curve consisting of three circular arcs (Figure 9.5).

Similarly, if C is any curve from (1, 0, 0) to (0, 1, 0), such as the arc C_1 in the xy-plane of Example 9.3, then by the conservative vector field theorem:

$$\int_C x\,dx + 2y\,dy + 3z\,dz = f(0, 1, 0) - f(1, 0, 0) = (0 + 1 + 0) - (\frac{1}{2} + 0 + 0) = \frac{1}{2}.$$

Again, this agrees with what we got before. More importantly, it is a much simpler way to evaluate the integral.

Example 9.7. Let G be the inverse square field:

$$\mathbf{G}(x, y, z) = \left(-\frac{x}{(x^2 + y^2 + z^2)^{3/2}}, -\frac{y}{(x^2 + y^2 + z^2)^{3/2}}, -\frac{z}{(x^2 + y^2 + z^2)^{3/2}} \right),$$

where (x, y, z)≠(0, 0, 0). Find $\int_C \mathbf{G} \cdot d\mathbf{s}$ if C is the portion of the helix parametrized by α(t) = (cos t, sin t, t), 0 ≤ t ≤ 3π.

We could calculate the integral explicitly using the definition and the parametrization, but, inspired by the previous example, we check first to see if G is conservative. After a little trial-and-error, we find that $g(x, y, z) = (x^2 + y^2 + z^2)^{-1/2} = \frac{1}{\|(x,y,z)\|}$ works as a potential function.

For instance, $\frac{\partial g}{\partial x} = -\frac{1}{2}(x^2 + y^2 + z^2)^{-3/2} \cdot 2x = -\frac{x}{(x^2+y^2+z^2)^{3/2}}$, and so on. Since C goes from p = α(0) = (1, 0, 0) to q = α(3π) = (-1, 0, 3π), the conservative vector field theorem implies:

$$\int_C \mathbf{G} \cdot d\mathbf{s} = g(-1, 0, 3\pi) - g(1, 0, 0) = \frac{1}{\sqrt{1 + 9\pi^2}} - 1.$$

The truth is that evaluating this integral directly using the parametrization is actually fairly straightforward. Try it—it turns out that it's not as bad as it might look. This is often the case with conservative fields. Instead, perhaps the most meaningful takeaway from the approach we took is that we can say that

$\int_C \mathbf{G} \cdot ds = \frac{1}{\sqrt{1+9\pi^2}} - 1$

for any piecewise smooth curve in the domain of G, $\mathbb{R}^3 - \{(0, 0, 0)\}$, from p = (1, 0, 0) to q = (-1, 0, 3π). That the given curve is a helix is irrelevant. Indeed, more generally, if C is any piecewise smooth curve in $\mathbb{R}^3 - \{(0, 0, 0)\}$ from a point p to a point q, then using the potential function $g(\mathbf{x}) = \frac{1}{\|\mathbf{x}\|}$ above gives:

$$\int_C \mathbf{G} \cdot ds = \frac{1}{\|\mathbf{q}\|} - \frac{1}{\|\mathbf{p}\|}.$$

In particular, the integral of the inverse square field around any piecewise smooth oriented closed curve that avoids the origin is 0.

Example 9.8. Let F be the counterclockwise circulation field F(x, y) = (-y, x) on $\mathbb{R}^2$ (Figure 9.2). Is F a conservative field?

This time, the answer is no, and we give a couple of ways of looking at it. First, if F were conservative, then its integral around any closed curve would be zero. But just by looking at the way the field circulates around the origin, one can see that the integral around a circle centered at the origin and oriented counterclockwise is positive. In fact, we calculated earlier in Example 9.1 that, if C is the unit circle traversed counterclockwise, then $\int_C -y\,dx + x\,dy = 2\pi$. Since C is closed and the integral is not zero, F is not conservative.

A more direct approach is to try to guess a potential function f. Here, we want ∇f = F = (-y, x), that is:

$$\begin{cases} \frac{\partial f}{\partial x} = -y \\ \frac{\partial f}{\partial y} = x. \end{cases} \tag{9.7}$$

To satisfy the first condition, we might try $f(x, y)$ = -xy, but then $\frac{\partial f}{\partial y} = -x$. not x, and there doesn't seem to be any way to fix it.

This shows that our first guess didn't work, but it's possible that we are simply bad guessers. To show that (9.7) is truly inconsistent, we bring in the second-order partials. For if there were a function f that satisfied (9.7), then:

$$\frac{\partial^2 f}{\partial y\,\partial x} = \frac{\partial}{\partial y}\left(\frac{\partial f}{\partial x}\right) = \frac{\partial}{\partial y}(-y) = -1 \quad \text{and} \quad \frac{\partial^2 f}{\partial x\,\partial y} = \frac{\partial}{\partial x}\left(\frac{\partial f}{\partial y}\right) = \frac{\partial}{\partial x}(x) = 1.$$

This contradicts the equality of mixed partials. Hence no such f exists.

Theorem 9.9 (Mixed partials theorem). Let F be a smooth vector field on an open set U in $\mathbb{R}^n$, written in terms of components as F = $(F_1, F_2, \ldots, F_n)$. If F is conservative, then:

$$\frac{\partial F_i}{\partial x_j} = \frac{\partial F_j}{\partial x_i} \text{ for all } i, j = 1, 2, \ldots, n.$$

Equivalently, if $\frac{\partial F_i}{\partial x_j} \neq \frac{\partial F_j}{\partial x_i}$ for some i, j, then F is not conservative.

Proof. Assume that F is conservative, and let f be a potential function. Then F = ∇f, that is, $(F_1, F_2, \ldots, F_n) = (\frac{\partial f}{\partial x_1}, \frac{\partial f}{\partial x_2}, \ldots, \frac{\partial f}{\partial x_n})$. Therefore $F_i = \frac{\partial f}{\partial x_i}$ and $F_j = \frac{\partial f}{\partial x_j}$, and so $\frac{\partial F_i}{\partial x_j} = \frac{\partial^2 f}{\partial x_j \partial x_i}$ and $\frac{\partial F_j}{\partial x_i} = \frac{\partial^2 f}{\partial x_i \partial x_j}$. By the equality of mixed partials, these last two partial derivatives are equal.

We look at what the mixed partials theorem says for $\mathbb{R}^2$ and $\mathbb{R}^3$. For a vector field F = (F_1, F_2) on a subset of $\mathbb{R}^2$, there is just one mixed partials pair: $\frac{\partial F_1}{\partial y}$ and $\frac{\partial F_2}{\partial x}$. Thus:

> In $\mathbb{R}^2$, if $\mathbf{F} = (F_1, F_2)$ is conservative, then $\frac{\partial F_1}{\partial y} = \frac{\partial F_2}{\partial x}$.

Next, for F = (F_1, F_2, F_3) on a subset of R3, there are three mixed partials pairs: (i) $\frac{\partial F_1}{\partial y}$ and $\frac{\partial F_2}{\partial x}$, (ii) $\frac{\partial F_1}{\partial z}$ and $\frac{\partial F_3}{\partial x}$, and (iii) $\frac{\partial F_2}{\partial z}$ and $\frac{\partial F_3}{\partial y}$.. We describe a slick way to compare the partial derivatives in each pair.

Let $\nabla = (\frac{\partial}{\partial x}, \frac{\partial}{\partial y}, \frac{\partial}{\partial z})$. This is not a legitimate vector, but we think of it as an "operator." For instance, it acts on a real-valued function f (x, y, z) by scalar multiplication: $\nabla f = (\frac{\partial}{\partial x}, \frac{\partial}{\partial y}, \frac{\partial}{\partial z}) f = (\frac{\partial f}{\partial x}, \frac{\partial f}{\partial y}, \frac{\partial f}{\partial z})$. This agrees with the usual notation for the gradient.

For a vector field F = (F_1, F_2, F_3), ∇ acts by the cross product:

$$\nabla \times \mathbf{F} = \det \begin{bmatrix} i & j & k \\ \frac{\partial}{\partial x} & \frac{\partial}{\partial y} & \frac{\partial}{\partial z} \\ F_1 & F_2 & F_3 \end{bmatrix}$$

$$= \left(\frac{\partial F_3}{\partial y} - \frac{\partial F_2}{\partial z}, -(\frac{\partial F_3}{\partial x} - \frac{\partial F_1}{\partial z}), \frac{\partial F_2}{\partial x} - \frac{\partial F_1}{\partial y} \right)$$

$$= \left(\frac{\partial F_3}{\partial y} - \frac{\partial F_2}{\partial z}, \frac{\partial F_1}{\partial z} - \frac{\partial F_3}{\partial x}, \frac{\partial F_2}{\partial x} - \frac{\partial F_1}{\partial y} \right).$$

This is called the curl of F. Note that its components are precisely the differences within the three mixed partials pairs. Hence:

> In $\mathbb{R}^3$, if $\mathbf{F} = (F_1, F_2, F_3)$ is conservative, then $\nabla \times \mathbf{F} = \mathbf{0}$.

Example 9.10. Let F(x, y, z) = $(2xy^3z^4 + x, 3x^2y^2z^4 + 3, 4x^2y^3z^3 + y^2)$. Is F conservative?

We begin by trying to guess a potential function. In order to get $\frac{\partial f}{\partial x} = 2xy^3z^4 + x$, we guess $f = x^2y^3z^4 + \frac{1}{2}x^2$. Then, to get $\frac{\partial f}{\partial y} = 3x^2y^2z^4 + 3$, we modify this to $f = x^2y^3z^4 + \frac{1}{2}x^2 + 3y$. Then, to get $\frac{\partial f}{\partial z} = 4x^2y^3z^3 + y^2$, we modify again to

$f = x^2y^3z^4 + \frac{1}{2}x^2 + 3y + y^2z$. Unfortunately, this spoils $\frac{\partial f}{\partial y}$. It seems unlikely that a potential function exists.

Having raised the suspicion, we compute the curl:

$$\nabla \times \mathbf{F} = \det \begin{bmatrix} \mathbf{i} & \mathbf{j} & \mathbf{k} \\ \frac{\partial}{\partial x} & \frac{\partial}{\partial y} & \frac{\partial}{\partial z} \\ 2xy^3z^4 + x & 3x^2y^2z^4 + 3 & 4x^2y^3z^3 + y^2 \end{bmatrix}$$

$$= \left(12x^2y^2z^3 + 2y - 12x^2y^2z^3,\ 8xy^3z^3 - 8xy^3z^3,\ 6xy^2z^4 - 6xy^2z^4\right)$$

$$= (2y, 0, 0).$$

This is not identically 0, so F is not conservative.

We remark that it is not true in general that, if $\frac{\partial F_i}{\partial x_j} = \frac{\partial F_j}{\partial x_i}$ for all i, j, then F is conservative.

In other words, the converse of the mixed partials theorem does not hold. This is a point to which we return in Sections 9.5 and 9.6.

9.4 GREEN'S THEOREM

We now discuss the first of three major theorems of multivariable calculus, each of which is named after somebody. The one at hand applies to smooth vector fields F = (F_1, F_2) on subsets of $\mathbb{R}^2$.

Suppose that such a vector field F is defined on a rectangle R = [a, b] × [c, d] and that we integrate F around the boundary of R, oriented counterclockwise. Let's call this oriented curve C. We break up C into the four sides C_1, C_2, C_3, C_4 of the rectangle but orient the horizontal sides C_1 and C_3 to the right and the vertical sides C_2 and C_4 upward. See Figure 9.8.

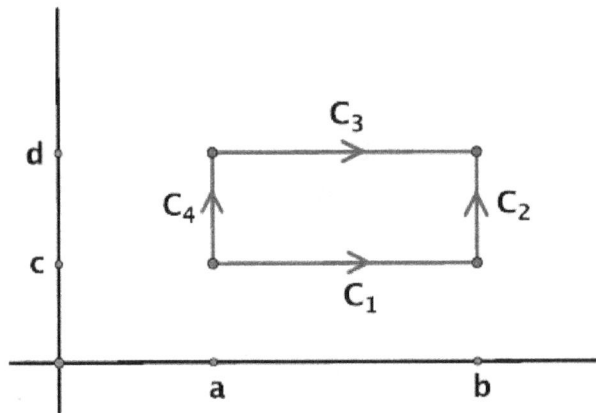

Figure 9.8: The boundary of a rectangle R = [a, b] × [c, d]

Taking into account the orientations, we write $C = C_1 \cup C_2 \cup (-C_3) \cup (-C_4)$, where the minus signs indicate that the segment has been given the orientation opposite the direction in which it is actually traversed in C. Then the line integral over C can be written as:

$$\int_C F_1 \, dx + F_2 \, dy = \left(\int_{C_1} + \int_{C_2} - \int_{C_3} - \int_{C_4} \right) F_1 \, dx + F_2 \, dy. \tag{9.8}$$

Hopefully, the notation, where the integral signs are meant to distribute over the integrand, is self-explanatory.

For the horizontal segment C_1, we use x as parameter, keeping y = c fixed: $\alpha_1(x) = (x, c)$, $a \le x \le b$. Then, along C_1. $\frac{dx}{dx} = 1$ and $\frac{dy}{dx} = \frac{d}{dx}(c) = 0$. so:

$$\int_{C_1} F_1 \, dx + F_2 \, dy = \int_a^b (F_1(x, c) \cdot 1 + F_2(x, c) \cdot 0) \, dx = \int_a^b F_1(x, c) \, dx$$

Similarly, for C3:

$$\int_{C_3} F_1 \, dx + F_2 \, dy = \int_a^b F_1(x, d) \, dx.$$

For the vertical segments C_2 and C_4, we use y as parameter, $c \le y \le \quad$ d, keeping x fixed. Hence $\frac{dx}{dy} = 0$ and $\frac{dy}{dy} = 1$, which gives:

$$\int_{C_2} F_1 \, dx + F_2 \, dy = \int_c^d F_2(b, y) \, dy \quad \text{and} \quad \int_{C_4} F_1 \, dx + F_2 \, dy = \int_c^d F_2(a, y) \, dy.$$

Substituting these four calculations into equation (9.8) results in:

$$\int_C F_1 \, dx + F_2 \, dy = \int_a^b F_1(x, c) \, dx + \int_c^d F_2(b, y) \, dy - \int_a^b F_1(x, d) \, dx - \int_c^d F_2(a, y) \, dy$$

$$= \int_c^d (F_2(b, y) - F_2(a, y)) \, dy - \int_a^b (F_1(x, d) - F_1(x, c)) \, dx.$$

Somewhat remarkably, the fundamental theorem of calculus applies to each of these terms:

$$\int_C F_1 \, dx + F_2 \, dy = \int_c^d \left(F_2(x, y) \Big|_{x=a}^{x=b} \right) dy - \int_a^b \left(F_1(x, y) \Big|_{y=c}^{y=d} \right) dx$$

$$= \int_c^d \left(\int_a^b \frac{\partial F_2}{\partial x}(x, y) \, dx \right) dy - \int_a^b \left(\int_c^d \frac{\partial F_1}{\partial y}(x, y) \, dy \right) dx$$

$$= \iint_R \left(\frac{\partial F_2}{\partial x} - \frac{\partial F_1}{\partial y} \right) dx \, dy. \tag{9.9}$$

This expresses the line integral around the boundary of the rectangle as a certain double integral over the rectangle itself.

Next, suppose that D is a rectangle with a rectangular hole, as on the left of Figure 9.9. We examine the relationship (9.9), working from right to left, for this new domain. We write D as the union of four rectangles R_1, R_2, R_3, R_4 that overlap at most along portions of their boundaries, as on the right of Figure 9.9, so that the double integral over D may be decomposed:

$$\iint_D \left(\frac{\partial F_2}{\partial x} - \frac{\partial F_1}{\partial y}\right) dx\, dy = \left(\iint_{R_1} + \iint_{R_2} + \iint_{R_3} + \iint_{R_4}\right) \left(\frac{\partial F_2}{\partial x} - \frac{\partial F_1}{\partial y}\right) dx\, dy.$$

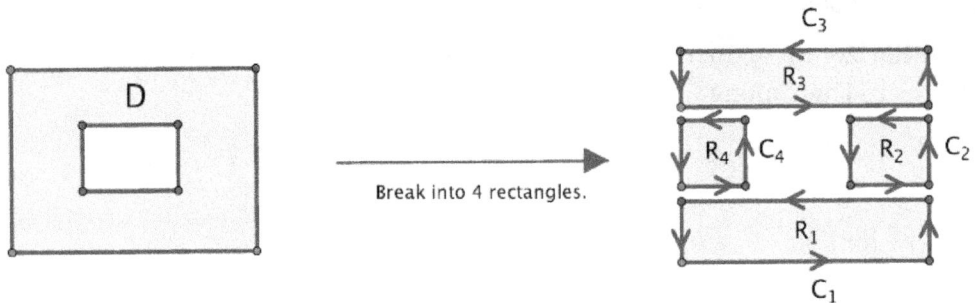

Figure 9.9: A rectangle with a rectangular hole

Let C_j be the boundary of R_j, j = 1, 2, 3, 4, oriented counterclockwise. Equation (9.9) applies to each of the preceding integrals over the rectangles R_1, R_2, R_3, R_4, which yields:

$$\iint_D \left(\frac{\partial F_2}{\partial x} - \frac{\partial F_1}{\partial y}\right) dx\, dy = \left(\int_{C_1} + \int_{C_2} + \int_{C_3} + \int_{C_4}\right) F_1\, dx + F_2\, dy.$$

Note that, for the boundaries of R_1, R_2, R_3, R_4, the portions in the interior of D appear as part of two different C_j traversed in opposite directions. Thus these portions of the line integrals cancel out, leaving only the parts left exposed around the boundary of the original region D. Hence, if C denotes the boundary of D, we have:

$$\iint_D \left(\frac{\partial F_2}{\partial x} - \frac{\partial F_1}{\partial y}\right) dx\, dy = \int_C F_1\, dx + F_2\, dy.$$

This is the same relation as we obtained for the filled-in rectangle in equation (9.9). As the argument shows, it is valid provided that C is oriented so that the outer perimeter is traversed counterclockwise and the inner perimeter is traversed clockwise, as shown in Figure 9.10.

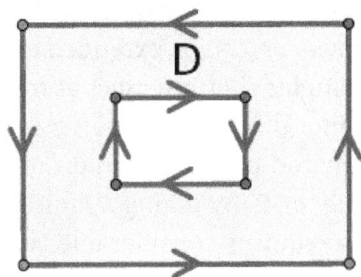

Figure 9.10: The oriented boundary of a rectangle with a rectangular hole

Before generalizing, we need a further definition.

Definition. We have said that a curve C is called closed if it starts and ends at the same point.

In addition, C is said to be **simple** if, apart from this, it does not intersect itself.

For example, a circle or ellipse is simple, but a figure eight is not.

Now, let D be a subset of $\mathbb{R}^2$ whose boundary consists of a finite number of simple closed curves. We adopt the following notation:

∂D denotes the boundary of D oriented so that, as

you traverse the boundary, D stays on the left.

For instance, this describes the orientation of the boundary of the rectangle-with-rectangular-hole in Figure 9.10. Note that oriented boundaries are completely different from partial derivatives, even though the same symbol "∂" happens to be used for both.

The relation we found for line integrals around the oriented boundaries of rectangles or rectangles with holes holds in greater generality.

Theorem 9.11 (Green's theorem). Let U be an open set in $\mathbb{R}^2$, and let D be a bounded subset of U such that the boundary of D consists of a finite number of piecewise smooth simple closed curves. If F = (F_1, F_2) is a smooth vector field on U, then:

$$\int_{\partial D} F_1\, dx + F_2\, dy = \iint_D \left(\frac{\partial F_2}{\partial x} - \frac{\partial F_1}{\partial y} \right) dx\, dy.$$

For example, consider the special case that F happens to be a conservative field. Then $\int_{\partial D} F_1\, dx + F_2\, dy = 0$, since ∂D consists of closed curves and the integral of a conservative field around any closed curve is 0. On the other hand, by the mixed partials theorem, the mixed partials $\frac{\partial F_2}{\partial x}$ and $\frac{\partial F_1}{\partial y}$ are equal, so $\iint_D (\frac{\partial F_2}{\partial x} - \frac{\partial F_1}{\partial y})\, dx\, dy = 0$ as well. Thus, for conservative fields, Green's theorem confirm the generally accepted belief that 0 = 0.

We have proven Green's theorem only in very special cases. The argument given for a rectangle with a hole, however, can be extended easily to any domain D that is a union of finitely many rectangles that intersect at most along common portions of their boundaries. For a general region D, one strategy for giving a proof is to show that D can be approximated by such a union of rectangles so that Green's theorem is approximately true for D. By taking a limit of such approximations, the theorem can be proven. This requires considerable technical expertise which we leave for another course.

Example 9.12. Let F(x, y) = (x - y², xy). $\int_C$ **F** · *ds* if C is the oriented simple closed curve consisting of the semicircular arc in the upper half-plane from (2, 0) to (-2, 0) followed by the line segment from (-2, 0) to (2, 0). See Figure 9.11.

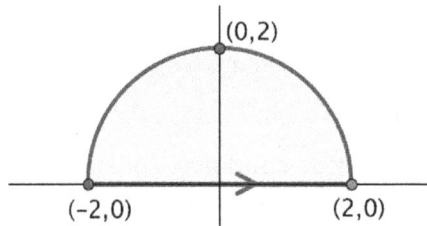

Figure 9.11: An integral around the boundary of a half-disk

Let D be the filled-in half-disk bounded by C so that C = ∂D. By Green's theorem:

$$\int_{C=\partial D} (x - y^2)\, dx + xy\, dy = \iint_D \left(\frac{\partial}{\partial x}(xy) - \frac{\partial}{\partial y}(x - y^2)\right) dx\, dy$$

$$= \iint_D (y - (-2y))\, dx\, dy$$

$$= \iint_D 3y\, dx\, dy.$$

This last double integral can be evaluated using polar coordinates:

$$\int_C (x - y^2)\, dx + xy\, dy = \int_0^\pi \left(\int_0^2 3r\sin\theta \cdot r\, dr\right) d\theta$$

$$= 3\left(\int_0^2 r^2\, dr\right)\left(\int_0^\pi \sin\theta\, d\theta\right)$$

$$= 3\left(\frac{1}{3}r^3\Big|_0^2\right)\left(-\cos\theta\Big|_0^\pi\right)$$

$$= 3 \cdot \frac{8}{3} \cdot 2$$

$$= 16.$$

Example 9.13. Once again, consider the vector field F(x, y) = (-y, x) that circulates counterclockwise about the origin. Let C be a piecewise smooth simple closed curve in $\mathbb{R}^2$, oriented counterclockwise. For example, it could be the one shown in Figure 9.12. Can we say anything about the integral of F over C?

There are portions of C that go counterclockwise about the origin, for instance, the parts furthest from the origin. The tangential component of F along these portions is positive. But there may also be places where C moves in the clockwise direction around the origin. For instance, if C does not encircle the origin, this happens at points nearest the origin. The tangental component is negative along such places. This means that there may be two types of contributions to the integral that counteract one another. The magnitude of F increases as you move away from the origin, however, so we might expect the positive contribution to be greater. In fact, using Green's theorem, we can measure precisely how much greater it is.

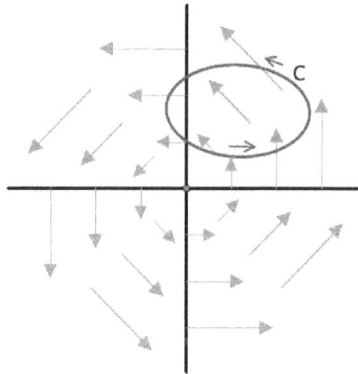

Figure 9.12: Integrating the circulating vector field F(x, y) = (-y, x) around a simple closed curve C

Let D be the filled-in region bounded by C. Then, by Green's theorem:

$$\int_C -y\,dx + x\,dy = \iint_D \left(\frac{\partial}{\partial x}(x) - \frac{\partial}{\partial y}(-y)\right) dx\,dy = \iint_D 2\,dx\,dy = 2\,\text{Area}\,(D).$$

In general, the same reasoning shows the following fact.

Corollary 9.14. Let D be a subset of $\mathbb{R}^2$ that satisfies the assumptions of Green's theorem. Then:

$$\text{Area}\,(D) = \frac{1}{2}\int_{\partial D} -y\,dx + x\,dy.$$

This illustrates the striking principle that information about all of D can be recovered from measurements taken only along its boundary.

9.5 THE VECTOR FIELD W

One of the early examples we looked at when we started studying vector fields was the vector field $W : \mathbb{R}^2 - \{(0,0)\} \to \mathbb{R}^2$ given by:

$$\mathbf{W}(x,y) = \left(-\frac{y}{x^2+y^2}, \frac{x}{x^2+y^2}\right).$$

This was Example 8.3 in Chapter 8. Like our usual circulating vector field, W circulates counterclockwise around the origin, but, due to the factor of $\frac{1}{x^2+y^2}$, the arrows get shorter as you move away (Figure 9.13). We are about to see that W is an instructive example to keep in mind.

First, we integrate W around the circle C_a of radius a centered at the origin and oriented counterclockwise. We evaluate the integral in two different ways.

Approach A. We use Green's theorem. Let Da be the filled-in disk of radius a so that $C_a = \partial D_a$. See Figure 9.14.

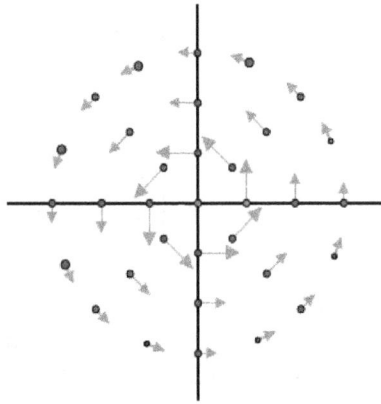

Figure 9.13: The vector field W

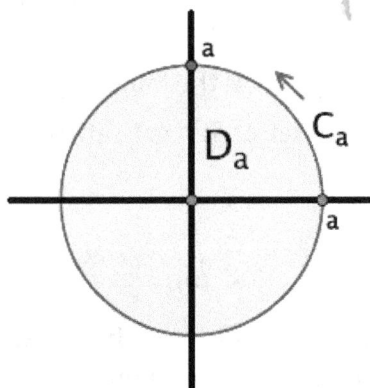

Figure 9.14: The circle and disk of radius a: $C_a = \partial D_a$

By the quotient rule:

$$\frac{\partial W_2}{\partial x} = \frac{\partial}{\partial x}\left(\frac{x}{x^2 + y^2}\right) = \frac{(x^2 + y^2) \cdot 1 - x \cdot 2x}{(x^2 + y^2)^2} = \frac{y^2 - x^2}{(x^2 + y^2)^2}$$

and $\quad \dfrac{\partial W_1}{\partial y} = \dfrac{\partial}{\partial y}\left(-\dfrac{y}{x^2 + y^2}\right) = -\dfrac{(x^2 + y^2) \cdot 1 - y \cdot 2y}{(x^2 + y^2)^2} = \dfrac{y^2 - x^2}{(x^2 + y^2)^2}.$

Hence $\frac{\partial W_2}{\partial x} = \frac{\partial W_1}{\partial y}$. so Green's theorem gives:

$$\int_{C_a} -\frac{y}{x^2 + y^2}\, dx + \frac{x}{x^2 + y^2}\, dy = \iint_{D_a}\left(\frac{\partial W_2}{\partial x} - \frac{\partial W_1}{\partial y}\right) dx\, dy = \iint_{D_a} 0\, dx\, dy = 0.$$

Approach B. We calculate the line integral directly using a parametrization of C_a, say $\alpha(t) = (a \cos t, a \sin t), 0 \leq t \leq 2\pi$. This gives:

$$\int_{C_a} -\frac{y}{x^2 + y^2}\, dx + \frac{x}{x^2 + y^2}\, dy$$

$$= \int_0^{2\pi}\left(-\frac{a \sin t}{a^2 \cos^2 t + a^2 \sin^2 t} \cdot (-a \sin t) + \frac{a \cos t}{a^2 \cos^2 t + a^2 \sin^2 t} \cdot (a \cos t)\right) dt$$

$$= \int_0^{2\pi} \frac{a^2 \sin^2 t + a^2 \cos^2 t}{a^2}\, dt$$

$$= \int_0^{2\pi} \frac{a^2}{a^2}\, dt$$

$$= \int_0^{2\pi} 1\, dt$$

$$= 2\pi.$$

In other words, $0 = 2\pi$.

Evidently, there's been a mistake. The truth is that Approach A is flawed. The disk D_a contains the origin, and the origin is not in the domain of W. Thus Green's theorem does not apply to W and D_a. Instead, the correct answer is B :

$$\int_{C_a} -\frac{y}{x^2 + y^2}\, dx + \frac{x}{x^2 + y^2}\, dy = 2\pi. \tag{9.10}$$

This calculation has some noteworthy consequences.

- ◉ The value of the integral is 2π regardless of the radius of the circle.
- ◉ W is not conservative, as integrating it around a closed curve does not always give 0.
- ◉ Nevertheless, all possible mixed partials pairs are equal: $\frac{\partial W_2}{\partial x} = \frac{\partial W_1}{\partial y}$. Hence the converse of the mixed partials theorem can fail.

We broaden the analysis by considering any piecewise smooth simple closed curve C in the plane, oriented counterclockwise, that does not pass through the origin, meaning that $(0, 0) \in C$. This doesn't seem like very much to go on, but it turns out that we can say a lot about the integral of W over C.

There are two cases, corresponding to the fact that C separates the plane into two regions, which we refer to as the interior and the exterior.

Case 1. $(0, 0)$ lies in the exterior of C. See Figure 9.15.

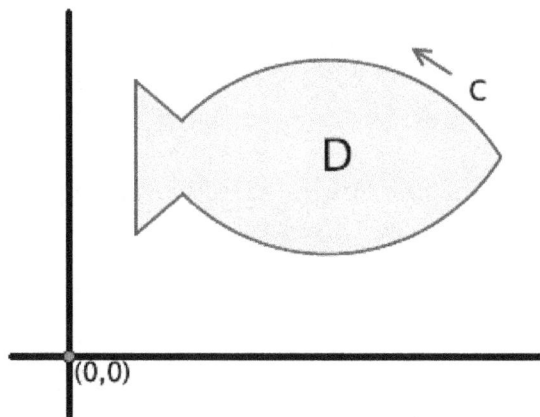

Figure 9.15: The origin lies in the exterior of C: ∂D = C.

Then the region D bounded by C does not contain the origin, so Approach A above is valid and Green's theorem applies with $C = \partial D$. Since the mixed partials of W are equal, it follows that:

$$\int_C -\frac{y}{x^2 + y^2}\, dx + \frac{x}{x^2 + y^2}\, dy = \iint_D \left(\frac{\partial W_2}{\partial x} - \frac{\partial W_1}{\partial y} \right) dx\, dy = 0.$$

Case 2. $(0, 0)$ lies in the interior of C. See Figure 9.16.

As we saw with the flaw in Approach A, we can no longer fill in C and use Green's theorem. Instead, let C_ϵ be a circle of radius ϵ centered at the origin and oriented counterclockwise, where ϵ is chosen small enough that C_ϵ is contained entirely in the interior of C.

Let D be the region inside of C but outside of C_ϵ. The origin does not belong to D, so Green's theorem applies. Since the mixed partials of W are equal, this gives:

$$\int_{\partial D} -\frac{y}{x^2 + y^2}\, dx + \frac{x}{x^2 + y^2}\, dy = \iint_D \left(\frac{\partial W_2}{\partial x} - \frac{\partial W_1}{\partial y} \right) dx\, dy = 0.$$

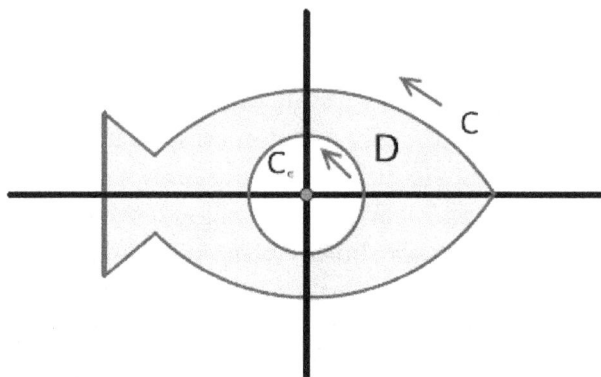

Figure 9.16: The origin lies in the interior of C: ∂D = C ∪ (-C_ε).

On the other hand, taking orientation into account, ∂D = C ∪ (-C_ε), so:

$$\int_{\partial D} -\frac{y}{x^2+y^2}\,dx + \frac{x}{x^2+y^2}\,dy = \int_C -\frac{y}{x^2+y^2}\,dx + \frac{x}{x^2+y^2}\,dy - \int_{C_\epsilon} -\frac{y}{x^2+y^2}\,dx + \frac{x}{x^2+y^2}\,dy.$$

Combining these results gives:

$$\int_C -\frac{y}{x^2+y^2}\,dx + \frac{x}{x^2+y^2}\,dy = \int_{C_\epsilon} -\frac{y}{x^2+y^2}\,dx + \frac{x}{x^2+y^2}\,dy.$$

But we showed in equation (9.10) that the integral of W around any counterclockwise circle about the origin, such as C_ϵ, is 2π. Hence the value around C must be the same.

To summarize, we have proven:

$$\int_C -\frac{y}{x^2+y^2}\,dx + \frac{x}{x^2+y^2}\,dy = \begin{cases} 0 & \text{if } (0,0) \text{ lies in the exterior of } C, \\ 2\pi & \text{if } (0,0) \text{ lies in the interior of } C. \end{cases} \tag{9.11}$$

If C is oriented clockwise instead, then the change in orientation reverses the sign of the integral, so the corresponding values in the two cases are 0 and -2π, respectively. That there are exactly three possible values of the integral of W over all piecewise smooth oriented simple closed curves in $\mathbb{R}^2 - \{(0,0)\}$ may be somewhat surprising.

9.6 THE CONVERSE OF THE MIXED PARTIALS THEOREM

We now discuss briefly and informally conditions regarding when a vector field $F = (F_1, F_2, \ldots, F_n)$ is conservative. On the one hand, we have seen that, if F is conservative, then:

C1. the integral of F over any piecewise smooth oriented curve C depends only on the endpoints of C,

C2. the integral of F over any piecewise smooth oriented closed curve is 0, and

C3. all mixed partial pairs are equal: $\frac{\partial F_i}{\partial x_j} = \frac{\partial F_j}{\partial x_i}$ for all i, j.

So these conditions are necessary for F to be conservative. Are any of them sufficient? It turns out that the converses corresponding to C1 and C2 are true, but we know from the example of W in the previous section that the converse of C3 is not. Computing the mixed partials and seeing if they are equal is the simplest of the three conditions to check, however, so we could ask if there are special circumstances under which the converse of C3 does hold. We investigate this next and toss in a few remarks about the other two conditions along the way.

Definition. A subset U of $\mathbb{R}^n$ is called **simply connected** if:

- every pair of points in U can be joined by a piecewise smooth curve in U and

- every closed curve in U can be continuously contracted within U to a point.

For example, $\mathbb{R}^n$ itself is simply connected. So are rectangles and balls. A sphere in R3 is simply connected as well. A closed curve in any of them can be collapsed to a point within the set rather like a rubber band shrinking in on itself. See Figure 9.17.

Figure 9.17: Examples of simply connected sets: a disk, a rectangle, and a sphere

The punctured plane $\mathbb{R}^2 - \{(0, 0)\}$ is not simply connected, however. A closed curve that goes around the origin cannot be shrunk to a point without passing through the origin, i.e., without leaving the set. Similarly, an annulus in $\mathbb{R}^2$, that is, the ring between two concentric circles, is not simply connected. An example of a non-simply connected surface is the surface of a donut, which is called a **torus**. A loop that goes through the hole cannot be contracted to a point while remaining on the surface. These examples are shown in Figure 9.18.

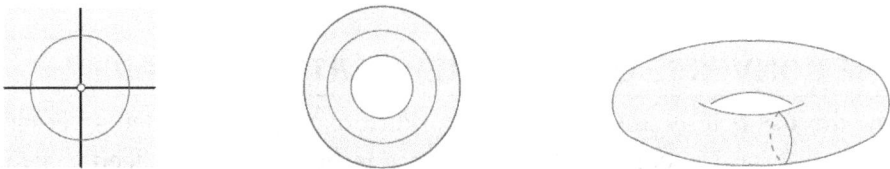

Figure 9.18: Examples of non-simply connected sets and closed curves that cannot be contracted within the set: the punctured plane, an annulus, and a torus

Theorem 9.15 (Converse of the mixed partials theorem). Let $F = (F_1, F_2, \ldots, F_n)$ be a smooth vector field on an open set U in $\mathbb{R}^n$. If

⦿ $\frac{\partial F_i}{\partial x_j} = \frac{\partial F_j}{\partial x_i}$ for all $i, j = 1, 2, \ldots, n$ and

⦿ U is simply connected,

then F is a conservative vector field on U.

A rigorous proof of this theorem is quite hard, so we postpone even any discussion of it until after we look at an example.

Example 9.16. Consider again the vector field $W(x, y) = \left(-\frac{y}{x^2+y^2}, \frac{x}{x^2+y^2}\right)$, $(x, y) \neq (0, 0)$, of Section 9.5. We established that W is not conservative on $\mathbb{R}^2 - \{(0, 0)\}$. On the other hand, the mixed partials of W are equal; therefore Theorem 9.15 implies that W is conservative when restricted to any simply connected subset of $\mathbb{R}^2 - \{(0, 0)\}$.

For example, let U be the right half-plane, $U = \{(x, y) \in \mathbb{R}^2 : x > 0\}$. This is a simply connected set, so a potential function for W must exist on U . Indeed, one can check that $w(x, y) = \arctan\left(\frac{y}{x}\right)$ works. For instance, $\frac{\partial w}{\partial x} = \frac{1}{1+\left(\frac{y}{x}\right)^2} \cdot \left(-\frac{y}{x^2}\right) = -\frac{y}{x^2+y^2}$.

To illustrate how this might be used, suppose that C is a piecewise smooth curve in the right half-plane that goes from a point p, say on the line y = -x in the fourth quadrant, to a point q on the line y = x in the first quadrant, as in Figure 9.19.

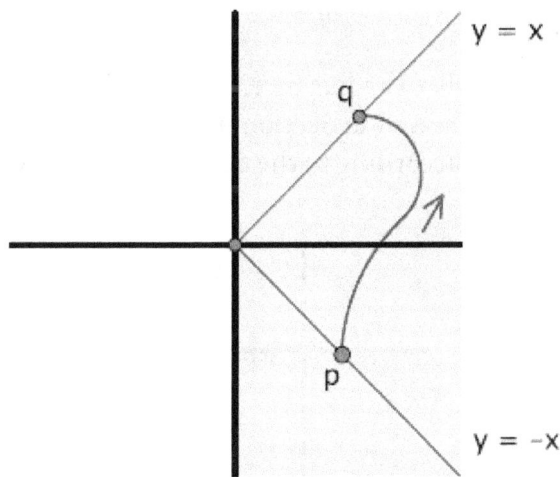

Figure 9.19: A curve in the right half-plane from the line y = -x to the line y = x

Using w as potential function:

$$\int_C -\frac{y}{x^2+y^2}\,dx + \frac{x}{x^2+y^2}\,dy = w(\mathbf{q}) - w(\mathbf{p}) = \arctan(1) - \arctan(-1) = \frac{\pi}{4} - \left(-\frac{\pi}{4}\right) = \frac{\pi}{2}.$$

Note that this difference in the arctangent measures the angle about the origin swept out by C.

The same potential function works in the left half-plane x < 0. Similarly, on the upper half-plane y > 0 and the lower half-plane y < 0, $v(x, y) = -\arctan\left(\frac{x}{y}\right)$ works as a potential function for W. Integrating W over a curve in any of these half-planes amounts to finding the counterclockwise change in angle traced out by the curve.

We can use these observations to go beyond the result of (9.11) and give a more complete description of the integral of W over curves in $\mathbb{R}^2 - \{(0, 0)\}$. Any such curve C can be divided into pieces, each of which lies in one of the four half-planes above. By summing the integrals over each of these pieces, we find that the integral of W over all of C is the total counterclockwise angle about the origin swept out by C from start to finish.

If C is a closed curve so that it comes back to where it started, the total angle is an integer multiple of 2π: $\int_C -\frac{y}{x^2+y^2}\,dx + \frac{x}{x^2+y^2}\,dy = 2\pi n$. This conclusion does not require that C be simple. The integer n represents the net number of times that C goes around the origin in the counterclockwise sense and is given a special name.

Definition. Let C be a piecewise smooth oriented closed curve in $\mathbb{R}^2 - \{(0, 0)\}$. Then the winding number of C with respect to $(0, 0)$ is the integer defined by the equation:

$$\text{winding } \# = \frac{1}{2\pi}\int_C -\frac{y}{x^2+y^2}\,dx + \frac{x}{x^2+y^2}\,dy.$$

See Figure 9.20 for some examples.

For instance, if C_a is the circle of radius a centered at the origin, traversed once counterclockwise, then we showed in Section 9.5 that $\int_{C_a} -\frac{y}{x^2+y^2}\,dx + \frac{x}{x^2+y^2}\,dy = 2\pi$ (see equation (9.10)). Hence, according to the definition, Ca has winding number 1.

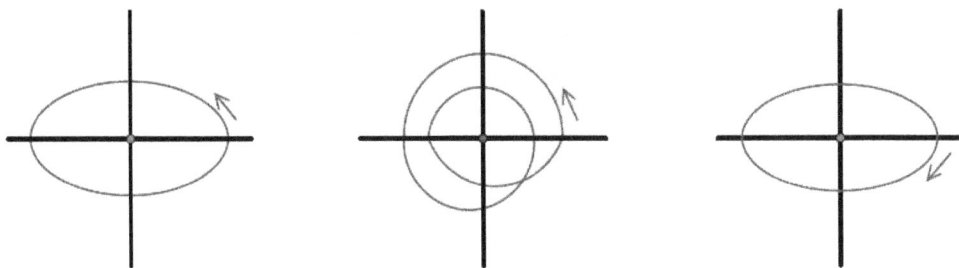

Figure 9.20: Oriented closed curves with winding number 1 (left), 2 (middle), and −1 (right)

One reason that this discussion is interesting is that, in first-year calculus, we tend to focus on the integral of a function as telling us mainly about the function. The winding number, on the other hand, is an instance where the integral is telling us about the geometry of the domain, i.e., the curve, over which the integral is taken.

We close by going over some of the ideas behind the proof of Theorem 9.15, the converse of the mixed partials theorem. We do this only in the case of a vector field $F = (F_1, F_2)$ on an open set U in $\mathbb{R}^2$. Thus we assume that $\frac{\partial F_2}{\partial x} = \frac{\partial F_1}{\partial y}$ and that U is simply connected, and we want to prove that F is conservative on U , that is, that it has a potential function f. We think of f as an "anti-gradient" of F. Then the method of finding f is a lot like what is done in first-year calculus for the fundamental theorem, where an antiderivative is constructed by integrating.

Choose a point a of U . If x is any point of U , then, as part of the definition of simply connected, there is a piecewise smooth curve C in U from a to x. The idea is to show that the line integral $\int_C F \cdot ds = \int_C F_1\, dx + F_2\, dy$ is the same regardless of which curve C is chosen. For suppose that C_1 is another piecewise smooth curve in U from a to x. See Figure 9.21. We want to prove that:

$$\int_C F_1\, dx + F_2\, dy = \int_{C_1} F_1\, dx + F_2\, dy.$$

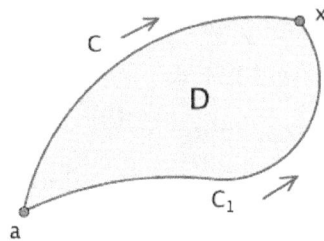

Figure 9.21: Two curves C and C_1 from a to x and the region D that fills in the closed curve $C_2 = C \cup (-C_1)$

By reversing the orientation on C_1 and appending it to C, we obtain a closed curve $C_2 = C \cup (-C_1)$ in U . Here's the hard part. Since U is simply connected, a closed curve such as C_2 can be shrunk within U to a point. The idea is that, in the process of the contraction, C_2 sweeps out a filled-in region D such that $C_2 = \pm \partial D$. It is difficult to turns this intuition into a rigorous proof. Assuming it has been done, we apply Green's theorem together with the assumption of equal mixed partials:

$$\int_{C_2} F_1\, dx + F_2\, dy = \pm \iint_D \left(\frac{\partial F_2}{\partial x} - \frac{\partial F_1}{\partial y} \right) dx\, dy = \pm \iint_D 0\, dx\, dy = 0. \qquad (9.12)$$

At the same time, since $C_2 = C \cup (-C_1)$, we have:

$$\int_{C_2} F_1\, dx + F_2\, dy = \int_C F_1\, dx + F_2\, dy - \int_{C_1} F_1\, dx + F_2\, dy. \qquad (9.13)$$

By combining equations (9.12) and (9.13), we find that

$\int_C F_1\,dx + F_2\,dy = \int_{C_1} F_1\,dx + F_2\,dy,$ as desired.

As a result, if $x \in U$, we may define $f(x) = \int_C F_1\,dx + F_2\,dy,$ where C is any piecewise smooth curve in U from a to x. We write this more suggestively as:

$$f(x) = \int_a^x F_1\,dx + F_2\,dy. \tag{9.14}$$

Note that, whether U is simply connected or not, as long as condition C1 above holds—namely, the line integral depends only on the endpoints of a curve—then the formula in equation (9.14) describes a well-defined function f. The reasoning above also contains the argument that, if condition C2 holds—that the integral around any closed curve is 0—that too is enough to show that the function f in (9.14) is well-defined. This basically follows from equation (9.13).

Thus, to show that any of C1, C2, or, on simply connected domains, C3 suffice to imply that F is conservative, it remains only to confirm that f in (9.14) is a potential function for F: $\frac{\partial f}{\partial x} = F_1$ and $\frac{\partial f}{\partial y} = F_2$. This part is actually not so bad, and we leave it for the exercises (Exercise 6.4).

9.7 EXERCISES FOR CHAPTER 9

Section 1 Definitions and examples

1.1. Let F be the vector field on $\mathbb{R}^2$ given by F(x, y) = (x, y).
   a. Sketch the vector field.
   b. Based on your drawing, describe some paths over which you expect the integral of F to be positive. Find an example of such a path, and calculate the integral over it.
   c. Describe some paths over which you expect the integral to be 0. Find an example of such a path, and calculate the integral over it.

1.2. Find $\int_C F \cdot ds$ if F(x, y) = (-y, x) and C is the circle $x^2 + y^2 = a^2$ of radius a, traversed counterclockwise.

1.3. Find $\int_C F \cdot ds$ if F(x, y) = (cos^2 x, sin y cos y) and C is the curve parametrized by $\alpha(t) = (t, t^2)$, $0 \le t \le \pi$.

1.4. Find $\int_C F \cdot ds$ if F(x, y, z) = (x + y, y + z, x + z) and C is the curve parametrized by $\alpha(t) = (t, t^2, t^3)$, $0 \le t \le 1$.

1.5. Find $\int_C F \cdot ds$ if F(x, y, z) = (-y, x, z) and C is the curve parametrized by $\alpha(t) =$ (cos t, sin t, t), $0 \le t \le 2\pi$.

1.6. Consider the line integral $\int_C (x + y)\,dx + (y - x)\,dy.$
   a. What is the vector field that is being integrated?
   b. Evaluate the integral if C is the line segment parametrized by $\alpha(t) = (t, t)$, 0

Line integrals

$\leq t \leq 1$, from $(0, 0)$ to $(1, 1)$.

    c. Evaluate the integral if C consists of the line segment from $(0, 0)$ to $(1, 0)$ followed by the line segment from $(1, 0)$ to $(1, 1)$.

    d. Evaluate the integral if C consists of the line segment from $(0, 0)$ to $(0, 1)$ followed by the line segment from $(0, 1)$ to $(1, 1)$.

1.7. Consider the line integral $\int_C xy\,dx + yz\,dy + xz\,dz.$

    a. What is the vector field that is being integrated?

    b. Evaluate the integral if C is the line segment from $(1, 2, 3)$ to $(4, 5, 6)$.

    c. Evaluate the integral if C is the curve from $(1, 2, 3)$ to $(4, 5, 6)$ consisting of three consec- utive line segments, each parallel to one of the coordinate axes: from $(1, 2, 3)$ to $(4, 2, 3)$ to $(4, 5, 3)$ to $(4, 5, 6)$.

1.8. Find $\int_C (x - y^2)\,dx + 3xy\,dy$ if C is the portion of the parabola $y = x^2$ from $(-1, 1)$ to $(2, 4)$.

1.9. Find $\int_C y\,dx + z\,dy + x\,dz$ if C is the curve of intersection in $\mathbb{R}^3$ of the cylinder $x^2 + y^2 = 1$ and the plane $z = x + y$, oriented counterclockwise as viewed from high above the xy-plane, looking down.

For the line integrals in Exercises 1.10–1.11, (a) sketch the curve C over which the integral is taken and (b) evaluate the integral. (Hint: Look for ways to reduce, or eliminate, messy calculation.)

1.10. $\int_C 8z\,e^{(x+y)^2}\,dx + (2x^2 + y^2)^2\,dy + 6xy^2z\,dz,$ where C is parametrized by:

$$\alpha(t) = ((\sin t)^{t+12}, (\sin t)^{t+12}, (\sin t)^{t+12}), \quad 0 \leq t \leq \pi/2$$

1.11. $\int_C e^{8xy}\,dx - \ln(\cos^2(x + y) + \pi^x y^{1,000,000})\,dy.$ where C is parametrized by:

$$\alpha(t) = \begin{cases} (\cos t, \sin t) & \text{if } 0 \leq t \leq \pi, \\ (\cos t, -\sin t) & \text{if } \pi \leq t \leq 2\pi \end{cases}$$

1.12. If $\alpha : [a, b] \to \mathbb{R}^n$ is a smooth parametrization of a curve C and f is a continuous real-valued function defined on C, let us write $\int_\alpha f\,ds$ to denote the quantity $\int_a^b f(\alpha(t))\,\|\alpha'(t)\|\,dt$ that appears in the definition of the integral with respect to arclength $\int_C f\,ds.$ Write out the details that the integral is independent of the parametrization in the sense that, if $\beta : [c, d] \to \mathbb{R}^n$ is another smooth parametrization of the same curve C, then:

$$\int_\alpha f\,ds = \int_\beta f\,ds.$$

Consider both cases that α and β traverse C in the same or opposite direction. You may assume α and β are one-to-one, except possibly at the endpoints of their respective domains.

1.13. In this exercise, we return to the geometry of curves in $\mathbb{R}^3$ and study the role of the parametrization in defining curvature. Recall that, if C is a curve in $\mathbb{R}^3$ with a smooth parametrization $\alpha : I \to \mathbb{R}^3$, then the curvature is defined by the equation:

$$\kappa_\alpha(t) = \frac{\|T'_\alpha(t)\|}{v_\alpha(t)},$$

where $T_\alpha(t)$ is the unit tangent vector and $v_\alpha(t)$ is the speed. We have embellished our original notation with the subscript α to emphasize our interest in the effect of the parametrization. We assume that $v_\alpha(t) \neq 0$ for all t so that $\kappa\alpha(t)$ is defined.

Let $\beta : J \to \mathbb{R}^3$ be another such parametrization of C. As in the text, we assume that α and β are one-to-one, except possibly at the endpoints of their respective domains. As in equation (9.3), there is a function $g : J \to I$, assumed to be smooth, such that $\beta(u) = \alpha(g(u))$ for all u in J. In other words, g(u) is the value of t such that $\beta(u) = \alpha(t)$.

    a. Show that the velocities of α and β are related by $v_\beta(u) = v_\alpha g(u) g'(u)$.
    b. Show that the unit tangents are related by $T_\beta(u) = \pm T_\alpha (g(u))$.
    c. Show that $\kappa (u) = \kappa_\beta g(u) \kappa_\alpha(g(u))$.

In other words, if $x = \beta(u) = \alpha(t)$ is any point of C other than an endpoint, then $\kappa_\beta(u) = \kappa_\alpha(t)$. We denote this common value by $\kappa(x)$, i.e., $\kappa(x)$ is defined to be $\kappa\alpha(t)$ for any smooth parametrization α of C, where t is the value of the parameter such that $\alpha(t) = x$. In other words, the curvature at x depends only on the point x and not on how C is parametrized.

Section 3 Conservative fields

3.1 Let $F : \mathbb{R}^2 \to \mathbb{R}^2$ be the vector field $F(x, y) = (y, x)$.

Show that F is not conservative.

Find an oriented closed curve C in $\mathbb{R}^2$ such that $\int_C F \cdot ds \neq 0$.

In Exercises 3.3–3.4, (a) find the curl of F and (b) determine if F is conservative. If it is conservative, find a potential function for it, and, if not, explain why not.

3.3. $F(x, y, z) = (3y^4z^2, 4x^3z^2, -3x^2y^2)$

3.4. $F(x, y, z) = (y - z, z - x, y - x)$

In Exercises 3.5–3.8, determine whether the vector field F is conservative. If it is, find a potential function for it. If not, explain why not.

3.5. $F(x, y) = (x^2 - y^2, 2xy)$

3.6. $F(x, y) = (x^2 + y^2, 2xy)$

3.7. $F(x, y, z) = (x + yz, y + xz, z + xy)$

3.8. $F(x, y, z) = (x - yz, -y + xz, z - xy)$

3.9. Let p be a positive real number, and let F be the vector field on $\mathbb{R}^3 - \{(0, 0, 0)\}$ given by:

$$F(x, y, z) = \left(-\frac{x}{(x^2 + y^2 + z^2)^p}, -\frac{y}{(x^2 + y^2 + z^2)^p}, -\frac{z}{(x^2 + y^2 + z^2)^p} \right).$$

For instance, the inverse square field is the case p = 3/2.

     a. The norm of F is given by $\|F(x, y, z)\| = \|(x, y, z)\|q$ for some exponent q. Find q in terms of p.

     b. For which values of p is F a conservative vector field? The case p = 1 may require special attention.

3.10. Find $\int_C e^y \, dx + xe^y \, dy$ if C is the curve parametrized by $\alpha(t) = (e^{t^2}, t^3), 0 \le t \le 1$.

3.11. Let a and b be real numbers. If C is a piecewise smooth oriented closed curve in the punctured plane $\mathbb{R}^2 - \{(0, 0)\}$, show that:

$$\int_C \frac{ax}{x^2 + y^2} \, dx + \frac{by}{x^2 + y^2} \, dy = \int_C \frac{(b - a)y}{x^2 + y^2} \, dy.$$

3.12. Let $G(x, y, z) = \left(-\frac{x}{(x^2+y^2+z^2)^{3/2}}, -\frac{y}{(x^2+y^2+z^2)^{3/2}}, -\frac{z}{(x^2+y^2+z^2)^{3/2}} \right)$ be the inverse square field. Find the integral of G over the curve parametrized by $\alpha(t)$ = $(1 + t + t^2, t + t^2 + t^3, t^2 + t^3 + t^4), 0 \le t \le 1$.

3.13. Find $\int_C (y + z) \, dx + (x + z) \, dy + (x + y) \, dz$ if C is the curve in $\mathbb{R}^3$ from the origin to (1, 1, 1) that consists of the sequence of line segments, each parallel to one of the coordinate axes, from (0, 0, 0) to (1, 0, 0) to (1, 1, 0) and finally to (1, 1, 1).

3.14. Let C be a piecewise smooth oriented curve in $\mathbb{R}^3$ from a point p = (x1, y1, z1) to a point q = (x_2, y_2, z_2). Show that $\int_C 1 \, dx = x_2 - x_1$. (Similar formulas apply to $\int_C 1 \, dy$ and $\int_C 1 \, dz$.)

3.15.   a. Find a vector field F = (F_1, F_2) on $\mathbb{R}^2$ with the property that, whenever C is a piecewise smooth oriented curve in $\mathbb{R}^2$, then:

$$\int_C F_1 \, dx + F_2 \, dy = \|q\|^2 - \|p\|^2.$$

     where p and q are the starting and ending points, respectively, of C.

     b. Sketch the vector field F that you found in part (a).

3.16. Does there exist a vector field F on $\mathbb{R}^n$ with the property that, whenever C is a piecewise smooth oriented curve in $\mathbb{R}^n$, then:

$$\int_C F \cdot ds = p \cdot q.$$

where p and q are the starting and ending points, respectively, of C? Either describe such a vector field as precisely as you can, or explain why none exists.

3.17. Let $f, g : \mathbb{R}^n \to \mathbb{R}$ be smooth real-valued functions. Prove that, for all piecewise smooth oriented closed curves C in $\mathbb{R}^n$:

$$\int_C (f\,\nabla g) \cdot d\mathbf{s} = -\int_C (g\,\nabla f) \cdot d\mathbf{s}.$$

(Hint: See Exercise 1.19 in Chapter 4.)

3.18. For a vector field $F = (F_1, F_2, F_3)$ on an open set in $\mathbb{R}^3$, is the cross product $\nabla \times F$, i.e., the curl, necessarily orthogonal to F? Either prove that it is, or find an example where it isn't.

3.19. Newton's second law of motion, F = ma, relates the force F acting on an object to the object's mass m and acceleration a. Let F be a smooth force field on an open set of $\mathbb{R}^3$, and assume that, under the influence of F, an object of mass m travels so that its motion is described by a smooth path $\alpha : [a, b] \to \mathbb{R}^3$ with velocity v(t) and acceleration a(t).

  a. The quantity $K(t) = \frac{1}{2}m\|\mathbf{v}(t)\|^2$ is called the **kinetic energy** of the object. Use Newton's second law to show that:

$$\int_C \mathbf{F} \cdot d\mathbf{s} = K(b) - K(a),$$

  where C is the oriented curve parametrized by α. In other words, the work done by the force equals the change in the object's kinetic energy.

  b. Assume, in addition, that F is a conservative field with potential function f. Show that the function is constant. (The function -f is called a **potential energy** for F. The fact that, for a conservative force field F, the sum of the kinetic and potential energies along an object's path is constant is known as the principle of **conservation of energy**.)

$$E(t) = \frac{1}{2}m\|\mathbf{v}(t)\|^2 - f(\alpha(t))$$

3.20. Let F be a smooth vector field on an open set U in $\mathbb{R}^n$. A smooth path $\alpha : I \to U$ defined on an interval I is called an integral path of F if $\alpha'(t) = F(\alpha(t))$ for all t. In other words, at time t, when the position is the point $\alpha(t)$, the velocity equals the value of the vector field at that point. (See page 201.)

  a. If $\alpha : [a, b] \to U$ is a nonconstant integral path of F and C is the oriented curve parametrized by α, show that $\int_C F \cdot ds > 0$. (Hint: The assumption that α is not constant means that there is some point t in [a, b] where $\alpha'(t) \neq 0$.)

  b. Let $f : U \to \mathbb{R}$ be a smooth real-valued function on U, and let $F = \nabla f$. If there is a nonconstant integral path of F going from a point p to a point q, show that $f(q) > f(p)$.

c. Show that a conservative vector field F cannot have a nonconstant closed integral path.

Let U be an open set in $\mathbb{R}^n$ with the property that every pair of points in U can be joined by a piecewise smooth curve in U . Let $f : U \to R$ be a smooth real-valued function defined on U . If $\nabla f (x) = 0$ for all x in U , use line integrals to give a simple proof that f is a constant function.

Section 4 Green's theorem

4.1. Find $\int_C (x^2 - y^2)\, dx + 2xy\, dy$ if C is the boundary of the square with vertices $(0, 0)$, $(1, 0)$, $(1, 1)$, and $(0, 1)$, oriented counterclockwise.

4.2. Find $\int_C (2xy^2 - y^3)\, dx + (x^2 y + x^3)\, dy$ if C is the unit circle $x^2 + y^2 = 1$, oriented counter- clockwise.

4.3. Find $\int_C -y \cos x\, dx + xy\, dy$ if C is the boundary of the parallelogram with vertices $(0, 0)$, $(2\pi, \pi)$, $(3\pi, 3\pi)$, and $(\pi, 2\pi)$, oriented counterclockwise.

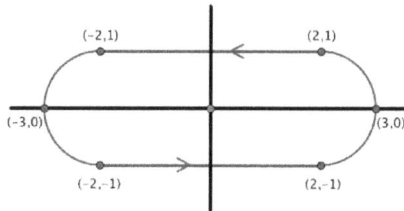

Figure 9.22: An oval track

4.4. Find $\int_C (e^{x+2y} - 3y)\, dx + (4x + 2e^{x+2y})\, dy$ if C is the track shown in Figure 9.22 consisting of two straightaways joined by semicircles at each end, oriented counterclockwise.

4.5.
a. Use the parametrization $\alpha(t) = (a \cos t, a \sin t)$, $0 \le t \le 2\pi$, and Corollary 9.14 to verify that the area of the disk $x^2 + y^2 \le a^2$ of radius a is $\pi a2$.

b. By modifying the parametrization in part (a), use Corollary 9.14 to find the area of the region $\frac{x^2}{a^2} + \frac{y^2}{b^2} \le 1$ inside an ellipse.

In Exercises 4.6–4.11, evaluate the line integral using whatever methods seem best.

4.6. $\int_C (x + y)\, dx + (x - y)\, dy$. where C is the curve in $\mathbb{R}^2$ consisting of the line segment from $(0, 1)$ to $(1, 0)$ followed by the line segment from $(1, 0)$ to $(2, 1)$

4.7. $\int_C \frac{e^x}{x^2+y^2}\, dx - \frac{x}{x^2+y^2}\, dy$. where C is the unit circle $x^2 + y^2 = 1$ in $\mathbb{R}^2$, oriented counterclockwise

4.8. $\int_C (3x^2 y - xy^2 - 3x^2 y^2)\, dx + (x^3 - 2x^3 y + x^2 y)\, dy$. where C is the closed triangular curve in $\mathbb{R}^2$ with vertices $(0, 0)$, $(1, 1)$, and $(0, 1)$, oriented counterclockwise

4.9. $\int_C yz^2\, dx + x^4 z\, dy + x^2 y^2\, dz$. where C is the curve parametrized by $\alpha(t) = (t, t^2, t^3)$, $0 \le t \le 1$

4.10. $\int_C (6x^2 - 4y + 2xy)\,dx + (2x - 2\sin y + 3x^2)\,dy$, where C is the diamond-shaped curve $|x| + |y| = 1$ in $\mathbb{R}^2$, oriented counterclockwise

4.11. $\int_C yz\,dx + xz\,dy + xy\,dz$, where C is the curve of intersection in $\mathbb{R}^3$ of the cylinder $x^2 + y^2 = 1$ and the saddle surface $z = x^2 - y^2$, oriented counterclockwise as viewed from high above the xy-plane, looking down

Let U be an open set in $\mathbb{R}^2$. A smooth real-valued function h : U → R is called **harmonic** on U if:

$$\frac{\partial^2 h}{\partial x^2} + \frac{\partial^2 h}{\partial y^2} = 0 \quad \text{at all points of } U.$$

Exercises 4.12–4.15 concern harmonic functions.

4.12. Show that $h(x, y) = e^{-x} \sin y$ is a harmonic function on $\mathbb{R}^2$.

4.13. Let D be a bounded subset of U whose boundary consists of a finite number of piecewise smooth simple closed curves. If h is harmonic on U , prove that:

$$\int_{\partial D} -h\frac{\partial h}{\partial y}\,dx + h\frac{\partial h}{\partial x}\,dy = \iint_D \|\nabla h(x, y)\|^2\,dx\,dy.$$

4.14. Let D be a region as in the preceding exercise, and suppose that every pair of points in D can be joined by a piecewise smooth curve in D. If h is harmonic on U and if h = 0 at all points of ∂D, prove that h = 0 on all of D. (Hint: In addition to the preceding exercise, see Exercises 2.4 of Chapter 5 and 3.21 of this chapter.)

4.15. Continuing with the previous exercise, prove that a harmonic function is completely deter- mined on D by its values on the boundary in the sense that, if h_1 and h_2 are harmonic on U and if $h_1 = h_2$ at all points of ∂D, then $h_1 = h_2$ on all of D.

Section 5 The vector field W

5.1. Let $F = (F_1, F_2)$ be a smooth vector field that:

 ⊙ is defined everywhere in $\mathbb{R}^2$ except at three points **p, q, r** and

 ⊙ satisfies the mixed partials condition $\frac{\partial F_2}{\partial x} = \frac{\partial F_1}{\partial y}$ everywhere in its domain.

Let C_1, C_2, C_3 be small circles centered at **p, q, r**, respectively, oriented counterclockwise as shown in Figure 9.23. Assume that:

$$\int_{C_1} F_1\,dx + F_2\,dy = 12, \quad \int_{C_2} F_1\,dx + F_2\,dy = 10, \text{ and } \int_{C_3} F_1\,dx + F_2\,dy = 15.$$

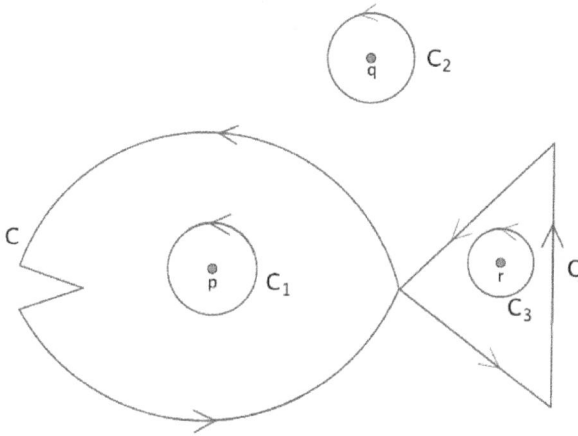

Figure 9.23: Three circles C_1, C_2, and C_3 and a fish-shaped curve C

a. Find $\int_C F_1\,dx + F_2\,dy$ is the oriented fish-shaped curve shown in the figure.
b. Draw a piecewise smooth oriented closed curve K such that $\int_K F_1\,dx + F_2\,dy = 1$. Your curve need not be simple. Include in your drawing the direction in which K is traversed.

Section 6 The converse of the mixed partials theorem

6.1.   a. Let C be the curve in $\mathbb{R}^2 - \{(0,0)\}$ parametrized by $\alpha(t) = (\cos t, -\sin t)$, $0 \le t \le 2\pi$. Describe C geometrically, and calculate its winding number using the definition.

b. More generally, let n be an integer (positive, negative, or zero allowed), and let C_n be the curve parametrized by $\alpha(t) = (\cos nt, \sin nt)$, $0 \le t \le 2\pi$. Give a geometric description of C_n, and calculate its winding number using the definition.

6.2. Let C be a piecewise smooth simple closed curve in $\mathbb{R}^2$, oriented counterclockwise, such that the origin lies in the exterior of C. What are the possible values of the winding number of C?

6.3. Let $F = (F_1, F_2)$ be a smooth vector field on the punctured plane $U = \mathbb{R}^2 - \{(0,0)\}$ such that $\frac{\partial F_2}{\partial x} = \frac{\partial F_1}{\partial y}$ on U , and let C be a piecewise smooth oriented closed curve in U.

a. If C does not intersect the x-axis, explain why $\int_C F \cdot ds = 0$.
b. Suppose instead that C may intersect the positive x-axis but not the negative x-axis. Is it necessarily still true that $\int_C F \cdot ds = 0$? Explain.

6.4.   a. Let $c = (c, d)$ be a point of $\mathbb{R}^2$, and let $B = B(c, r)$ be an open ball centered at c. For the moment, we think of $\mathbb{R}^2$ as the uv-plane to avoid having x and y mean too many different things later on. Let $F(u, v) = (F_1(u, v), F_2(u, v))$ be a continuous vector field on B.

Given any point x = (x, y) in B, let C_1 be the curve from (c, d) to (x, y) consisting of a vertical line segment followed by a horizontal line segment, as indicated on the left of Figure 9.24. Define a function $f_1 : B \to R$ by:

$$f_1(x, y) = \int_{C_1} F_1 \, du + F_2 \, dv.$$

Show that $\frac{\partial f_1}{\partial x}(x, y) = F_1(x, y)$ for all (x, y) in B. (Hint: Parametrize C_1 and use the fundamental theorem of calculus, keeping in mind what a partial derivative is.)

Figure 9.24: The curves of integration C_1 and C_2 for functions f_1 and f_2, respectively

  b. Similarly, let C_2 be the curve from (c, d) to (x, y) consisting of a horizontal segment followed by a vertical segment, as shown on the right of Figure 9.24, and define:

$$f_2(x, y) = \int_{C_2} F_1 \, du + F_2 \, dv.$$

Show that $\frac{\partial f_2}{\partial y}(x, y) = F_2(x, y)$ for all (x, y) in B.

  c. Now, let U be an open set in $\mathbb{R}^2$ such that every pair of points in U can be joined by a piecewise smooth curve in U . Let a be a point of U . Assume that F is a smooth vector field on U such that the function $f: U \to R$ given in equation (9.14) is well-defined. That is:

$$f(\mathbf{x}) = \int_a^\mathbf{x} F_1 \, du + F_2 \, dv,$$

where the notation means $\int_C F_1 \, du + F_2 \, dv$ for any piecewise smooth curve C in U from a to x.

Show that f is a potential function for F on U , i.e., $\frac{\partial f}{\partial x} = F_1$ and $\frac{\partial f}{\partial y} = F_2$. In particular, F is a conservative vector field. (Hint: Given a point c in U , choose an open ball B = B(c, r) that is contained in U . Define functions f_1 and f_2 as in parts (a) and (b). For x in B, how is f (x) related to $f(c) + f_1(x)$ and $f(c) + f_2(x)$?)

10 SURFACE INTEGRALS

Chapter

We next study how to integrate vector fields over surfaces. One important caveat: our discussion applies only to surfaces in $\mathbb{R}^3$. For surfaces in $\mathbb{R}^n$ with n > 3, the expressions that one integrates are more complicated than vector fields.

It is easy to get lost in the weeds with the details of surface integrals, so, in the first section of the chapter, we just try to build some intuition about what the integral measures and how to compute that measurement. The formal definition of the integral appears in the second section. Calculating surface integrals is not necessarily difficult but it can be messy, so having a range of options is especially welcome. In addition to the definition and the intuitive approach, sometimes a surface integral can be converted to a different type of integral altogether. For example, one of the theorems of the chapter, Stokes's theorem, relates surface integrals to line integrals, and another, Gauss's theorem, relates them to triple integrals. It would be a mistake to think of these results primarily as computational tools, however. Their theoretical consequences are at least as important, and we try to give a taste of some of the new lines of reasoning about integrals that they open up. The chapter closes with a couple of sections in which we tie up some loose ends. This also puts us in position to wrap things up in the next, and final, chapter.

10.1 WHAT THE SURFACE INTEGRAL MEASURES

If S is a surface in $\mathbb{R}^3$, we learned in Section 5.5 how to integrate a continuous real-valued function $f : S \to R$ over S. By definition:

$$\iint_S f\, dS = \iint_D f(\sigma(s,t)) \left\| \frac{\partial \sigma}{\partial s} \times \frac{\partial \sigma}{\partial t} \right\| ds\, dt, \tag{10.1}$$

where $\sigma : D \to \mathbb{R}^3$ is a smooth parametrization of S defined on a subset D of the st-plane, as in Figure 10.1. This is called the integral with respect to surface area. As before, we assume now and in the future that D is the closure of a bounded open set in $\mathbb{R}^2$, that is, a bounded open set together with all its boundary points, such as a closed disk or rectangle. As a matter of full disclosure, we did not show that the value of the integral is independent of the particular parametrization σ, a point that we address at the end of the chapter.

Now, let $F : U \to \mathbb{R}^3$ be a vector field on an open set U in $\mathbb{R}^3$ that contains S. The idea is that the integral of F over S measures the amount that F flows through S. This quantity is called the flux. For example, if F points in a direction normal to the surface, then it is flowing through effectively, whereas, if it points in a tangent direction, it is not flowing through at all. This stands in contrast with the interpretation of the line integral, where we were interested in the degree to which the vector field flowed along a curve.

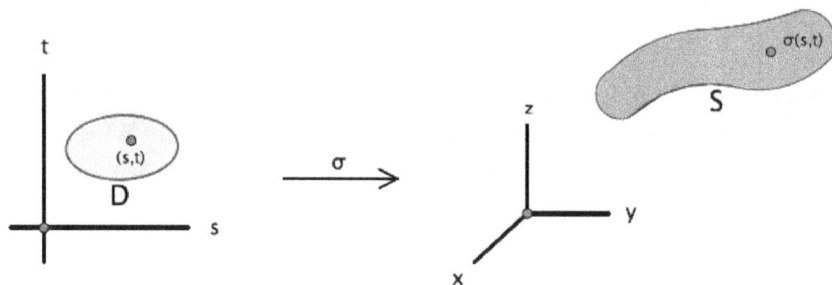

Figure 10.1: A parametrization σ of a surface S in $\mathbb{R}^3$

To begin, we need to designate a direction of flow through S that is considered to be positive.

Definition. An **orientation** of a smooth surface S in $\mathbb{R}^3$ is a continuous vector field $n : S \to \mathbb{R}^3$ such that, for every x in S, n(x) is a unit normal vector to S at x. See Figure 10.2. An **oriented surface** is a surface S together with a specified orientation n.

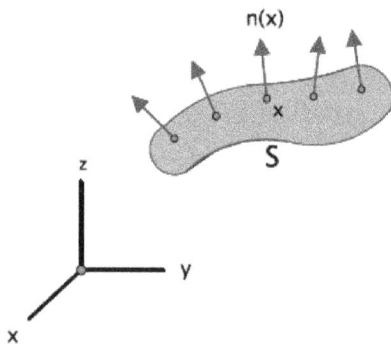

Figure 10.2: A surface S with orienting unit normal vector field n

Example 10.1. Suppose that S is a level set defined by $f(x, y, z) = c$ for some smooth real-valued function f of three variables. If $x \in S$ then by Proposition 4.17, $\nabla f(x)$ is a normal vector to S at x, so the formula $n(x) = \frac{1}{\|\nabla f(x)\|} \nabla f(x)$ describes an orientation of S, provided that $\nabla f(x) \neq 0$ for all x in S.

For instance, let S be the sphere $x^2 + y^2 + z^2 = a^2$ of radius a. Here, $f(x, y, z) = x^2 + y^2 + z^2$, so $\nabla f = (2x, 2y, 2z) = 2(x, y, z)$. As a result, an orientation of the sphere is given by:

$$\mathbf{n}(x, y, z) = \frac{1}{2\sqrt{x^2 + y^2 + z^2}} 2(x, y, z) = \frac{1}{\sqrt{a^2}}(x, y, z) = \frac{1}{a}(x, y, z)$$

for all points (x, y, z) on the sphere. This vector points away from the origin, so we say that it orients S with the outward normal. See Figure 10.3, left.

We could equally well orient the sphere with the inward unit normal. This would be given by $\mathbf{n}(x, y, z) = -\frac{1}{a}(x, y, z)$.

Or let S be the circular cylinder $x^2 + y^2 = a^2$. This is a level set of $f(x, y, z) = x^2 + y^2$. Then $\nabla f = (2x, 2y, 0) = 2(x, y, 0)$, and an orientation is:

$$\mathbf{n}(x, y, z) = \frac{1}{2\sqrt{x^2 + y^2}} 2(x, y, 0) = \frac{1}{a}(x, y, 0)$$

for all (x, y, z) on the cylinder. Once again, this is the outward normal (Figure 10.3, right).

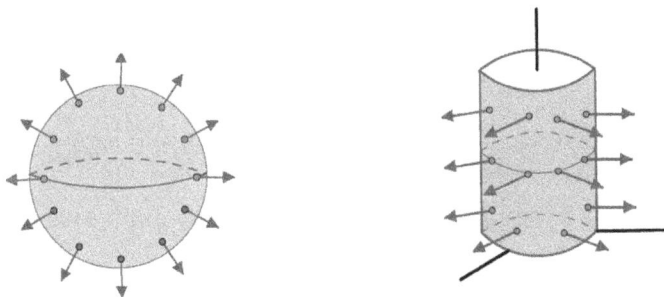

Figure 10.3: A sphere and cylinder oriented by the outward normal

Figure 10.4: A mammal with orientation[9]

In general, let S be an oriented surface with orienting normal n. To measure the flow of a vector field F = (F$_1$, F$_2$, F$_3$) through S, we find the scalar component F$_{norm}$ of F in the normal direction at every point of S and integrate it over S. This component is given by F$_{norm}$ = ||F|| cos θ, where θ is the angle between F and n, as shown in Figure 10.5.

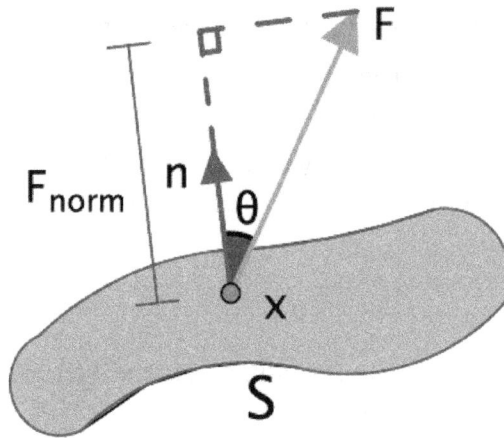

Figure 10.5: The component of a vector field in the normal direction at a point x

Hence:

$$F_{norm} = \|\mathbf{F}\| \cos \theta = \|\mathbf{F}\| \|\mathbf{n}\| \cos \theta = \mathbf{F} \cdot \mathbf{n}.$$

The integral of F over S, denoted by $\iint_S \mathbf{F} \cdot d\mathbf{S}$, can be expressed as:

$$\iint_S \mathbf{F} \cdot d\mathbf{S} = \iint_S F_{norm} \, dS = \iint_S \mathbf{F} \cdot \mathbf{n} \, dS.$$

As the integral of the real-valued function F · n, the expression on the right is an integral with respect to surface area of the type considered previously in (10.1). We take equation (10.2) as a tentative definition of the integral of F and use it to work through some examples.

Example 10.2. Let S be the sphere x^2 + y^2 + z^2 = a^2 of radius a, oriented by the outward normal. Find $\iint_S \mathbf{F} \cdot d\mathbf{S}$ if:

a. $\mathbf{F}(x, y, z) = -\frac{1}{(x^2+y^2+z^2)^{3/2}} (x, y, z)$ (i.e., the inverse square field),

b. F(x, y, z) = (0, 0, z),

c. F(x, y, z) = (0, 0, –1).

As shown in Example 10.1, the outward unit normal to S is given by $\mathbf{n} = \frac{1}{a}(x, y, z)$.

The inverse square field, illustrated in Figure 10.6, points directly inwards towards the origin, so F · n is negative at every point of S. In other words, if the positive direction is taken to be outward, then F is flowing in the negative direction. Thus we expect the integral to be negative.

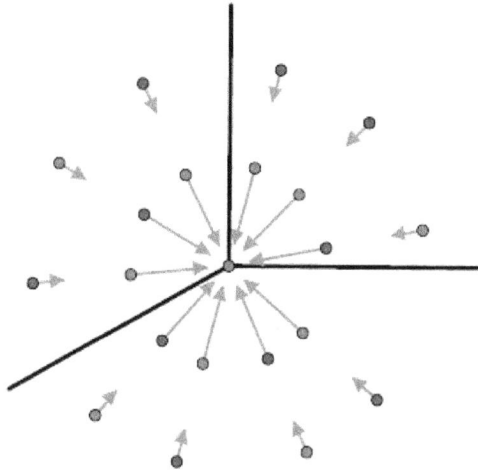

Figure 10.6: The inverse square field

To calculate it precisely, note that on

$$S, \ \mathbf{F}(x, y, z) = -\frac{1}{(a^2)^{3/2}}(x, y, z) = -\frac{1}{a^3}(x, y, z), \ \text{so:}$$

$$\mathbf{F} \cdot \mathbf{n} = -\frac{1}{a^3}(x, y, z) \cdot \frac{1}{a}(x, y, z) = -\frac{1}{a^4}(x^2 + y^2 + z^2) = -\frac{1}{a^4} \cdot a^2 = -\frac{1}{a^2}.$$

Thus:

$$\iint_S \mathbf{F} \cdot d\mathbf{S} = \iint_S \mathbf{F} \cdot \mathbf{n} \, dS = \iint_S -\frac{1}{a^2} \, dS = -\frac{1}{a^2} \, \text{Area}\,(S) = -\frac{1}{a^2} \cdot 4\pi a^2 = -4\pi.$$

This uses the formula that the surface area is $4\pi a^2$ from Example 5.19 in Chapter 5. We shall refer to this surface integral later, so to repeat for emphasis: the integral of the inverse square field over a sphere of radius a centered at the origin and with the outward orientation is -4π, regardless of the radius of the sphere.

The vector field F(x, y, z) = (0, 0, z) points straight upward when z > 0 and straight downward when z < 0 (Figure 10.7). In either region, the component in the outward direction of S is positive. Thus we expect the integral to be positive.

In fact, $\mathbf{F} \cdot \mathbf{n} = (0, 0, z) \cdot \frac{1}{a}(x, y, z) = \frac{1}{a}z^2$, so:

$$\iint_S \mathbf{F} \cdot d\mathbf{S} = \frac{1}{a} \iint_S z^2 \, dS.$$

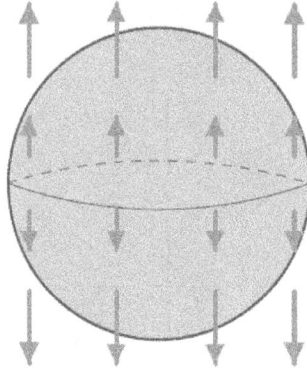

Figure 10.7: The vector field F(x, y, z) = (0, 0, z) flowing through a sphere

This last integral is precisely one of the examples we calculated using a parametrization when we studied integrals with respect to surface area (Example 5.19 again). The answer is $\iint_S z^2 \, dS = \frac{4}{3}\pi a^4$. (We also gave a symmetry argument that made use of the facts that $\iint_S x^2 \, dS = \iint_S y^2 \, dS = \iint_S z^2 \, dS$ and $\iint_S (x^2 + y^2 + z^2) \, dS = \iint_S a^2 \, dS = a^2 \cdot (4\pi a^2) = 4\pi a^4$.) **Thus:**

$$\iint_S \mathbf{F} \cdot d\mathbf{S} = \frac{1}{a} \cdot \frac{4}{3}\pi a^4 = \frac{4}{3}\pi a^3.$$

We shall take another look at this example later as well.

Here, F(x, y, z) = (0, 0, -1) is a constant vector field consisting of unit vectors that point straight down everywhere (Figure 10.8). Roughly speaking, F points inward on the upper hemisphere, i.e., the component in the outward direction is negative, and outward on the lower hemisphere. From the symmetry of the situation, we might expect these two contributions to cancel out exactly, giving an integral of 0. Informally, the flow in equals the flow out, so the net flow is 0.

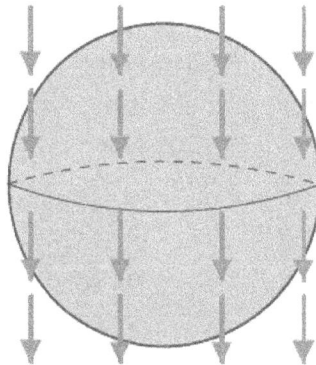

Figure 10.8: The vector field F(x, y, z) = (0, 0, -1) flowing through a sphere

To verify this, note that $\mathbf{F} \cdot \mathbf{n} = (0, 0, -1) \cdot \frac{1}{a}(x, y, z) = -\frac{1}{a}z$. Thus:

$$\iint_S \mathbf{F} \cdot d\mathbf{S} = \iint_S -\frac{1}{a}z \, dS = -\frac{1}{a} \iint_S z \, dS.$$

By symmetry, this last integral is indeed 0. More precisely, S is symmetric in the xy-plane, i.e., (x, y, -z) is in S whenever (x, y, z) is, and the integrand $f(x, y, z) = z$ satisfies $f(x, y, -z) = -z = -f(x, y, z)$, so the contributions of the northern and southern hemispheres cancel. Thus $\iint_S \mathbf{F} \cdot d\mathbf{S} = 0$.

In this last example, if we had integrated instead only over the northern hemisphere, the same line of thought would lead us to expect that the integral is negative. As a test of your intuition about what the integral represents, you might think about whether this is related to the following example.

Example 10.3. Let S be the disk of radius a in the xy-plane described by oriented by the upward normal, and let F(x, y, z) = (0, 0, -1) as in Example 10.2(c) above.

$$x^2 + y^2 \leq a^2, \quad z = 0,$$

See Figure 10.9. Find $\iint_S \mathbf{F} \cdot d\mathbf{S}$.

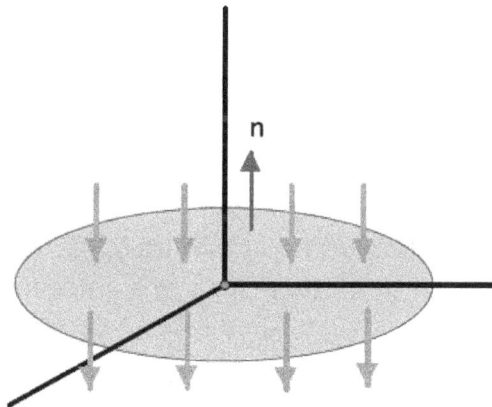

Figure 10.9: The vector field F(x, y, z) = (0, 0, -1) flowing through a disk in the xy-plane

The upward unit normal to S is n = k = (0, 0, 1), so F · n = (0, 0, -1) · (0, 0, 1) = -1. Therefore:

$$\iint_S \mathbf{F} \cdot d\mathbf{S} = \iint_S -1 \, dS = -\text{Area}\,(S) = -\pi a^2.$$

Continuing with the comments right before the example, the result turnsed out be negative, but is the flow of F = (0, 0, -1) through the disk really related to the flow through the northe$\mathbb{R}^n$ hemisphere? We shall be able to give a precise answer based on general principles once we learn a little more about surface integrals (see Exercises 4.4 and 5.5 at the end of the chapter).

10.2 THE DEFINITION OF THE SURFACE INTEGRAL

We now present the actual definition of the vector field surface integral $\iint_S \mathbf{F} \cdot d\mathbf{S}$. This follows a strategy we have used before: to define this new type of integral, we parametrize the surface, substitute for x, y, and z in terms of the two parameters, and then pull back to a double integral in the parameter plane. Of course, we need to figure out what the pullback should be.

Let S be an oriented surface in $\mathbb{R}^3$ with orienting normal n, and let $\sigma : D \to \mathbb{R}^3$ be a smooth parametrization of S with domain D in $\mathbb{R}^2$:

$$\sigma(s,t) = (x(s,t), y(s,t), z(s,t)), \quad (s,t) \in D.$$

We first describe how σ can be used to give an independent orientation of S. Recall from Section 5.5 that the vectors $\frac{\partial \sigma}{\partial s} = \left(\frac{\partial x}{\partial s}, \frac{\partial y}{\partial s}, \frac{\partial z}{\partial s}\right)$ and $\frac{\partial \sigma}{\partial t} = \left(\frac{\partial x}{\partial t}, \frac{\partial y}{\partial t}, \frac{\partial z}{\partial t}\right)$ are "velocity" vectors with respect to s and t, respectively, and thus are tangent to S. As a result, $\frac{\partial \sigma}{\partial s} \times \frac{\partial \sigma}{\partial t}$ is normal to S. and $\frac{\frac{\partial \sigma}{\partial s} \times \frac{\partial \sigma}{\partial t}}{\|\frac{\partial \sigma}{\partial s} \times \frac{\partial \sigma}{\partial t}\|}$ is a unit normal, i.e., an orientation of S, provided that $\frac{\partial \sigma}{\partial s} \times \frac{\partial \sigma}{\partial t} \neq 0$. Since the orientation n that is given for S is also a unit normal, it must be true that:

$$\mathbf{n} = \pm \frac{\frac{\partial \sigma}{\partial s} \times \frac{\partial \sigma}{\partial t}}{\|\frac{\partial \sigma}{\partial s} \times \frac{\partial \sigma}{\partial t}\|}. \tag{10.3}$$

We say that σ **preserves orientation** if the sign is + and that it **reverses orientation** if -.

Now, to formulate the definition of the surface integral, recall that the idea was that $\iint_S \mathbf{F} \cdot d\mathbf{S}$ should equal $\iint_S \mathbf{F} \cdot \mathbf{n}\, dS$. Using (10.3) to substitute for n, this becomes $\pm \iint_S \mathbf{F} \cdot \frac{\frac{\partial \sigma}{\partial s} \times \frac{\partial \sigma}{\partial t}}{\|\frac{\partial \sigma}{\partial s} \times \frac{\partial \sigma}{\partial t}\|} \, dS$.

Then the definition of the integral with respect to surface area (10.1) with $f = \mathbf{F} \cdot \frac{\frac{\partial \sigma}{\partial s} \times \frac{\partial \sigma}{\partial t}}{\|\frac{\partial \sigma}{\partial s} \times \frac{\partial \sigma}{\partial t}\|}$ gives:

$$\iint_S \mathbf{F} \cdot \mathbf{n} \, dS = \pm \iint_D \left(\mathbf{F}(\sigma(s,t)) \cdot \frac{\frac{\partial \sigma}{\partial s} \times \frac{\partial \sigma}{\partial t}}{\|\frac{\partial \sigma}{\partial s} \times \frac{\partial \sigma}{\partial t}\|} \right) \left\| \frac{\partial \sigma}{\partial s} \times \frac{\partial \sigma}{\partial t} \right\| \, ds\, dt$$

$$= \pm \iint_D \mathbf{F}(\sigma(s,t)) \cdot \left(\frac{\partial \sigma}{\partial s} \times \frac{\partial \sigma}{\partial t} \right) \, ds\, dt.$$

This is the thinking behind the following definition.

Definition. Let U be an open set in $\mathbb{R}^3$, and let $\mathbf{F} : U \to \mathbb{R}^3$ be a continuous vector field. Let S be an oriented surface contained in U , and let $\sigma : D \to \mathbb{R}^3$ be a smooth parametrization of S such that $\frac{\partial \sigma}{\partial s} \times \frac{\partial \sigma}{\partial t}$ is never 0. Then the integral of F over S, denoted by $\iint_S \mathbf{F} \cdot d\mathbf{S}$. is defined to be:

$$\iint_S \mathbf{F} \cdot d\mathbf{S} = \pm \iint_D \mathbf{F}(\sigma(s,t)) \cdot \left(\frac{\partial \sigma}{\partial s} \times \frac{\partial \sigma}{\partial t} \right) ds\,dt, \qquad (10.4)$$

where + is used if σ is orientation-preserving and − if orientation-reversing.

We defer the discussion that the definition is independent of the parametrization σ until the end of the chapter after we have gotten used to working with the definition and there are fewer new things to absorb. Also, the definition extends in the obvious way to surfaces that are piecewise smooth. That is, if a surface S is a union $S = S_1 \cup S_2 \cup \cdots \cup S_k$ of oriented smooth surfaces that intersect at most in pairs along portions of their boundaries, we define:

$$\iint_S \mathbf{F} \cdot d\mathbf{S} = \iint_{S_1} \mathbf{F} \cdot d\mathbf{S} + \iint_{S_2} \mathbf{F} \cdot d\mathbf{S} + \cdots + \iint_{S_k} \mathbf{F} \cdot d\mathbf{S}.$$

As a first example, we recalculate an integral that we evaluated before using the tentative definition in Example 10.2(b).

Example 10.4. Let S be the sphere $x^2 + y^2 + z^2 = a^2$, oriented by the outward normal. Find $\iint_S \mathbf{F} \cdot d\mathbf{S}$ if $F(x, y, z) = (0, 0, z)$.

We apply the definition using the standard parametrization of S with spherical coordinates φ and θ as parameters and ρ = a fixed:

$$\sigma(\phi, \theta) = (a \sin \phi \cos \theta, \, a \sin \phi \sin \theta, \, a \cos \phi), \quad 0 \le \phi \le \pi, \quad 0 \le \theta \le 2\pi.$$

The domain of the parametrization is the rectangle D = [0, π] × [0, 2π] in the φθ-plane. See Figure 10.10.

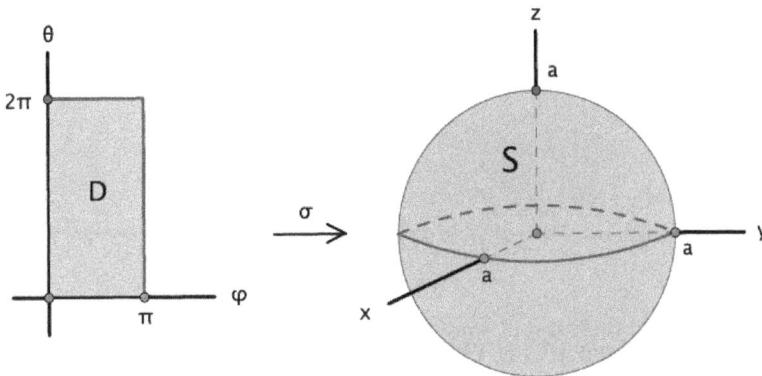

Figure 10.10: A parametrization of the sphere of radius a

Substituting in terms of the parameters gives $F(\sigma(\phi, \theta)) = (0, 0, z(\phi, \theta)) = (0, 0, a \cos \phi)$. In addition:

$$\frac{\partial \sigma}{\partial \phi} \times \frac{\partial \sigma}{\partial \theta} = \det \begin{bmatrix} \mathbf{i} & \mathbf{j} & \mathbf{k} \\ a\cos\phi\cos\theta & a\cos\phi\sin\theta & -a\sin\phi \\ -a\sin\phi\sin\theta & a\sin\phi\cos\theta & 0 \end{bmatrix}$$

$$= \vdots \quad \text{(we calculated this before in Example 5.19)}$$

$$= a^2 \sin\phi \left(\sin\phi\cos\theta, \ \sin\phi\sin\theta, \ \cos\phi\right). \tag{10.5}$$

This vector is guaranteed to be normal to S, but, to determine whether σ is orientation-preserving or reversing, we must check if it agrees with the given orientation, which is outward. Note that, in the first octant, where φ and θ are both between 0 and $\frac{\pi}{2}$, all components of $\frac{\partial \sigma}{\partial \phi} \times \frac{\partial \sigma}{\partial \theta}$ in equation (10.5) are positive, hence $\frac{\partial \sigma}{\partial \phi} \times \frac{\partial \sigma}{\partial \theta}$.points outward. Thus σ is orientation-preserving.

This can also be seen geometrically. At any point x of S, $\frac{\partial \sigma}{\partial \phi}$ is a tangent vector in the direction of increasing φ. Similarly, $\frac{\partial \sigma}{\partial \theta}$ is tangent in the direction of increasing θ. These vectors are shown in Figure 10.11. By the right-hand rule, $\frac{\partial \sigma}{\partial \phi} \times \frac{\partial \sigma}{\partial \theta}$ points in the outward direction.

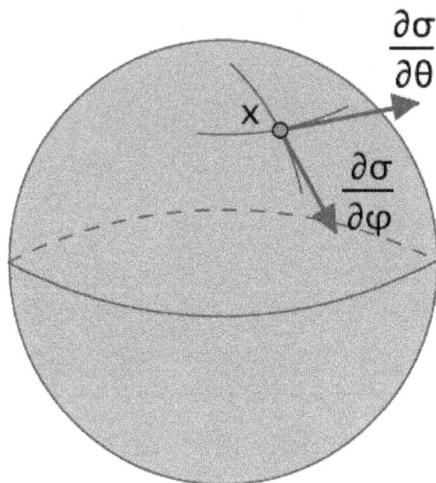

Figure 10.11: The tangent vectors $\frac{\partial \sigma}{\partial \phi}$ and $\frac{\partial \sigma}{\partial \theta}$

To use the definition of the integral (10.4), we integrate the function:

$$\mathbf{F}(\sigma(\phi,\theta)) \cdot \left(\frac{\partial \sigma}{\partial \phi} \times \frac{\partial \sigma}{\partial \theta}\right) = (0, 0, a\cos\phi) \cdot \left(a^2 \sin\phi \left(\sin\phi\cos\theta, \ \sin\phi\sin\theta, \ \cos\phi\right)\right)$$

$$= a^2 \sin\phi \left(0 + 0 + a\cos^2\phi\right)$$

$$= a^3 \sin\phi\cos^2\phi.$$

Thus:

$$\iint_S \mathbf{F} \cdot d\mathbf{S} = +\iint_D \mathbf{F}(\sigma(\phi,\theta)) \cdot \left(\frac{\partial \sigma}{\partial \phi} \times \frac{\partial \sigma}{\partial \theta}\right) d\phi\, d\theta$$

$$= \iint_D a^3 \sin \phi \cos^2 \phi \, d\phi\, d\theta$$

$$= \int_0^{2\pi} \left(\int_0^{\pi} a^3 \sin \phi \cos^2 \phi \, d\phi\right) d\theta$$

$$= a^3 \left(\int_0^{\pi} \sin \phi \cos^2 \phi \, d\phi\right)\left(\int_0^{2\pi} 1 \, d\theta\right)$$

$$= a^3 \left(-\frac{1}{3}\cos^3 \phi \Big|_0^{\pi}\right)\left(\theta \Big|_0^{2\pi}\right)$$

$$= a^3 \cdot \left(-\frac{1}{3}(-1-1)\right) \cdot 2\pi$$

$$= a^3 \cdot \frac{2}{3} \cdot 2\pi$$

$$= \frac{4}{3}\pi a^3.$$

This agrees with the answer we got earlier in Example 10.2(b), but the calculation here was at least as messy as the one obtained there by integrating the normal component of F. In other words, the definition may not always be the most efficient way to go. We'll return to this example one more time later and apply a method that's simplest of all.

Example 10.5. Let F(x, y, z) = (0, 1, -3) for all (x, y, z) in $\mathbb{R}^3$. This is a constant vector $\iint_S \mathbf{F} \cdot d\mathbf{S}$ e think of as "rain" falling through space. See Figure 10.12, left. Find if:

S is the cone $z = \sqrt{x^2 + y^2}$, where x² + y² ≤ 4, oriented by the upward normal (Figure 10.12, middle),

S is the disk that caps the cone in part (a), i.e., the disk x² + y² ≤ 4 in the plane z = 2, oriented by the upward normal (Figure 10.12, right).

As we saw in Section 5.4, there are several ways to parametrize the cone, but it's the graph of a function of x and y so one way is to use x and y as parameters:

$$\sigma(x, y) = (x, y, \sqrt{x^2 + y^2}), \quad x^2 + y^2 \le 4.$$

The domain D in the xy-plane is the disk x² + y² ≤ 4 of radius 2. Moreover:

$$\frac{\partial \sigma}{\partial x} \times \frac{\partial \sigma}{\partial y} = \det \begin{bmatrix} \mathbf{i} & \mathbf{j} & \mathbf{k} \\ 1 & 0 & \frac{x}{\sqrt{x^2+y^2}} \\ 0 & 1 & \frac{y}{\sqrt{x^2+y^2}} \end{bmatrix} = \left(-\frac{x}{\sqrt{x^2+y^2}}, -\frac{y}{\sqrt{x^2+y^2}}, 1\right).$$

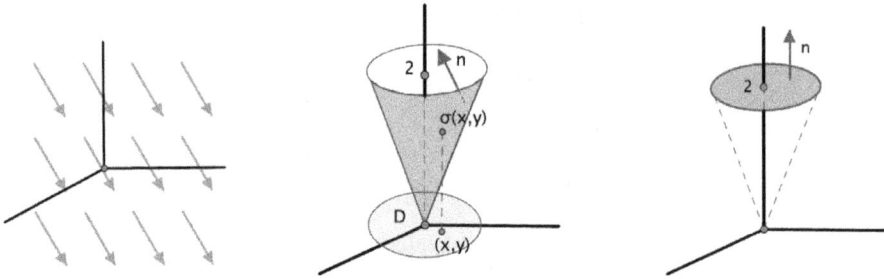

Figure 10.12: Rain (left), the cone $z = \sqrt{x^2 + y^2}$, $x^2 + y^2 \leq 4$ **(middle), and the cap of the cone, x² + y² ≤ 4, z = 2 (right)**

The z-component is positive, so $\frac{\partial \sigma}{\partial x} \times \frac{\partial \sigma}{\partial y}$ points upward. Thus σ is orientation-preserving. In addition:

$$\mathbf{F}(\sigma(x,y)) \cdot \left(\frac{\partial \sigma}{\partial x} \times \frac{\partial \sigma}{\partial y}\right) = (0, 1, -3) \cdot \left(-\frac{x}{\sqrt{x^2+y^2}}, -\frac{y}{\sqrt{x^2+y^2}}, 1\right)$$

$$= -\frac{y}{\sqrt{x^2+y^2}} - 3.$$

Thus, by definition:

$$\iint_S \mathbf{F} \cdot d\mathbf{S} = \iint_D \left(-\frac{y}{\sqrt{x^2+y^2}} - 3\right) dx\, dy$$

$$= -\iint_D \frac{y}{\sqrt{x^2+y^2}}\, dx\, dy - 3\iint_D 1\, dx\, dy.$$

The first term is 0 using the symmetry of D in the x-axis, and the second is 3 times the area of D. Therefore $\iint_S \mathbf{F} \cdot d\mathbf{S} = -3\,\text{Area}(D) = -3 \cdot (\pi \cdot 2^2) = -12\pi$.

The negative answer makes sense, since S is oriented so that the upward direction is considered positive whereas the rain is falling downward.

To integrate over the oriented disk on the right of Figure 10.12, we won't parametrize at all but rather go back to integrating the normal component of F over S, which is what the actual definition formalized. The upward unit normal to S is n = (0, 0, 1), so F· n = (0, 1, -3)· (0, 0, 1) = -3. Hence:

$$\iint_S \mathbf{F} \cdot d\mathbf{S} = \iint_S \mathbf{F} \cdot \mathbf{n}\, dS = \iint_S -3\, dS = -3\,\text{Area}\,(S) = -3 \cdot (\pi \cdot 2^2) = -12\pi.$$

We close by picking up the thread with which we ended the previous section, which is to note that the answers to parts (a) and (b) of the last example are the same. We can interpret this to say that the flow through the top of the cone equals the flow through the sides. If we could have been sure that this was true in advance, then we could have calculated the integral over the cone by replacing it with the much simpler calculation over the capping disk. The next three sections have something to say about why this would have been valid from a couple of different perspectives. They also contain the last two big theorems of introductory multivariable calculus.

10.3 STOKES'S THEOREM

There are two types of points on a surface. At some points, one can approach along the surface from any direction and remain on the surface, at least for a little while, after continuing on through the point. These are sometimes called **interior points**. Other points lie on the edge of the surface in the sense that, after approaching the point from certain directions, one falls off the surface immediately after continuing through. These are called **boundary points**. The set of all boundary points is called the **boundary** of the surface.

For instance, a hemisphere has a boundary consisting of the equatorial circle, and a cylinder has a boundary consisting of two pieces, namely, the circles at each end. On the other hand, a sphere and a torus have no boundary points. See Figure 10.13. In general, the boundary of a surface is either empty or consists of one or more simple closed curves.

Figure 10.13: The boundary of a hemisphere is a circle and that of a cylinder is two circles. The boundaries of a sphere and torus are empty.

Stokes's theorem relates a line integral $\int_B \mathbf{F} \cdot d\mathbf{s}$ around the boundary B of a surface in $\mathbb{R}^3$ to a vector field integral over the surface itself. It can be seen as an extension of Green's theorem from double integrals to surface integrals. The theorem is not obvious, and we shall try to show where it comes from, though in a very special case similar to the one we used to explain Green's theorem. Namely, we assume that S is an oriented surface with a smooth orientation-preserving parametrization σ(s, t) = (x(s, t), y(s, t), z(s, t) such that:

- ⊙ the domain of σ is a rectangle R = [a, b] × [c, d] and
- ⊙ σ is one-to-one and, in particular, transforms the boundary of R precisely onto the boundary of S.

The situation is shown in Figure 10.14.

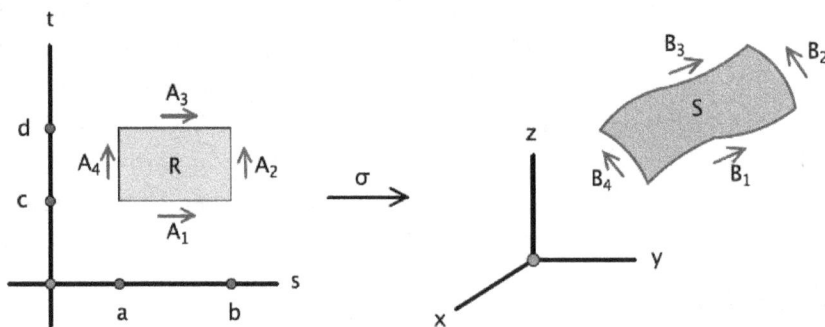

Figure 10.14: A magic carpet: a surface S with parametrization σ defined on a rectangle R. The sides of R are oriented so that ∂R = A$_1$ ∪ A$_2$ ∪ (-A$_3$) ∪ (-A$_4$).

Let A denote the boundary of R and B the boundary of S. We orient A counterclockwise and break it into its four sides A$_1$, A$_2$, A$_3$, A$_4$, where the horizontal sides A$_1$ and A$_3$ are oriented to the right and the vertical sides A$_2$ and A$_4$ are oriented upward. Taking orientation into account, we have A = A$_1$ ∪ A$_2$ ∪ (-A$_3$) ∪ (-A$_4$). Applying σ then breaks B up into four curves B$_1$, B$_2$, B$_3$, B$_4$, where B$_j$ = σ(A$_j$) for j = 1, 2, 3, 4. B inherits the orientation B = B$_1$ ∪ B$_2$ ∪ (-B$_3$) ∪ (-B$_4$).

Let F = (F$_1$, F$_2$, F$_3$) be a smooth vector field defined on an open set in R3 that contains S. We consider the line integral of F around B, the boundary of S:

$$\int_B \mathbf{F} \cdot d\mathbf{s} = \left(\int_{B_1} + \int_{B_2} - \int_{B_3} - \int_{B_4} \right) \mathbf{F} \cdot d\mathbf{s}.$$

We use differential form notation: $\int_B \mathbf{F} \cdot d\mathbf{s} = \int_B F_1\,dx + F_2\,dy + F_3\,dz$. To cut down on the amount of notation in any one place, we find $\int_B F_1\,dx,\ \int_B F_2\,dy,\ \int_B F_3\,dz$ separately and add the results at the end.

We can parametrize each of the four pieces of B by applying σ to the corresponding side of A, fixing one of s or t and using the other as parameter.

For B$_1$: α1(s) = σ(s, c), a ≤ s ≤ b.

For B$_2$: α2(t) = σ(b, t), c ≤ t ≤ d.

For B$_3$: α3(s) = σ(s, d), a ≤ s ≤ b.

For B$_4$: α4(t) = σ(a, t), c ≤ t ≤ d.

So for $\int_B F_1 \, dx$, we integrate over each of B1 through B4 and take the sum with appropriate signs. For example, for B1, the parametrization is $\alpha_1(s) = \sigma(s, c) = (x(s, c), y(s, c), z(s, c))$, where $t = c$ is held fixed. Hence the derivatives with respect to the parameter are derivatives with respect to s, partial derivatives where appropriate. For instance, $\alpha_1'(s) = \frac{\partial \sigma}{\partial s}(s, c) = (\frac{\partial x}{\partial s}(s, c), \frac{\partial y}{\partial s}(s, c), \frac{\partial z}{\partial s}(s, c))$. As a result, we obtain:

$$\int_{B_1} F_1 \, dx = \int_a^b F_1(\sigma(s, c)) \frac{\partial x}{\partial s}(s, c) \, ds.$$

Similarly, over B_2 with parameter t:

$$\int_{B_2} F_1 \, dx = \int_c^d F_1(\sigma(b, t)) \frac{\partial x}{\partial t}(b, t) \, dt.$$

Continuing in this way with B3 and B4 and combining the results gives:

$$\int_B F_1 \, dx = \int_a^b F_1(\sigma(s, c)) \frac{\partial x}{\partial s}(s, c) \, ds + \int_c^d F_1(\sigma(b, t)) \frac{\partial x}{\partial t}(b, t) \, dt$$
$$- \int_a^b F_1(\sigma(s, d)) \frac{\partial x}{\partial s}(s, d) \, ds - \int_c^d F_1(\sigma(a, t)) \frac{\partial x}{\partial t}(a, t) \, dt. \qquad (10.6)$$

Upon closer examination, each of the terms in this last expression is a line integral over one of the sides of the boundary of R back in the st-plane. For instance, the first term is $\int_{A_1}(F_1 \circ \sigma) \frac{\partial x}{\partial s} \, ds$ and the second is $\int_{A_2}(F_1 \circ \sigma) \frac{\partial x}{\partial t} \, dt$. line integrals over A_1 and A_2, respectively. In fact, since t is constant on A_1 and s is constant on A_2, we can give these terms a more uniform appearance by writing them as:

$$\int_{A_1}(F_1 \circ \sigma) \frac{\partial x}{\partial s} \, ds + (F_1 \circ \sigma) \frac{\partial x}{\partial t} \, dt \quad \text{and} \quad \int_{A_2}(F_1 \circ \sigma) \frac{\partial x}{\partial s} \, ds + (F_1 \circ \sigma) \frac{\partial x}{\partial t} \, dt.$$

respectively. The extra summands are superfluous to the integrals, since the derivatives of the extra coordinates are zero on the corresponding segment. After converting the remaining two terms to line integrals over A_3 and A_4, equation (10.6) becomes a line integral over all of A:

$$\int_B F_1 \, dx = \int_A (F_1 \circ \sigma) \frac{\partial x}{\partial s} \, ds + (F_1 \circ \sigma) \frac{\partial x}{\partial t} \, dt.$$

But now that we are back to a line integral in the plane, Green's theorem applies, and, as $A = \partial R$, we can replace the line integral over A with a double integral over the rectangle R. Thus:

$$\int_B F_1 \, dx = \iint_R \left(\frac{\partial}{\partial s}((F_1 \circ \sigma) \frac{\partial x}{\partial t}) - \frac{\partial}{\partial t}((F_1 \circ \sigma) \frac{\partial x}{\partial s}) \right) ds \, dt. \qquad (10.7)$$

We work with the integrand of the double integral. Using the product rule to calculate the partial derivatives with respect to s and t, the integrand becomes:

$$\frac{\partial}{\partial s}(F_1 \circ \sigma) \cdot \frac{\partial x}{\partial t} + (F_1 \circ \sigma) \cdot \frac{\partial^2 x}{\partial s \, \partial t} - \frac{\partial}{\partial t}(F_1 \circ \sigma) \cdot \frac{\partial x}{\partial s} - (F_1 \circ \sigma) \cdot \frac{\partial^2 x}{\partial t \, \partial s}.$$

Thanks to the equality of the mixed partials $\frac{\partial^2 x}{\partial s \, \partial t} = \frac{\partial^2 x}{\partial t \, \partial s}$, the second and fourth terms cancel, leaving:

$$\frac{\partial}{\partial s}(F_1 \circ \sigma) \cdot \frac{\partial x}{\partial t} - \frac{\partial}{\partial t}(F_1 \circ \sigma) \cdot \frac{\partial x}{\partial s}.$$

So after substituting into equation (10.7), at this point, we have:

$$\int_B F_1 \, dx = \iint_R \left(\frac{\partial}{\partial s}(F_1 \circ \sigma) \cdot \frac{\partial x}{\partial t} - \frac{\partial}{\partial t}(F_1 \circ \sigma) \cdot \frac{\partial x}{\partial s} \right) ds \, dt. \tag{10.8}$$

We compute the partials of $F_1 \circ \sigma$ using the chain rule. Fans of dependence diagrams will welcome Figure 10.15. By the chain rule, the integrand in equation (10.8) becomes:

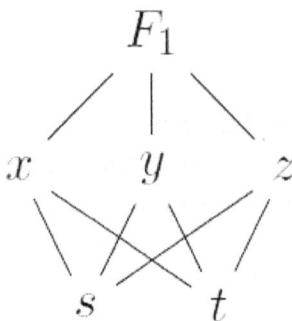

$$F_1$$
$$x \qquad y \qquad z$$
$$s \qquad t$$

Figure 10.15: A dependence diagram for the composition $F_1 \circ \sigma$

$$\left(\frac{\partial F_1}{\partial x} \frac{\partial x}{\partial s} + \frac{\partial F_1}{\partial y} \frac{\partial y}{\partial s} + \frac{\partial F_1}{\partial z} \frac{\partial z}{\partial s} \right) \frac{\partial x}{\partial t} - \left(\frac{\partial F_1}{\partial x} \frac{\partial x}{\partial t} + \frac{\partial F_1}{\partial y} \frac{\partial y}{\partial t} + \frac{\partial F_1}{\partial z} \frac{\partial z}{\partial t} \right) \frac{\partial x}{\partial s}$$

$$= \frac{\partial F_1}{\partial z} \left(\frac{\partial z}{\partial s} \frac{\partial x}{\partial t} - \frac{\partial x}{\partial s} \frac{\partial z}{\partial t} \right) - \frac{\partial F_1}{\partial y} \left(\frac{\partial x}{\partial s} \frac{\partial y}{\partial t} - \frac{\partial y}{\partial s} \frac{\partial x}{\partial t} \right).$$

where $\frac{\partial F_1}{\partial x}$, $\frac{\partial F_1}{\partial y}$, and $\frac{\partial F_1}{\partial z}$ are evaluated at $\sigma(s, t) = x(s, t), y(s, t), z(s, t))$. Substituting this into (10.8) yields:

$$\int_B F_1 \, dx = \iint_R \left(\frac{\partial F_1}{\partial z} \left(\frac{\partial z}{\partial s} \frac{\partial x}{\partial t} - \frac{\partial x}{\partial s} \frac{\partial z}{\partial t} \right) - \frac{\partial F_1}{\partial y} \left(\frac{\partial x}{\partial s} \frac{\partial y}{\partial t} - \frac{\partial y}{\partial s} \frac{\partial x}{\partial t} \right) \right) ds \, dt.$$

The calculations of $\int_B F_2 \, dy$ and $\int_B F_3 \, dz$ are similar and end up giving:

$$\int_B F_2 \, dy = \iint_R \left(-\frac{\partial F_2}{\partial z} \left(\frac{\partial y}{\partial s} \frac{\partial z}{\partial t} - \frac{\partial z}{\partial s} \frac{\partial y}{\partial t} \right) + \frac{\partial F_2}{\partial x} \left(\frac{\partial x}{\partial s} \frac{\partial y}{\partial t} - \frac{\partial y}{\partial s} \frac{\partial x}{\partial t} \right) \right) ds \, dt$$

and $$\int_B F_3 \, dz = \iint_R \left(\frac{\partial F_3}{\partial y} \left(\frac{\partial y}{\partial s} \frac{\partial z}{\partial t} - \frac{\partial z}{\partial s} \frac{\partial y}{\partial t} \right) - \frac{\partial F_3}{\partial x} \left(\frac{\partial z}{\partial s} \frac{\partial x}{\partial t} - \frac{\partial x}{\partial s} \frac{\partial z}{\partial t} \right) \right) ds \, dt.$$

Adding the three calculations and suitably regrouping the terms results in the formula:

$$\int_B F_1 \, dx + F_2 \, dy + F_3 \, dz = \iint_R \left(\left(\frac{\partial F_3}{\partial y} - \frac{\partial F_2}{\partial z} \right) \left(\frac{\partial y}{\partial s} \frac{\partial z}{\partial t} - \frac{\partial z}{\partial s} \frac{\partial y}{\partial t} \right) \right.$$
$$+ \left(\frac{\partial F_1}{\partial z} - \frac{\partial F_3}{\partial x} \right) \left(\frac{\partial z}{\partial s} \frac{\partial x}{\partial t} - \frac{\partial x}{\partial s} \frac{\partial z}{\partial t} \right)$$
$$\left. + \left(\frac{\partial F_2}{\partial x} - \frac{\partial F_1}{\partial y} \right) \left(\frac{\partial x}{\partial s} \frac{\partial y}{\partial t} - \frac{\partial y}{\partial s} \frac{\partial x}{\partial t} \right) \right) ds \, dt. \qquad (10.9)$$

This looks awful, but actually the ungainly double integral can be expressed much more concisely. The integrand is the dot product of the vectors:

$$\mathbf{v} = \left(\frac{\partial F_3}{\partial y} - \frac{\partial F_2}{\partial z}, \ \frac{\partial F_1}{\partial z} - \frac{\partial F_3}{\partial x}, \ \frac{\partial F_2}{\partial x} - \frac{\partial F_1}{\partial y} \right)$$

and $$\mathbf{w} = \left(\frac{\partial y}{\partial s} \frac{\partial z}{\partial t} - \frac{\partial z}{\partial s} \frac{\partial y}{\partial t}, \ \frac{\partial z}{\partial s} \frac{\partial x}{\partial t} - \frac{\partial x}{\partial s} \frac{\partial z}{\partial t}, \ \frac{\partial x}{\partial s} \frac{\partial y}{\partial t} - \frac{\partial y}{\partial s} \frac{\partial x}{\partial t} \right).$$

For v, the components are the differences within the mixed partial pairs of F, and in fact v = ∇×F, the curl of F. Likewise, w is also a cross product:

$$\mathbf{w} = \det \begin{bmatrix} \mathbf{i} & \mathbf{j} & \mathbf{k} \\ \frac{\partial x}{\partial s} & \frac{\partial y}{\partial s} & \frac{\partial z}{\partial s} \\ \frac{\partial x}{\partial t} & \frac{\partial y}{\partial t} & \frac{\partial z}{\partial t} \end{bmatrix} = \frac{\partial \sigma}{\partial s} \times \frac{\partial \sigma}{\partial t}.$$

Thus, after all this work, equation (10.9) becomes:

$$\int_B F_1 \, dx + F_2 \, dy + F_3 \, dz = \iint_R (\nabla \times \mathbf{F}) \cdot \left(\frac{\partial \sigma}{\partial s} \times \frac{\partial \sigma}{\partial t} \right) ds \, dt.$$

This last expression is precisely the definition of the surface integral of the vector field ∇ × F over S. We have arrived at the final formula:

$$\int_B \mathbf{F} \cdot d\mathbf{s} = \int_B F_1 \, dx + F_2 \, dy + F_3 \, dz = \iint_S (\nabla \times \mathbf{F}) \cdot d\mathbf{S}. \qquad (10.10)$$

This was a long calculation. To recap the strategy, to evaluate the line integral over the boundary B of S, we used a parametrization of S to pull back to a line integral in the plane, applied Green's theorem there, and lastly identified the resulting double integral (10.9) as the pullback of an integral over the surface.

The argument behind (10.10) assumed that the domain of the parametrization is a rectangle, but the conclusion remains true for surfaces more broadly. In order to formulate a general statement, we need to say something about the relation between the orientation of a surface and the orientation of its boundary. Let S be a smooth oriented surface in $\mathbb{R}^3$ with orienting normal n. We think of n as defining "up" and imagine walking around S with our feet on the surface and our heads in the direction of n. Then:

∂S denotes the boundary of S oriented so that, as

you traverse the boundary, S stays on the left.

If S happens to be contained in the plane $\mathbb{R}^2$, this is consistent with the way we oriented its boundary in Green's theorem, assuming that S is oriented by the upward normal $n = (0, 0, 1)$.

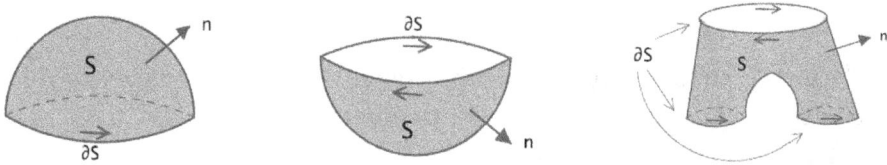

Figure 10.16: Three oriented surfaces and their oriented boundaries

For example, let S be a sphere that is oriented by the outward normal. Then the boundary of the northern hemisphere is the equatorial circle oriented counterclockwise, while the boundary of the southern hemisphere is the same circle but oriented clockwise. If S is a pair of pants oriented by the outward normal, then ∂S consists of three closed curves: the waist oriented one way—let's call it clockwise—and the bottoms of the two legs oriented counterclockwise. See Figure 10.16.

For a piecewise smooth surface $S = S_1 \cup S_2 \cup \cdots \cup S_k$, we require that the smooth pieces $S_1, S_2, \ldots, S_k$ are oriented compatibly in the sense that the oriented boundaries of adjacent pieces are traversed in opposite directions wherever they intersect. These intersections are not considered to be part of the boundary of the whole surface S. This is illustrated in Figure 10.17.

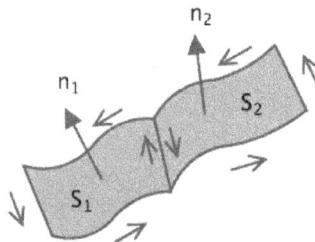

Figure 10.17: Compatible orienting normal n_1, n_2 for adjacent surfaces S_1, S_2, respectively, with corresponding orientations of ∂S_1, ∂S_2

For the extension of (10.10) beyond surfaces whose parameter domains are rectangles, one approach might be to mimic the strategy we outlined informally for Green's theorem. Namely, first verify it in the case where the parameter domain is a union of rectangles that overlap at most along parts of their boundaries and then argue that the domain of any parametrization can be approximated arbitrarily well by such a union. We won't give the details.

With these preliminaries out of the way, we can state the theorem.

Theorem 10.6 (Stokes's theorem). Let S be a piecewise smooth oriented surface in $\mathbb{R}^3$ that is bounded as a subset of $\mathbb{R}^3$, and let F be a smooth vector field defined on an open set containing S. Then:

$$\int_{\partial S} \mathbf{F} \cdot ds = \iint_S (\nabla \times \mathbf{F}) \cdot d\mathbf{S}.$$

In the form stated, Stokes's theorem converts a line integral to a surface integral. Since surface integrals are usually messier than line integrals, this is not the usual way in which the theorem is applied, at least as far as calculating specific examples goes. (See Exercises 3.3–3.6 for some exceptions, however.) Sometimes though, the theorem can be used to go the other way, converting a surface integral to a line integral. This possibility has implications for the general properties of surface integrals.

$$x2 + y2, x2 + y2 \le 4,$$

Example 10.7. Let F be the "falling rain" vector field F(x, y, z) = (0, 1, -3), and let S be the cone $z = \sqrt{x^2 + y^2}$, $x^2 + y^2 \le 4$, oriented by the upward normal. In Example 10.5, we integrated F over S using a parametrization. Since we evaluated the integral before, there seems no harm in giving away a spoiler: the answer is $\iint_S \mathbf{F} \cdot d\mathbf{S} = -12\pi$.

Could this surface integral have been computed using Stokes's theorem? The answer is yes provided that there is a vector field G such that F = ∇ × G, for then Stokes's theorem gives $\iint_S \mathbf{F} \cdot d\mathbf{S} = \iint_S (\nabla \times \mathbf{G}) \cdot d\mathbf{S} = \int_{\partial S} \mathbf{G} \cdot ds$. In other words, we could find the surface integral of F over S by computing the line integral of G around the boundary.

We try to find such a G = (G₁, G₂, G₃) by setting F = ∇ × G and guessing:

$$\mathbf{F} = (0, 1, -3) = \det \begin{bmatrix} \mathbf{i} & \mathbf{j} & \mathbf{k} \\ \frac{\partial}{\partial x} & \frac{\partial}{\partial y} & \frac{\partial}{\partial z} \\ G_1 & G_2 & G_3 \end{bmatrix}.$$

This leads to the system:

$$
\begin{cases}
0 &= \frac{\partial G_3}{\partial y} - \frac{\partial G_2}{\partial z} \\
1 &= \frac{\partial G_1}{\partial z} - \frac{\partial G_3}{\partial x} \\
-3 &= \frac{\partial G_2}{\partial x} - \frac{\partial G_1}{\partial y}.
\end{cases}
$$

From the second equation, we guess G_1 = z, from the third G_2 = -3x, and from the first G_3 = 0. One can check that these guesses are actually consistent with all three equations. Thus G(x, y, z) = (z, -3x, 0) works! Our virtuous lifestyle has paid off.

This gives:

$$
\iint_S \mathbf{F}\cdot d\mathbf{S} = \int_{\partial S} \mathbf{G}\cdot ds = \int_{\partial S} z\,dx - 3x\,dy + 0\,dz = \int_{\partial S} z\,dx - 3x\,dy.
$$

We calculate the line integral by parametrizing C = ∂S, which is a circle of radius 2 in the plane z = 2, oriented counterclockwise (see Figure 10.12, middle, or, if you're willing to peek ahead, Figure 10.18, left). The circle can be parametrized by α(t) = (2 cos t, 2 sin t, 2), $0 \le t \le 2\pi$. Thus:

$$
\begin{aligned}
\iint_S \mathbf{F}\cdot d\mathbf{S} = \int_C z\,dx - 3x\,dy &= \int_0^{2\pi} \left(2\cdot(-2\sin t) - 6\cos t\cdot 2\cos t\right) dt \\
&= \int_0^{2\pi} (-4\sin t - 12\cos^2 t)\,dt \\
&= 4\cos t\Big|_0^{2\pi} - 12\cdot\frac{2\pi}{2} \qquad (10.11) \\
&= 0 - 12\pi \\
&= -12\pi,
\end{aligned}
$$

where, in step (10.11), we used our trick to integrate cos2 t (Exercise 1.1, Chapter 7).

10.4 CURL FIELDS

In the last example, we got the same answer of -12π using Stokes's theorem as we did in Example 10.5 using a parametrization, but it's not clear that the Stokes's theorem approach was any simpler. Actually, it seemed a little roundabout. The argument has some important consequences, however.

Continuing with the example, let C denote the oriented boundary of the cone S, and suppose that $\widetilde{S}$ is any piecewise smooth oriented surface having C as its oriented boundary. For instance, $\widetilde{S}$ could be the disk in the plane z = 2 that caps off the cone, oriented by the upward normal, which was also considered in Example 10.5. See Figure 10.18.

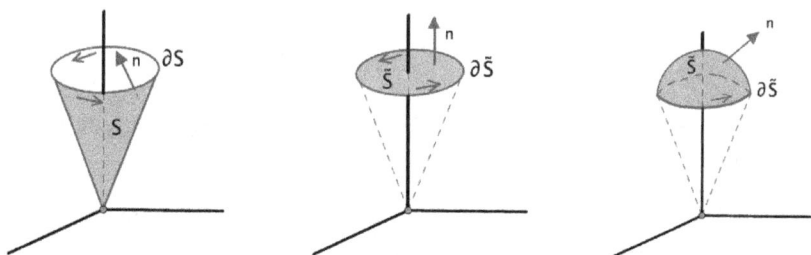

Figure 10.18: An oriented cone S and two other oriented surfaces $\widetilde{S}$, all three having the same oriented boundary

If we integrate the falling rain vector field F over the new surface $\widetilde{S}$, we obtain:

$$\iint_{\widetilde{S}} \mathbf{F} \cdot d\mathbf{S} = \iint_{\widetilde{S}} (\nabla \times \mathbf{G}) \cdot d\mathbf{S} \quad \text{(where } \mathbf{G} = (z, -3x, 0) \text{ as before)}$$

$$= \int_{\partial \widetilde{S}} \mathbf{G} \cdot ds \quad \text{(by Stokes's theorem)}$$

$$= \int_{C} \mathbf{G} \cdot ds$$

$$= -12\pi \quad \text{(by the calculation in the previous example).}$$

In other words, the integral of F is -12π over all oriented surfaces whose boundary is C. This idea can be formulated more generally.

Definition. Let U be an open set in $\mathbb{R}^3$. A vector field $F : U \to \mathbb{R}^3$ is called a curl field on U if there exists a smooth vector field $G : U \to \mathbb{R}^3$ such that $F = \nabla \times G$ at all points of U .

For instance, we saw in Example 10.7 that F = (0, 1, –3) is a curl field on R3 with "anti-curl" G = (z, –3x, 0).

One perspective on curl fields is the following general working principle: curl fields are to surface integrals as conservative fields are to line integrals. We consider a couple of illustrations of this.

Example 10.8. For line integrals: The integral of a conservative field over an oriented curve depends only on the endpoints of the curve.

For surface integrals: The integral of a curl field over an oriented surface depends only on the oriented boundary of the surface. More precisely:

Theorem 10.9. Let S_1 and S_2 be piecewise smooth oriented surfaces in an open set U in $\mathbb{R}^3$ that have the same oriented boundary, i.e., $\partial S_1 = \partial S_2$. If F is a curl field on U, then:

$$\iint_{S_1} \mathbf{F} \cdot d\mathbf{S} = \iint_{S_2} \mathbf{F} \cdot d\mathbf{S}.$$

Proof. Suppose that F = ∇ × G. Then by Stokes's theorem, both sides of the equation are equal to $\int_C \mathbf{G} \cdot ds$, where $C = \partial S_1 = \partial S_2$.

Example 10.10. For line integrals: The integral of a conservative field over a piecewise smooth oriented closed curve is 0.

For surface integrals:

Definition. Let S be a piecewise smooth surface in $\mathbb{R}^n$ such that every pair of points in S can be joined by a piecewise smooth curve in S. Then S is called closed if ∂S = ∅ (the empty set).

For example, spheres and tori are closed surfaces, but hemispheres and cylinders are not.

Theorem 10.11. If F is a curl field on an open set U in $\mathbb{R}^3$, then $\iint_S \mathbf{F} \cdot d\mathbf{S} = 0$ for any piecewise smooth oriented closed surface S in U.

Proof. Let S be a piecewise smooth oriented closed surface in U. The simplest argument would be to say that, if F = ∇ × G, then, by Stokes's theorem $\iint_S \mathbf{F} \cdot d\mathbf{S} = \int_{\partial S} \mathbf{G} \cdot ds = 0$ since ∂S = ∅.

If integrating over the empty set makes you nervous, then perhaps a more aboveboard alternative would be to cut S into two nonempty pieces S_1 and S_2 such that S = $S_1 \cup S_2$ and S_1 and S_2 intersect only along their common boundary. For instance, S_1 could be a small disklike piece of S, and S_2 could be what's left. See Figure 10.19. Then:

$$\iint_S \mathbf{F} \cdot d\mathbf{S} = \iint_{S_1} \mathbf{F} \cdot d\mathbf{S} + \iint_{S_2} \mathbf{F} \cdot d\mathbf{S}$$
$$= \int_{\partial S_1} \mathbf{G} \cdot ds + \int_{\partial S_2} \mathbf{G} \cdot ds. \qquad (10.12)$$

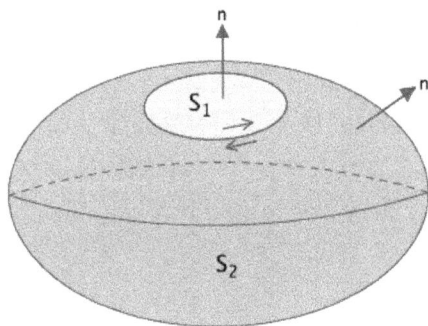

Figure 10.19: Dividing S into two pieces, S = $S_1 \cup S_2$

S_1 and S_2 inherit the orienting unit normal from S, so, in order to keep them on the left, their boundaries must be traversed in opposite directions. This is consistent with our requirement for orientations of piecewise smooth surfaces. In other words, ∂S_1 and ∂S_2 are the same curve but with opposite orientations. Hence the line integrals in equation (10.12) cancel each other out.

Lastly, we look for an analogue for curl fields of the mixed partials theorem for conservative fields. In $\mathbb{R}^3$, the mixed partials theorem takes the form that, if F is conservative, then $\nabla \times F = 0$.

Now, let F be a curl field, say $F = \nabla \times G$, so $(F_1, F_2, F_3) = \det \begin{bmatrix} i & j & k \\ \frac{\partial}{\partial x} & \frac{\partial}{\partial y} & \frac{\partial}{\partial z} \\ G_1 & G_2 & G_3 \end{bmatrix}$. Hence:

$$\begin{cases} F_1 &= \frac{\partial G_3}{\partial y} - \frac{\partial G_2}{\partial z} \\ F_2 &= \frac{\partial G_1}{\partial z} - \frac{\partial G_3}{\partial x} \\ F_3 &= \frac{\partial G_2}{\partial x} - \frac{\partial G_1}{\partial y}. \end{cases}$$

Taking certain partial derivatives of F_1, F_2, F_3 introduces the second-order mixed partials of G_1, G_2, G_3:

$$\begin{cases} \frac{\partial F_1}{\partial x} &= \frac{\partial^2 G_3}{\partial x\, \partial y} - \frac{\partial^2 G_2}{\partial x\, \partial z} \\ \frac{\partial F_2}{\partial y} &= \frac{\partial^2 G_1}{\partial y\, \partial z} - \frac{\partial^2 G_3}{\partial y\, \partial x} \\ \frac{\partial F_3}{\partial z} &= \frac{\partial^2 G_2}{\partial z\, \partial x} - \frac{\partial^2 G_1}{\partial z\, \partial y}. \end{cases}$$

The terms on the right side are mixed partial pairs that appear with opposite signs. Thus, by the equality of mixed partials, taking the sum gives:

$$\frac{\partial F_1}{\partial x} + \frac{\partial F_2}{\partial y} + \frac{\partial F_3}{\partial z} = 0. \tag{10.13}$$

The sum can also be written in terms of the ∇ operator as $\frac{\partial F_1}{\partial x} + \frac{\partial F_2}{\partial y} + \frac{\partial F_3}{\partial z} = (\frac{\partial}{\partial x}, \frac{\partial}{\partial y}, \frac{\partial}{\partial z}) \cdot (F_1, F_2, F_3) = \nabla \cdot F$, where the product is dot product. This quantity has a name.

Definition. If $F = (F_1, F_2, F_3)$ is a smooth vector field on an open set in $\mathbb{R}^3$, then:

$$\nabla \cdot F = \frac{\partial F_1}{\partial x} + \frac{\partial F_2}{\partial y} + \frac{\partial F_3}{\partial z}$$

is called the divergence of F.

Note that, while the curl of a vector field is again a vector field, the divergence is a real-valued function.

From (10.13), we have proven the following.

Theorem 10.12. Let U be an open set in $\mathbb{R}^3$, and let F be a smooth vector field on U. If F is a curl field on U, then:

$$\nabla \cdot F = 0$$

at all points of U. Equivalently, given a smooth vector field F on U, if $\nabla \cdot F \neq 0$ at some point, then F is not a curl field.

Example 10.13. The vector field $F(x, y, z) = (2xy + z^2, 2yz + x^2, 2xz + y^2)$ is conservative on $\mathbb{R}^3$. It has $f(x, y, z) = x^2y + y^2z + z^2x$ as a potential function. Is F also a curl field on $\mathbb{R}^3$?

We could write down the conditions for being a curl field and try to guess a vector field G such that $F = \nabla \times G$ the way we)did be fore. It i)s easy, however, to compute the divergence:

$$\nabla \cdot \mathbf{F} = \tfrac{\partial}{\partial x}\left(2xy + z^2\right) + \tfrac{\partial}{\partial y}\left(2yz + x^2\right) + \tfrac{\partial}{\partial z}\left(2xz + y^2\right) = 2y + 2z + 2x.$$

Since this is not identically 0, F is not a curl field.

10.5 GAUSS'S THEOREM

Gauss's theorem relates the integral of a vector field over a closed surface to a triple integral over the three-dimensional region that the surface encloses. As with Green's and Stokes's theorems, we indicate where the theorem comes from by deriving it in a very special case, then hint at how it might be extended to more general situations.

Let $F = (F_1, F_2, F_3)$ be a smooth vector field defined on an open set U in R3. We consider the case of a surface S that is the boundary of a rectangular box W = [a, b] × [c, d] × [e,f] contained in U . Thus S consists of the six rectangular faces of the box, each parallel to one of the coordinate planes. We orient each face with the unit normal pointing away from the box, as shown for three of the faces in Figure 10.20. This makes S into a piecewise smooth oriented closed surface.

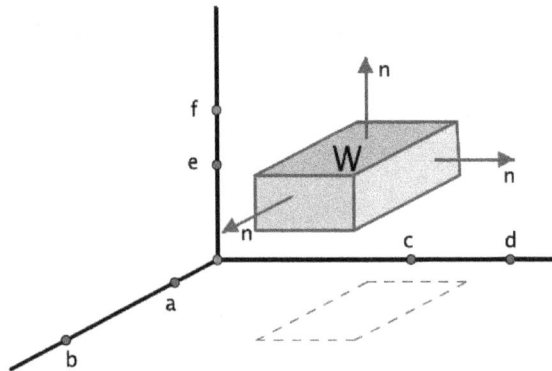

Figure 10.20: The rectangular box W = [a, b]×[c, d]×[e, f] in $\mathbb{R}^3$ and the orientation of its boundary surface S

We integrate F over S. To do so, we consider each face separately and add the results:

$$\iint_S \mathbf{F} \cdot d\mathbf{S} = \left(\iint_{\text{top}} + \iint_{\text{bottom}} + \iint_{\text{front}} + \iint_{\text{back}} + \iint_{\text{left}} + \iint_{\text{right}} \right) \mathbf{F} \cdot d\mathbf{S}. \qquad (10.14)$$

For instance, the top may be parametrized using x and y as parameters with z = f fixed: σ(x, y) = (x, y, f), a ≤ x ≤ b, c ≤ y ≤ d. Then:

$$\frac{\partial \sigma}{\partial x} \times \frac{\partial \sigma}{\partial y} = (1, 0, 0) \times (0, 1, 0) = \mathbf{i} \times \mathbf{j} = \mathbf{k} = (0, 0, 1).$$

This points upward, hence away from the box W , so σ is orientation-preserving. Therefore:

$$
\begin{aligned}
\iint_{\text{top}} \mathbf{F} \cdot d\mathbf{S} &= \iint_{[a,b] \times [c,d]} \mathbf{F}(\sigma(x, y)) \cdot \left(\frac{\partial \sigma}{\partial x} \times \frac{\partial \sigma}{\partial y} \right) dx\, dy \\
&= \iint_{[a,b] \times [c,d]} (F_1, F_2, F_3) \cdot (0, 0, 1)\, dx\, dy \\
&= \iint_{[a,b] \times [c,d]} F_3(\sigma(x, y))\, dx\, dy \\
&= \iint_{[a,b] \times [c,d]} F_3(x, y, f)\, dx\, dy.
\end{aligned}
$$

For the face on the bottom, the similar parametrization σ(x, y) = (x, y, e) works, only now σ is orientation-reversing since the orientation of S on the bottom points downward in order to point away from W . This gives:

$$\iint_{\text{bottom}} \mathbf{F} \cdot d\mathbf{S} = - \iint_{[a,b] \times [c,d]} F_3(x, y, e)\, dx\, dy.$$

We add these results and apply the fundamental theorem of calculus:

$$
\begin{aligned}
\iint_{\text{top}} \mathbf{F} \cdot d\mathbf{S} + \iint_{\text{bottom}} \mathbf{F} \cdot d\mathbf{S} &= \iint_{[a,b] \times [c,d]} (F_3(x, y, f) - F_3(x, y, e))\, dx\, dy \\
&= \iint_{[a,b] \times [c,d]} \left(F_3(x, y, z) \Big|_{z=e}^{z=f} \right) dx\, dy \\
&= \iint_{[a,b] \times [c,d]} \left(\int_e^f \frac{\partial F_3}{\partial z}(x, y, z)\, dz \right) dx\, dy \\
&= \iiint_W \frac{\partial F_3}{\partial z}(x, y, z)\, dx\, dy\, dz.
\end{aligned}
$$

The corresponding calculations for the remaining four faces yield:

$$\iint_{\text{front}} \mathbf{F}\cdot d\mathbf{S} + \iint_{\text{back}} \mathbf{F}\cdot d\mathbf{S} = \iiint_W \frac{\partial F_1}{\partial x}(x,y,z)\,dx\,dy\,dz$$

and

$$\iint_{\text{left}} \mathbf{F}\cdot d\mathbf{S} + \iint_{\text{right}} \mathbf{F}\cdot d\mathbf{S} = \iiint_W \frac{\partial F_2}{\partial y}(x,y,z)\,dx\,dy\,dz.$$

Adding up the results over all the faces as in equation (10.14) then gives:

$$\iint_S \mathbf{F}\cdot d\mathbf{S} = \iiint_W \left(\frac{\partial F_1}{\partial x} + \frac{\partial F_2}{\partial y} + \frac{\partial F_3}{\partial z}\right) dx\,dy\,dz = \iiint_W \nabla\cdot\mathbf{F}\,dx\,dy\,dz. \qquad (10.15)$$

One can extend this formula to the case that W is a union of rectangular boxes that intersect at most along their boundaries by applying equation (10.15) to the individual boxes and adding. A key observation is that when two neighboring boxes intersect in the interior of W on a portion of a common face, then the orienting normals of each box point in opposite directions in order to point away from the respective boxes. Hence the normal components $\mathbf{F}\cdot n$ are also opposite, so that the surface integrals over the overlaps cancel out in the sum. This leaves only the surface integral over the boundary of W. One then claims that equation (10.15) remains true for more general regions by some sort of limiting argument.

Before stating the theorem, we identify the orientation convention implicit in the argument given above, namely, if W is a bounded set in R3 whose boundary consists of one or more closed surfaces, then:

∂W denotes the boundary of W oriented so that

the orienting unit normal points away from W.

An example is shown in Figure 10.21.

The general version of equation (10.15) is then the following.

Theorem 10.14 (Gauss's theorem). Let U be an open set in $\mathbb{R}^3$, and let F be a smooth vector field on U. If W is a bounded subset of U whose boundary consists of a finite number of piecewise smooth closed surfaces, then:

$$\iint_{\partial W} \mathbf{F}\cdot d\mathbf{S} = \iiint_W \nabla\cdot\mathbf{F}\,dx\,dy\,dz.$$

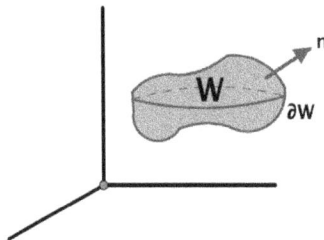

Figure 10.21: A solid region W in $\mathbb{R}^3$ and its oriented boundary ∂W

In the case that F is a curl field, the left side of Gauss's theorem is 0. This is because ∂W consists of closed surfaces and the integral of a curl field over a closed surface is 0 (Theorem 10.11). The right side is 0, too, because curl fields have divergence 0 (Theorem 10.12). Thus, for curl fields, Gauss's theorem reflects properties already known to us.

To illustrate Gauss's theorem, we begin by taking a new look at two surface integrals we evaluated earlier.

Example 10.15. Let Sa be the sphere $x^2 + y^2 + z^2 = a^2$ of radius a, oriented by the outward normal. Find $\iint_{S_a} \mathbf{F} \cdot d\mathbf{S}$ if:

    a. $F(x, y, z) = (0, 0, -1)$,
    b. $F(x, y, z) = (0, 0, z)$.

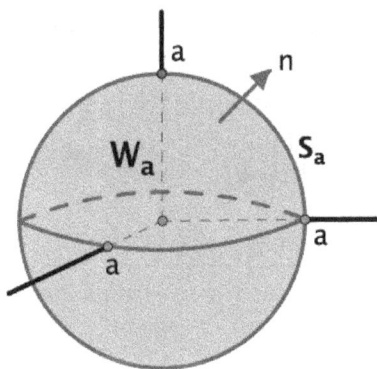

Figure 10.22: The solid ball W_a of radius a and its oriented boundary, the sphere S_a

This was Example 10.2(c). To answer it using Gauss's theorem, let W_a be the solid ball $x^2 + y^2 + z^2 \leq a^2$ so that $S_a = \partial W_a$, as in Figure 10.22. $\nabla \cdot \mathbf{F} = \frac{\partial}{\partial x}(0) + \frac{\partial}{\partial y}(0) + \frac{\partial}{\partial z}(-1) = 0$, Gauss's theorem says that:

$$\iint_{S_a} \mathbf{F} \cdot d\mathbf{S} = \iiint_{W_a} \nabla \cdot \mathbf{F}\, dx\, dy\, dz = \iiint_{W_a} 0\, dx\, dy\, dz = 0.$$

This is the third time we have evaluated this integral (see Examples 10.2(b) and 10.4). Here, $\nabla \cdot \mathbf{F} = 0 + 0 + 1 = 1$, so with W_a as above:

$$\iint_{S_a} \mathbf{F} \cdot d\mathbf{S} = \iiint_{W_a} 1\, dx\, dy\, dz = \text{Vol}(W_a) = \frac{4}{3}\pi a^3,$$

where we have used the formula for the volume of a three-dimensional ball from Example 7.8 in Chapter 7.

It seems odd to say that turning a problem into a triple integral makes it simpler, but, with the help of Gauss's theorem, finding the preceding two surface integrals became essentially trivial.

Example 10.16. Let F(x, y, z) = (x, y, z). This vector field radiates directly away from the origin at every point, and the arrows get longer the further out you go. Let W be a solid region in $\mathbb{R}^3$ as in Gauss's theorem. Then $\nabla \cdot \mathbf{F} = 1 + 1 + 1 = 3$, so $\iint_{\partial W} \mathbf{F} \cdot d\mathbf{S} = \iiint_W 3\, dx\, dy\, dz = 3\,\mathrm{Vol}\,(W)$. that is:

$$\mathrm{Vol}\,(W) = \frac{1}{3} \iint_{\partial W} (x, y, z) \cdot d\mathbf{S}.$$

This is another instance where information about a set can be gleaned from a calculation just involving its boundary. (Question: Thinking of the surface integral as the integral of the normal component and keeping in mind that ∂W is oriented by the outward normal, does it make sense that $\iint_{\partial W}(x, y, z) \cdot d\mathbf{S}$ is always positive? Note that the origin need not lie in W , so there may be places where the orienting normal n on the boundary points towards the origin and the normal component F · n = (x, y, z) · n is negative.)

10.6 THE INVERSE SQUARE FIELD

Throughout this section, G denotes the inverse square field G : $\mathbb{R}^3$ - {(0, 0, 0)} → $\mathbb{R}^3$, given by:

$$G(x, y, z) = \left(-\frac{x}{(x^2 + y^2 + z^2)^{3/2}}, -\frac{y}{(x^2 + y^2 + z^2)^{3/2}}, -\frac{z}{(x^2 + y^2 + z^2)^{3/2}} \right). \qquad (10.16)$$

We study G using Gauss's theorem, starting with a couple of preliminary facts.

Fact 1. Let S_a be the sphere $x^2 + y^2 + z^2 = a^2$, oriented by the outward normal. Then:

$$\iint_{S_a} \mathbf{G} \cdot d\mathbf{S} = -4\pi, \text{ independent of the radius } a.$$

Justification. We calculated and even emphasized this earlier in Example 10.2(a), which was the first surface integral of a vector field that we found. Note that this implies that G is not a curl field, since a curl field would integrate to 0 over the closed surface S_a.

Fact 2. $\nabla \cdot G = 0$.

Justification. This is a brute force calculation. For instance:

$$\frac{\partial G_1}{\partial x} = \frac{\partial}{\partial x}\left(-\frac{x}{(x^2+y^2+z^2)^{3/2}}\right)$$

$$= -\frac{(x^2+y^2+z^2)^{3/2}\cdot 1 - x\cdot\frac{3}{2}(x^2+y^2+z^2)^{1/2}\cdot 2x}{(x^2+y^2+z^2)^3}$$

$$= -\frac{(x^2+y^2+z^2) - 3x^2}{(x^2+y^2+z^2)^{5/2}}$$

$$= \frac{2x^2 - y^2 - z^2}{(x^2+y^2+z^2)^{5/2}}.$$

By symmetric calculations:

$$\frac{\partial G_2}{\partial y} = \frac{2y^2 - x^2 - z^2}{(x^2+y^2+z^2)^{5/2}} \quad \text{and} \quad \frac{\partial G_3}{\partial z} = \frac{2z^2 - x^2 - y^2}{(x^2+y^2+z^2)^{5/2}}.$$

Taking the sum gives $\nabla\cdot G = \frac{(2x^2-y^2-z^2)+(2y^2-x^2-z^2)+(2z^2-x^2-y^2)}{(x^2+y^2+z^2)^{5/2}} = 0$, as claimed.

This fact has an interesting implication, too, for we know that curl fields have divergence 0.

This shows that the converse need not hold, since $\nabla\cdot G = 0$ yet G is not a curl field by Fact 1.

Now, let S be a piecewise smooth closed surface in $\mathbb{R}^3 - \{(0, 0, 0)\}$, oriented by the outward normal. Is that enough for us to be able to say something about $\iint_S G\cdot dS$? Strikingly, the answer is yes. There are two cases.

Case 1. Suppose that the origin 0 lies in the exterior of S (Figure 10.23).

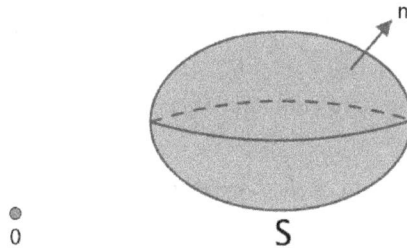

Figure 10.23: The origin lies exterior to S.

Let W be the three-dimensional solid inside of S so that $S = \partial W$. By assumption, $0 \in/ W$, so G is defined throughout W and we may apply Gauss's theorem. This gives:

$$\iint_S G\cdot dS = \iiint_W \nabla\cdot G\,dx\,dy\,dz$$

$$= \iiint_W 0\,dx\,dy\,dz \quad \text{(by Fact 2)}$$

$$= 0.$$

Case 2. Suppose that 0 lies in the interior of S.

The preceding argument is no longer valid, since filling in the interior of S would include the origin, which is excluded from the domain of G. Instead, we choose a very small sphere Sϵ of radius ϵ centered at the origin, oriented by the outward normal, and we let W be the region inside S but outside Sϵ. See Figure 10.24.

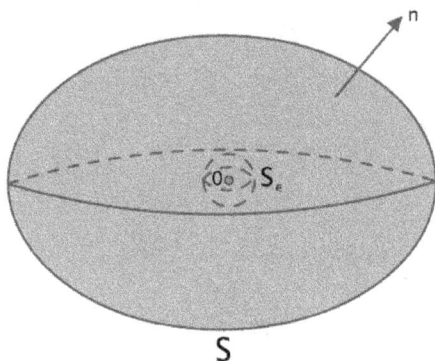

Figure 10.24: The origin lies in the interior of S.

Then Gauss's theorem applies to W , giving:

$$\iint_{\partial W} \mathbf{G} \cdot d\mathbf{S} = \iiint_W \nabla \cdot \mathbf{G} \, dx \, dy \, dz = 0, \qquad (10.17)$$

as before. This time, however, the boundary of W consists of S and Sϵ, oriented by the normal pointing away from W . This is the outward normal on S but points inward on Sϵ. In other words, $\partial W = S \, (\, S_\epsilon \,)$, and therefore $\iint_{\partial W} \mathbf{G} \cdot d\mathbf{S} = \iint_S \mathbf{G} \cdot d\mathbf{S} - \iint_{S_\epsilon} \mathbf{G} \cdot d\mathbf{S}$. By equation (10.17), the difference is 0, from which it follows that $\iint_S \mathbf{G} \cdot d\mathbf{S} = \iint_{S_\epsilon} \mathbf{G} \cdot d\mathbf{S}$. And, by Fact 1, the integral of G over any sphere centered at the origin is -4π. Hence:

$$\iint_S \mathbf{G} \cdot d\mathbf{S} = -4\pi.$$

After a slight rearrangement, we can summarize the discussion as follows. For a piecewise smooth closed surface S that doesn't pass through the origin and is oriented by the outward normal:

$$-\frac{1}{4\pi} \iint_S \mathbf{G} \cdot d\mathbf{S} = \begin{cases} 0 & \text{if } \mathbf{0} \text{ lies in the exterior of } S, \\ 1 & \text{if } \mathbf{0} \text{ lies in the interior of } S. \end{cases}$$

As we noted back in Chapter 8, the inverse square field describes the gravitational, or electro- static, force due to an object, or charged particle, located at the origin. If several objects of equal mass, or equal charge, are placed at the points $p_1, p_2, \ldots$

, p_k in $\mathbb{R}^3$, then the total force at a point x due to these objects is described by the vector field:

$$\mathbf{F}(\mathbf{x}) = \sum_{i=1}^{k} \mathbf{G}(\mathbf{x} - \mathbf{p}_i), \qquad (10.18)$$

where G is the same as above (10.16). It is not hard to verify that again $\nabla \cdot \mathbf{F} = 0$. A modification of the argument just given shows that, if S is a piecewise smooth closed surface contained in the set $\mathbb{R}^3 - \{p_1, p_2, \ldots, p_k\}$ and oriented by the outward normal, then:

$$-\frac{1}{4\pi} \iint_S \mathbf{F} \cdot d\mathbf{S} \text{ counts the number of the points } \mathbf{p}_1, \mathbf{p}_2, \ldots, \mathbf{p}_k \text{ that lie in the interior of } S.$$

This is an instance of what is known in physics as **Gauss's law**. In effect, the total mass, or charge, contained inside of S determines, and is determined by, the integral of F over S. An explicit formula or parametrization for S is not needed to obtain this information.

10.7 A MORE SUBSTITUTION-FRIENDLY NOTATION FOR SURFACE INTEGRALS

We now fill in a few details that we have been postponing. Some of the arguments are technical, but they also unveil themes that unify the theory behind what we have been doing.

In the case of line integrals, we have typically replaced the notation $\int_C \mathbf{F} \cdot d\mathbf{s}$ in favor of the more suggestive $\int_C F_1 \, dx_1 + F_2 \, dx_2 + \cdots + F_n \, dx_n$. The latter is useful in that it tells us more or less what we need to substitute in order to express the integral in terms of the parameter used to trace out C. We now introduce a corresponding notation for surface integrals.

Let $\mathbf{F} = (F_1, F_2, F_3)$ be a continuous vector field on an open set U in $\mathbb{R}^3$, and let S be a smooth oriented surface in U. To integrate F over S, we want to turn the calculation into a double integral in the two parameters of a parametrization. So let $\sigma : D \to \mathbb{R}^3$ be a smooth parametrization of S, where $\sigma(s, t) = (x(s, t), y(s, t), z(s, t))$. For any function of (x, y, z), such as the components $F_1, F_2,$ and F_3 of F, the parametrization tells us how to substitute for x, y, and z, resulting in a function of s and t.

For simplicity, let's assume that σ is orientation-preserving. By definition:

$$\iint_S \mathbf{F} \cdot d\mathbf{S} = \iint_D \mathbf{F}(\sigma(s,t)) \cdot \left(\frac{\partial \sigma}{\partial s} \times \frac{\partial \sigma}{\partial t} \right) ds\,dt$$

$$= \iint_D (F_1(\sigma(s,t)), F_2(\sigma(s,t)), F_3(\sigma(s,t))) \cdot \det \begin{bmatrix} \mathbf{i} & \mathbf{j} & \mathbf{k} \\ \frac{\partial x}{\partial s} & \frac{\partial y}{\partial s} & \frac{\partial z}{\partial s} \\ \frac{\partial x}{\partial t} & \frac{\partial y}{\partial t} & \frac{\partial z}{\partial t} \end{bmatrix} ds\,dt$$

$$= \iint_D \left(F_1(\sigma(s,t)) \cdot \det \begin{bmatrix} \frac{\partial y}{\partial s} & \frac{\partial z}{\partial s} \\ \frac{\partial y}{\partial t} & \frac{\partial z}{\partial t} \end{bmatrix} + F_2(\sigma(s,t)) \cdot \det \begin{bmatrix} \frac{\partial z}{\partial s} & \frac{\partial x}{\partial s} \\ \frac{\partial z}{\partial t} & \frac{\partial x}{\partial t} \end{bmatrix} \right.$$

$$\left. + F_3(\sigma(s,t)) \cdot \det \begin{bmatrix} \frac{\partial x}{\partial s} & \frac{\partial y}{\partial s} \\ \frac{\partial x}{\partial t} & \frac{\partial y}{\partial t} \end{bmatrix} \right) ds\,dt. \tag{10.19}$$

We adopt the notation:

$$\iint_S \mathbf{F} \cdot d\mathbf{S} = \iint_S F_1\, dy \wedge dz + F_2\, dz \wedge dx + F_3\, dx \wedge dy. \tag{10.20}$$

The "$\wedge$" symbol stands for a kind of multiplication known as the **wedge product**. We develop some of its properties along the way as we get more practice working with it.

For instance, if F(x, y, z) = (xyz, x² + y² + z², xy + yz), then $\iint_S \mathbf{F} \cdot d\mathbf{S}$ would be written $\iint_S xyz\, dy \wedge dz + (x^2 + y^2 + z^2)\, dz \wedge dx + (xy + yz)\, dx \wedge dy$. The order of the factors in the individual terms of the notation is important. The pattern is that the subscripts/variables appear in cyclic order, as indicated in Figure 10.25.

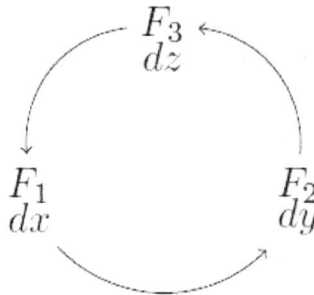

Figure 10.25: The cyclic ordering of coordinates in $\mathbb{R}^3$

In comparison with the expanded definition (10.19), the notation in (10.20) says that, as part of the substitutions for x, y, z in terms of s, t, one should make the substitution, for example:

$$dy \wedge dz = \det \begin{bmatrix} \frac{\partial y}{\partial s} & \frac{\partial z}{\partial s} \\ \frac{\partial y}{\partial t} & \frac{\partial z}{\partial t} \end{bmatrix} ds\,dt.$$

Actually, this may look more familiar if we take the transpose of the matrix, which does not affect the value of the determinant:

$$dy \wedge dz = \det \begin{bmatrix} \frac{\partial y}{\partial s} & \frac{\partial y}{\partial t} \\ \frac{\partial z}{\partial s} & \frac{\partial z}{\partial t} \end{bmatrix} ds\, dt. \tag{10.21}$$

This has the form of a change of variables, i.e., a substitution, for a transformation from the st- plane to the yz-plane. Similarly, to evaluate an integral presented in the notation of (10.20), one substitutes:

$$dz \wedge dx = \det \begin{bmatrix} \frac{\partial z}{\partial s} & \frac{\partial z}{\partial t} \\ \frac{\partial x}{\partial s} & \frac{\partial x}{\partial t} \end{bmatrix} ds\, dt$$

$$\text{and} \quad dx \wedge dy = \det \begin{bmatrix} \frac{\partial x}{\partial s} & \frac{\partial x}{\partial t} \\ \frac{\partial y}{\partial s} & \frac{\partial y}{\partial t} \end{bmatrix} ds\, dt. \tag{10.22}$$

which also have the form of certain changes of variables.

The integrand $\eta = F_1\, dy \wedge dz + F_2\, dz \wedge dx + F_3\, dx \wedge dy$ of (10.20) is called a **differential 2-form**. By comparison, the expressions $\omega = F_1\, dx + F_2\, dy + F_3\, dz$ that we integrated over curves in $\mathbb{R}^3$ are called differential 1-forms. To integrate the 2-form η over S, we substitute for x, y, z in terms of s, t in F_1, F_2, F_3 and replace the wedge products like $dy \wedge dz$ with expressions that come formally from the change of variables theorem. This converts the integral over S into a double integral over the parameter domain. The integrand of this double integral is called the pullback of η, denoted by $\sigma^*(\eta)$. From (10.19), it can be written:

$$\sigma^*(\eta) = \left((F_1 \circ \sigma) \cdot \det \begin{bmatrix} \frac{\partial y}{\partial s} & \frac{\partial y}{\partial t} \\ \frac{\partial z}{\partial s} & \frac{\partial z}{\partial t} \end{bmatrix} + (F_2 \circ \sigma) \cdot \det \begin{bmatrix} \frac{\partial z}{\partial s} & \frac{\partial z}{\partial t} \\ \frac{\partial x}{\partial s} & \frac{\partial x}{\partial t} \end{bmatrix} + (F_3 \circ \sigma) \cdot \det \begin{bmatrix} \frac{\partial x}{\partial s} & \frac{\partial x}{\partial t} \\ \frac{\partial y}{\partial s} & \frac{\partial y}{\partial t} \end{bmatrix} \right) ds \wedge dt.$$

With this notation, the definition of the surface integral takes the form:

$$\iint_{\sigma(D)} \eta = \iint_D \sigma^*(\eta). \tag{10.23}$$

where, on the right, an integral of the form $\iint_D f(s, t)\, ds \wedge dt$ is defined to be the usual double integral $\iint_D f(s, t)\, ds\, dt$. Equation 10.23 highlights that the definition has the same form as the change of variables theorem, continuing an emerging pattern (see (7.3) and (9.6)).

We say more in the next chapter about properties of differential forms, but our discussion so far can be used to motivate some elementary manipulations. The substitution formulas (10.21) and (10.22) for $dy \wedge dz$ and the like suggest a connection between wedge products and determinants..

To reinforce this, we adopt the following rules for working with differential forms:

⊙ $dz \wedge dy = -dy \wedge dz$,

$dx \wedge dz = -dz \wedge dx,$

$dy \wedge dx = -dx \wedge dy,$

⊙ $dx \wedge dx = 0,$

$dy \wedge dy = 0,$

$dz \wedge dz = 0.$

The first collection mirrors the property that interchanging two rows changes the sign of a determinant, and the second mirrors the property that two equal rows imply a determinantof 0. We postpone until the next chapter the discussion of how these properties may be used.

10.8 INDEPENDENCE OF PARAMETRIZATION

Finally, we address the long-standing question of the extent to which the official definition of the surface integral, $\iint_S \mathbf{F} \cdot d\mathbf{S} = \pm \iint_D \mathbf{F}(\sigma(s,t)) \cdot \left(\frac{\partial \sigma}{\partial s} \times \frac{\partial \sigma}{\partial t}\right) ds\,dt,$ depends on the smooth parametrization $\sigma : D \to \mathbb{R}^3$ of S, where D is a subset of the st-plane and, by assumption, $\frac{\partial \sigma}{\partial s} \times \frac{\partial \sigma}{\partial t} \neq 0$, except possibly on the boundary of D. To highlight the role of the parametrization σ, we write $\iint_\sigma \mathbf{F} \cdot d\mathbf{S}$ to denote the quantity $\iint_D \mathbf{F}(\sigma(s,t)) \cdot \left(\frac{\partial \sigma}{\partial s} \times \frac{\partial \sigma}{\partial t}\right) ds\,dt$ in the definition.

We continue to assume that σ is orientation-preserving. Let $\tau : D^* \to \mathbb{R}^3$ be another smooth parametrization of S. Call its parameters u and v so that $\tau(u,v) = (x(u,v), y(u,v), z(u,v))$. For simplicity, we assume that σ and τ are one-to-one, except possibly on the boundaries of their domains. The goal is to compare $\iint_\sigma \mathbf{F} \cdot d\mathbf{S}$ and $\iint_\tau \mathbf{F} \cdot d\mathbf{S}$. We use the formula for the surface integral after it has been expanded out in terms of coordinates as suggested by the differential form notation, that is:

$$\iint_\sigma \mathbf{F} \cdot d\mathbf{S} = \iint_D \left(F_1(\sigma(s,t)) \cdot \det \begin{bmatrix} \frac{\partial y}{\partial s} & \frac{\partial y}{\partial t} \\ \frac{\partial z}{\partial s} & \frac{\partial z}{\partial t} \end{bmatrix} + F_2(\sigma(s,t)) \cdot \det \begin{bmatrix} \frac{\partial z}{\partial s} & \frac{\partial z}{\partial t} \\ \frac{\partial x}{\partial s} & \frac{\partial x}{\partial t} \end{bmatrix} \right.$$
$$\left. + F_3(\sigma(s,t)) \cdot \det \begin{bmatrix} \frac{\partial x}{\partial s} & \frac{\partial x}{\partial t} \\ \frac{\partial y}{\partial s} & \frac{\partial y}{\partial t} \end{bmatrix} \right) ds\,dt, \quad (10.24)$$

as in equation (10.19), except that the 2 by 2 matrices have been transposed. The expression for $\iint_\tau \mathbf{F} \cdot d\mathbf{S}$ is similar, integrating over D^* and using u's and v's instead of s's and t's.

If $x \in S$, then $x = \sigma(s, t)$ for some (s, t) in D and $x = \tau(u, v)$ for some (u, v) in D^*. We write this as $(s, t) = T(u, v)$. This defines a function $T : D^* \to D$, where, given (u, v) in D^*, $T(u, v)$ is the point (s, t) in D such that $\sigma(s, t) = \tau(u, v)$. Then, by construction:

$$\tau(u,v) = \sigma(s,t) = \sigma(T(u,v)). \quad (10.25)$$

See Figure 10.26. (If σ or τ is not one-to-one, which is permitted only on the boundaries of their respective domains, there is some ambiguity on how to define

T there, but, as we have observed before, the integral can tolerate a certain amount of misbehavior on the boundary.)

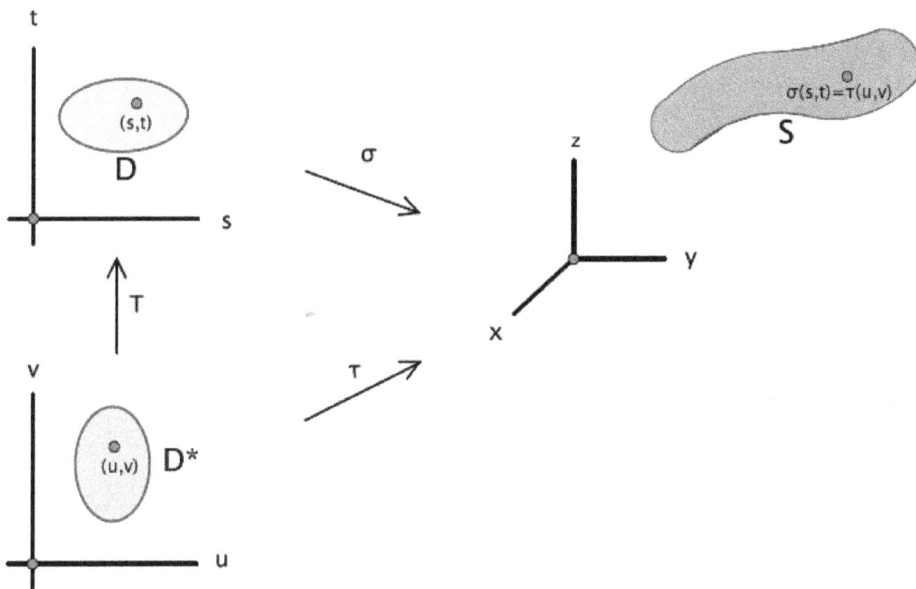

Figure 10.26: Two parametrizations σ and τ of the same surface S

By equation (10.25) and the chain rule, $D\tau(u, v) = D\sigma(T(u, v)) \cdot DT(u, v) = D\sigma(s, t) \cdot DT(u, v)$, or:

$$
\begin{bmatrix} \frac{\partial x}{\partial u} & \frac{\partial x}{\partial v} \\ \frac{\partial y}{\partial u} & \frac{\partial y}{\partial v} \\ \frac{\partial z}{\partial u} & \frac{\partial z}{\partial u} \end{bmatrix} = \begin{bmatrix} \frac{\partial x}{\partial s} & \frac{\partial x}{\partial t} \\ \frac{\partial y}{\partial s} & \frac{\partial y}{\partial t} \\ \frac{\partial z}{\partial s} & \frac{\partial z}{\partial s} \end{bmatrix} \begin{bmatrix} \frac{\partial s}{\partial u} & \frac{\partial s}{\partial v} \\ \frac{\partial t}{\partial u} & \frac{\partial t}{\partial v} \end{bmatrix}. \tag{10.26}
$$

We focus on the 2 by 2 blocks within the first two matrices of this equation. For instance, isolating the portions of the product involving the second and third rows, as indicated in red above, gives:

$$
\begin{bmatrix} \frac{\partial y}{\partial u} & \frac{\partial y}{\partial v} \\ \frac{\partial z}{\partial u} & \frac{\partial z}{\partial v} \end{bmatrix} = \begin{bmatrix} \frac{\partial y}{\partial s} & \frac{\partial y}{\partial t} \\ \frac{\partial z}{\partial s} & \frac{\partial z}{\partial t} \end{bmatrix} \begin{bmatrix} \frac{\partial s}{\partial u} & \frac{\partial s}{\partial v} \\ \frac{\partial t}{\partial u} & \frac{\partial t}{\partial v} \end{bmatrix} = \begin{bmatrix} \frac{\partial y}{\partial s} & \frac{\partial y}{\partial t} \\ \frac{\partial z}{\partial s} & \frac{\partial z}{\partial t} \end{bmatrix} \cdot DT(u, v).
$$

Since the determinants of a product is the product of the determinants—that is, $\det(AB) = (\det A)(\det B)$—we obtain:

$$
\det \begin{bmatrix} \frac{\partial y}{\partial u} & \frac{\partial y}{\partial v} \\ \frac{\partial z}{\partial u} & \frac{\partial z}{\partial v} \end{bmatrix} = \det \begin{bmatrix} \frac{\partial y}{\partial s} & \frac{\partial y}{\partial t} \\ \frac{\partial z}{\partial s} & \frac{\partial z}{\partial t} \end{bmatrix} \cdot \det DT(u, v). \tag{10.27}
$$

Referring to (10.24), the first determinant in the preceding equation appears in the definition of $\iint_\tau \mathbf{F} \cdot d\mathbf{S}$, while the second appears in the definition of $\iint_\sigma \mathbf{F} \cdot d\mathbf{S}$. By choosing different pairs of rows in the chain rule (10.26), the same reasoning applies to the other two determinants that appear in (10.24). In each case, the determinants for τ and σ differ by a factor of det DT (u, v).

In fact, these 2 by 2 determinants are precisely the components of the cross products $\frac{\partial \tau}{\partial u} \times \frac{\partial \tau}{\partial v}$ and $\frac{\partial \sigma}{\partial s} \times \frac{\partial \sigma}{\partial t}$, so this calculation shows that:

$$\frac{\partial \tau}{\partial u} \times \frac{\partial \tau}{\partial v} = \left(\frac{\partial \sigma}{\partial s} \times \frac{\partial \sigma}{\partial t}\right) \cdot \det DT(u, v).$$

where, contrary to standard practice, we have written the scalar factor det DT (u, v) on the right. The two cross products are normal vectors to S. Since we assumed that σ is orientation-preserving, we conclude that τ is orientation-preserving if det DT (u, v) is positive and orientation-reversing if it is negative.

We are ready to compare the integrals obtained using the two parametrizations. Beginning with the expansion (10.24) with τ as parametrization and then using (10.27) and its analogues to pull out a factor of det DT (u, v), we obtain:

$$\iint_\tau \mathbf{F} \cdot d\mathbf{S} = \iint_{D^*} \left((F_1 \circ \tau) \cdot \det \begin{bmatrix} \frac{\partial y}{\partial u} & \frac{\partial y}{\partial v} \\ \frac{\partial z}{\partial u} & \frac{\partial z}{\partial v} \end{bmatrix} + (F_2 \circ \tau) \cdot \det \begin{bmatrix} \frac{\partial z}{\partial u} & \frac{\partial z}{\partial v} \\ \frac{\partial x}{\partial u} & \frac{\partial x}{\partial v} \end{bmatrix} \right.$$
$$\left. + (F_3 \circ \tau) \cdot \det \begin{bmatrix} \frac{\partial x}{\partial u} & \frac{\partial x}{\partial v} \\ \frac{\partial y}{\partial u} & \frac{\partial y}{\partial v} \end{bmatrix} \right) du\, dv \qquad (10.28)$$

$$= \iint_{D^*} \left((F_1 \circ \tau) \cdot \det \begin{bmatrix} \frac{\partial y}{\partial s} & \frac{\partial y}{\partial t} \\ \frac{\partial z}{\partial s} & \frac{\partial z}{\partial t} \end{bmatrix} + (F_2 \circ \tau) \cdot \det \begin{bmatrix} \frac{\partial z}{\partial s} & \frac{\partial z}{\partial t} \\ \frac{\partial x}{\partial s} & \frac{\partial x}{\partial t} \end{bmatrix} \right.$$
$$\left. + (F_3 \circ \tau) \cdot \det \begin{bmatrix} \frac{\partial x}{\partial s} & \frac{\partial x}{\partial t} \\ \frac{\partial y}{\partial s} & \frac{\partial y}{\partial t} \end{bmatrix} \right) \cdot \det DT(u, v)\, du\, dv. \qquad (10.29)$$

Next, note that det DT (u, v) = ±| det DT (u, v)|, where the sign depends on whether det DT (u, v) is positive or negative, or equivalently, as noted above, whether τ is orientation-preserving or orientation-reversing. Keeping in mind that $\tau = \sigma \circ T$, we find:

$$\iint_\tau \mathbf{F} \cdot d\mathbf{S} = \pm \iint_{D^*} \left((F_1 \circ \sigma \circ T) \cdot \det \begin{bmatrix} \frac{\partial y}{\partial s} & \frac{\partial y}{\partial t} \\ \frac{\partial z}{\partial s} & \frac{\partial z}{\partial t} \end{bmatrix} + (F_2 \circ \sigma \circ T) \cdot \det \begin{bmatrix} \frac{\partial z}{\partial s} & \frac{\partial z}{\partial t} \\ \frac{\partial x}{\partial s} & \frac{\partial x}{\partial t} \end{bmatrix} \right.$$
$$\left. + (F_3 \circ \sigma \circ T) \cdot \det \begin{bmatrix} \frac{\partial x}{\partial s} & \frac{\partial x}{\partial t} \\ \frac{\partial y}{\partial s} & \frac{\partial y}{\partial t} \end{bmatrix} \right) \cdot | \det DT(u, v)|\, du\, dv.$$

This last integral is the result of a change of variables. It is an integral over D* that is the pullback of an integral over D using T as the change of variables transformation from the uv-plane to the st-plane. The integral over D to which the change of variables is applied is:

$$\pm \iint_D \left((F_1 \circ \sigma) \cdot \det \begin{bmatrix} \frac{\partial y}{\partial s} & \frac{\partial y}{\partial t} \\ \frac{\partial z}{\partial s} & \frac{\partial z}{\partial t} \end{bmatrix} + (F_2 \circ \sigma) \cdot \det \begin{bmatrix} \frac{\partial z}{\partial s} & \frac{\partial z}{\partial t} \\ \frac{\partial x}{\partial s} & \frac{\partial x}{\partial t} \end{bmatrix} + (F_3 \circ \sigma) \cdot \det \begin{bmatrix} \frac{\partial x}{\partial s} & \frac{\partial x}{\partial t} \\ \frac{\partial y}{\partial s} & \frac{\partial y}{\partial t} \end{bmatrix} \right) ds\, dt.$$

Going back to equation (10.24), this last integral is exactly what would be used to calculate the surface integral using σ as parametrization. That is, it is $\pm \sigma \mathbf{F} \cdot d\mathbf{S}$. We have attained our objective.

Proposition 10.17. Let $\sigma : D \to \mathbb{R}^3$ be a smooth orientation-preserving parametrization of an oriented surface S in $\mathbb{R}^3$. If $\tau : D* \to \mathbb{R}^3$ is another smooth parametrization of S, then:

$$\iint_{\tau} \mathbf{F} \cdot d\mathbf{S} = \pm \iint_{\sigma} \mathbf{F} \cdot d\mathbf{S},$$

where the sign depends on whether τ is orientation-preserving or orientation-reversing.

In particular, in order to compute $\iint_S \mathbf{F} \cdot d\mathbf{S} = \iint_{\sigma} \mathbf{F} \cdot d\mathbf{S}$. it does not matter which orientation- preserving parametrization σ is used.

For integrals of real-valued functions with respect to surface area, the corresponding assertion is that $\iint_{\sigma} f \, dS = \iint_{\tau} f \, dS$ for any two smooth parametrizations σ, τ of S. The proof combines the ideas just given above with those given in the analogous situation for integrals over curves with respect to arclength. We won't go through the details. This may be one of those rare occasions where it is all right to say that the authorities have weighed in and to let them have their way.

One last remark: We commented earlier that the definition of the vector field surface integral can be regarded as pulling back the integral of the 2–form $\eta = F_1 \, dy \wedge dz + F_2 \, dz \wedge dx + F_3 \, dx \wedge dy$ to a double integral in the parameter plane (see equation (10.23)). Using that language, our calculation of the independence of parametrization can be summarized in one line:

$$\iint_{D*} \tau^*(\eta) = \int_{D*} (\sigma \circ T)^*(\eta) = \iint_{D*} T^*(\sigma^*(\eta)) = \iint_{D} \sigma^*(\eta). \qquad (10.30)$$

In other words, using either σ or τ to pull back η gives the same value.

Don't panic! The first equation follows since T was defined so that $\tau = \sigma \circ T$. The second equation uses the fact that $(\sigma \circ T)^*(\eta) = T^*(\sigma*(\eta))$. This is essentially the chain rule-based calculation that brought out the factor of det DT (u, v) in going from (10.28) to (10.29), though we would need to develop the properties of pullbacks more carefully to explain this. (To get a taste of these properties, see Exercises 4.14– 4.18 in Chapter 11.) Finally, the step going from D* to D uses the pullback form of the change of variables theorem, which is a refinement that takes orientation into account and eliminates the absolute value around the determinant(see Exercise 4.20 in Chapter 11).

Though there are gaps that need to be filled in, the intended lesson here is that differential forms, properly developed, provide a structure in which the arguments can be presented with remarkable conciseness. Of course, the proper development may involve lengthy calculations similar to the ones we presented, but their validity is guided by and rooted in basic principles: the chain rule and change of variables.

10.9 EXERCISES FOR CHAPTER 10

Section 1 What the surface integral measures

1.1. Let S be the hemisphere $x_2 + y^2 + z^2 = a^2$, $z \geq 0$, oriented by the outward normal.

      a. Find $\iint_S \mathbf{F} \cdot d\mathbf{S}$ if $F(x, y, z) = (x, y, z)$.

      b. Find $\iint_S \mathbf{F} \cdot d\mathbf{S}$ if $F(x, y, z) = (y - z, z - x, x - y)$.

1.2. Let $F(x, y, z) = (z, x, y)$, and let S be the triangular surface in $\mathbb{R}^3$ with vertices (1, 0, 0), (0, 1, 0), and (0, 0, 1), oriented by the upward normal. Find $\iint_S \mathbf{F} \cdot d\mathbf{S}$.

1.3. Let F be a continuous vector field on $\mathbb{R}^3$ such that F(x, y, z) is always a scalar multiple of (0, 0, 1), i.e., F(x, y, z) = 0, 0, F3(x, y, z) , an d let S be the cylinder x2 + y2 = 1, $0 \leq z \leq 5$, oriented by the outward normal. Show that $\iint_S \mathbf{F} \cdot d\mathbf{S} = 0$.

1.4. Let U be an open set in $\mathbb{R}^3$, and let S be a smooth oriented surface in U with orienting normal vector field n. Let F be a continuous vector field on U . If F(x) = n(x) for all x in S, prove that $\iint_S \mathbf{F} \cdot d\mathbf{S} = \text{Area}\,(S)$.

1.5. Let $F : \mathbb{R}^3 \to \mathbb{R}^3$ be a vector field such that F(x) = F(x) for all x in $\mathbb{R}^3$, and let S be the sphere x2 + y2 + z2 = a2, oriented by the outward normal. Is it necessarily true that $\iint_S \mathbf{F} \cdot d\mathbf{S} = 0$? Explain.

Section 2 The definition of the surface integral

2.1. Let S be the portion of the plane x + y + z = 3 for which $0 \leq x \leq 1$ and $0 \leq y \leq 1$. Let $F(x, y, z) = (2y, x, z)$. If S is oriented by the upward normal, find $\iint_S \mathbf{F} \cdot d\mathbf{S}$.

2.2. Let $F(x, y, z) = (xz, yz, z^2)$.

      a. Let S be the cylinder $x^2 + y^2 = 4$, $1 \leq z \leq 3$, oriented by the outward normal. Find $\iint_S \mathbf{F} \cdot d\mathbf{S}$.

      b. Suppose that you "close up" the cylinder S in part (a) by adding the disks $x^2 + y^2 \leq 4$, z = 3, and $x^2 + y^2 \leq 4$, z = 1, at the top and bottom, respectively, where the disk at the top is oriented by the upward normal and the one at the bottom is oriented by the downward normal. Let S_1 be the resulting surface: cylinder, top, and bottom. Find $\iint_{S_1} \mathbf{F} \cdot d\mathbf{S}$.

2.3. Let S be the helicoid parametrized by:

$$\sigma(r, \theta) = (r\cos\theta, \ r\sin\theta, \ \theta), \qquad 0 \leq r \leq 1, \quad 0 \leq \theta \leq 4\pi,$$

oriented by the upward normal. Find $\iint_S \mathbf{F} \cdot d\mathbf{S}$. if $F(x, y, z) = (y, -x, z^2)$.

2.4. Let $F(x, y, z) = x, -z, 2y(x^2 - y^2)$. Find $\iint_S \mathbf{F} \cdot d\mathbf{S}$. if S is the saddle surface $z = x^2 - y^2$, where $x^2 + y^2 \leq 4$, oriented by the upward normal.

2.5. Let W be the solid region in $\mathbb{R}^3$ that lies inside the cylinders $x^2 + z^2 = 1$ and $y^2 + z^2 = 1$ and above the xy-plane. This solid appeared in Example 5.2 of Chapter 5, shown again in Figure 10.27. Let S be the piecewise smooth surface that is the exposed upper part of W, that is, the four pieces of the cylinders, but not the base, oriented by the upward normal.

Find $\iint_S \mathbf{F} \cdot d\mathbf{S}$. if F is the vector field $F(x, y, z) = (0, 0, z)$.

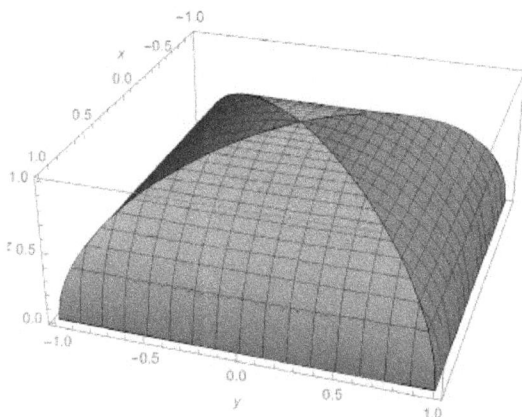

Figure 10.27: The surface S consisting of portions of the cylinders $x^2 + z^2 = 1$ and $y^2 + z^2 = 1$

Section 3 Stokes's theorem

3.1. Consider the vector field $F(x, y, z) = (2yz, y \sin z, 1 + \cos z)$.

  a. Find a vector field G whose curl is F.

  b. Let S be the half-ellipsoid $4x^2 + 4y^2 + z^2 = 4$, $z \geq 0$, oriented by the upward normal. Use Stokes's theorem to find $\iint_S \mathbf{F} \cdot d\mathbf{S}$..

  c. Find $\iint_S \mathbf{F} \cdot d\mathbf{S}$. if $\tilde{S}$ is the portion of the surface $z = 1 - x^2 - y^2$ above the xy-plane, oriented by the upward normal. (Hint: Take advantage of what you've already done.)

3.2. Consider the vector field $F(x, y, z) = (e^{x+y} - xe^{y+z}, e^{y+z} - e^{x+y} + ye^z, -e^z)$.

  a. Is F a conservative vector field? Explain.

  b. Find a vector field $G = (G_1, G_2, G_3)$ such that $G_2 = 0$ and the curl of G is F.

  c. Find $\iint_S \mathbf{F} \cdot d\mathbf{S}$. if S is the hemisphere $x^2 + y^2 + z^2 = 4$, $z \geq 0$, oriented by the outward normal.

  d. Find $\iint_S \mathbf{F} \cdot d\mathbf{S}$. if S is the hemisphere $x^2 + y^2 + z^2 = 4$, $z \leq 0$, oriented by the outward normal.

  e. Find $\iint_S \mathbf{F} \cdot d\mathbf{S}$. if S is the cylinder $x^2 + y^2 = 4$, $0 \leq z \leq 4$, oriented by the outward normal.

Exercises 3.3–3.6 illustrate situations where, by bringing in a surface integral, Stokes's theorem can be used to obtain information about line integrals that would be hard to find directly.

3.3. Let $F(x, y, z) = (xz + yz, 2xz, \frac{1}{2}y^2)$.

    a. Find $\nabla \times F$.

    b. Let C be the curve of intersection in $\mathbb{R}^3$ of the cylinder $x^2 + y^2 = 1$ and the saddle surface $z = x^2 - y^2$, oriented counterclockwise as viewed from high above the xy-plane, looking down. Use Stokes's theorem to find the line integral $\iint_S F \cdot dS$. (Hint: Find a surface whose boundary is C.)

3.4. Let P be a plane in $\mathbb{R}^3$ described by the equation $ax + by + cz = d$. Let C be a piecewise smooth oriented simple closed curve that lies in P, and let S be the subset of P that is enclosed by C, i.e., the region of P "inside" C. Show that the area of S is equal to the absolute value of:

$$\frac{1}{2\sqrt{a^2 + b^2 + c^2}} \int_C (bz - cy)\, dx + (cx - az)\, dy + (ay - bx)\, dz.$$

3.5.   a. Find an example of a smooth vector field F = (F1, F2, F3) defined on an open set U of $\mathbb{R}^3$ such that:

    ◉ its mixed partials are equal, i.e., $\frac{\partial F_1}{\partial y} = \frac{\partial F_2}{\partial x}$, $\frac{\partial F_1}{\partial z} = \frac{\partial F_3}{\partial x}$, and $\frac{\partial F_2}{\partial z} = \frac{\partial F_3}{\partial y}$. and

    ◉ there exists a piecewise smooth oriented simple closed curve C in U such that $\left| \int_C F_1\, dx + F_2\, dy + F_3\, dz \neq 0. \right|$

    b. On the other hand, show that, if F is any smooth vector field on an open set U in $\mathbb{R}^3$ whose mixed partials are equal, then $\int_C F_1\, dx + F_2\, dy + F_3\, dz \neq 0.$ for any piecewise smooth oriented simple closed curve C that is the boundary of an oriented surface S contained in U .

    c. What does part (b) tell you about the curve C you found in part (a)?

3.6. Let C_1 and C_2 be piecewise smooth simple closed curves contained in the cylinder $x^2 + y^2 = 1$ that do not intersect one another, both oriented counterclockwise when viewed from high above the xy-plane, looking down, as illustrated in Figure 10.28. If F is the vector field $F(x, y, z) = (z - 2y^3, 2yz + x^5, y^2 + x)$, show that

$$\int_{C_1} F \cdot ds = \int_{C_2} F \cdot ds.$$

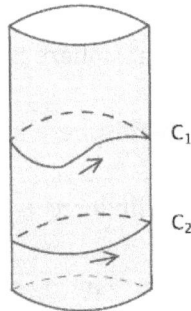

Figure 10.28: Two nonintersecting counterclockwise curves C_1 and C_2 in the cylinder $x^2 + y^2 = 1$

Section 4 Curl fields

In Exercises 4.1–4.3, determine whether the given vector field F is a curl field. If it is, find a vector field G whose curl is F. If not, explain why not.

4.1. $F(x, y, z) = (x, y, z)$

4.2. $F(x, y, z) = (y, z, x)$

4.3. $F(x, y, z) = (x + 2y + 3z, x^4 + y^5 + z^6, x^7y^8z^9)$

4.4. This exercise ties up some loose ends left dangling after Example 10.3 back in Section 10.1. Let S_1 be the hemisphere $x^2 + y^2 + z^2 = a^2$, $z \geq 0$, oriented by the outward normal, and let S2 be the disk $x^2 + y^2 \leq a^2$, $z = 0$, oriented by the upward normal.

Let F be the vector field $F(x, y, z) = (0, 0, -1)$.

    a. Show that F is a curl field by finding a vector field G whose curl is F.

    b. In Example 10.3, we showed that it is fairly easy to calculate that $\iint_{S_2} F \cdot dS = -\pi a^2$. Without any further calculation, explain why $\iint_{S_2} F \cdot dS = -\pi a^2$. as well.

4.5. Let F and G be vector fields on $\mathbb{R}^3$.

    a. True or false: If G = F + C for some constant vector C, then F and G have the same curl. Either prove that the statement is true, or find an example in which it is false.

    b. True or false: If F and G have the same curl, then G = F + C for some constant vector C. Either prove that the statement is true, or find an example in which it is false.

4.6. Does there exist a vector field F on $\mathbb{R}^3$, other than the constant vector field F = 0, with the property that:

    ⊙ $\int_C F \cdot ds = 0$ for every piecewise smooth oriented closed curve C in $\mathbb{R}^3$
    and

    ⊙ $\int_C F \cdot ds = 0$ for every piecewise smooth oriented closed surface S in $\mathbb{R}^3$?

If so, find such a vector field, and justify why it works. If no such vector field exists, explain why not.

Section 5 Gauss's theorem

5.1. Let $F(x, y, z) = (x3, y3, z3)$. Use Gauss's theorem to find $\iint_S F \cdot dS$ if S is the sphere $x^2 + v^2 + z^2 = 4$, oriented by the outward normal.

5.2. Find $\iint_S F \cdot dS$ if $F(x, y, z) = (x^3y^3z^4, x^2y^4z^4, x^2y^3z^5)$ and S is the boundary of the cube W = [0, 1] × [0, 1] × [0, 1], oriented by the normal pointing away from W .

5.3. Let S be the silo-shaped closed surface consisting of:

    ⊙ the cylinder $x^2 + y^2 = 4$, $0 \leq z \leq 3$,

- a hemispherical cap of radius 2 centered at $(0, 0, 3)$, i.e., $x^2 + y^2 + (z - 3)2 = 4$, $z \geq 3$, and

- the disk $x^2 + y^2 \leq 4$, $z = 0$, at the base,

- all oriented by the normal pointing away from the silo. Find $\iint_S \mathbf{F} \cdot d\mathbf{S}$ if $F(x, y, z) = (3xz^2, 2y, x + y^2 - z^3)$.

5.4. One way to produce a torus is to take the circle C in the xz-plane of radius b and center $(a, 0, 0)$ and rotate it about the z-axis. We assume that $a > b$. The surface that is swept out is a torus. The circle C can be parametrized by $\alpha(\psi) = (a + b \cos \psi, 0, b \sin \psi)$, $0 \leq \psi \leq 2\pi$.

The distance of $\alpha(\psi)$ to the z-axis is $a + b \cos \psi$, so rotating $\alpha(\psi)$ counterclockwise by θ about the z-axis brings it to the point $(a + b \cos \psi) \cos \theta$, $(a + b \cos \psi) \sin \theta$, $b \sin \psi)$. This gives the following parametrization of the torus with ψ and θ as parameters, illustrated in Figure 10.29:

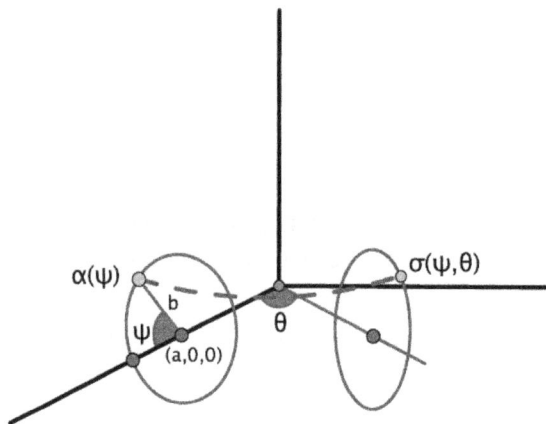

Figure 10.29: Parametrizing a torus

$$\sigma(\psi, \theta) = \big((a + b \cos \psi) \cos \theta, \ (a + b \cos \psi) \sin \theta, \ b \sin \psi\big), \qquad 0 \leq \psi \leq 2\pi, \quad 0 \leq \theta \leq 2\pi.$$

Let S denote the resulting torus, and assume that it is oriented by the unit normal vector that points towards the exterior of the torus.

   a. Show that:

$$\frac{\partial \sigma}{\partial \psi} \times \frac{\partial \sigma}{\partial \theta} = -b(a + b \cos \psi)\left(\cos \psi \cos \theta, \ \cos \psi \sin \theta, \ \sin \psi\right).$$

   b. Is σ orientation-preserving or orientation-reversing?

c. Find the surface area of the torus.

d. Let F(x, y, z) = (x, y, z). Use the parametrization of S and the definition of the surface integral to find $\iint_S F \cdot dS$. Then, use your answer and the result of Example 10.16 to find the volume of the solid torus, that is, the three-dimensional region enclosed by the torus.

e. Find the integral of the vector field F(x, y, z) = (0, 0, 1) over S. Use whatever method seems best. What does your answer say about the flow of F through S?

5.5. We describe an alternative approach to Exercise 4.4. Let F(x, y, z) = (0, 0, -1). The main point of the aforementioned exercise is that:

$$\iint_{S_1} F \cdot dS = \iint_{S_2} F \cdot dS, \qquad (10.31)$$

where S_1 is the hemisphere $x^2 + y^2 + z^2 = a^2$, $z \geq 0$, oriented by the outward normal, and S2 is the disk $x^2 + y^2 \leq a^2$, $z = 0$, oriented by the upward normal. The argument was based on Stokes's theorem.

Show that the same conclusion (10.31) can be reached using Gauss's theorem.

5.6. A bounded solid W in $\mathbb{R}^3$ is contained in the region $a \leq z \leq b$, where $a < b$. Its boundary is the union of three nonempty parts, as shown on the left of Figure 10.30:

⊙ a top surface T that is a subset of the plane z = b,

⊙ a bottom surface B that is a subset of the plane z = a, and

⊙ a surface S in between that connects the bottom and the top.

Let F(x, y, z) = (0, 0, 1). If the third piece above, S, is oriented by the normal pointing away from W , express $\iint_S F \cdot dS$ in terms of Area (T) and Area (B).

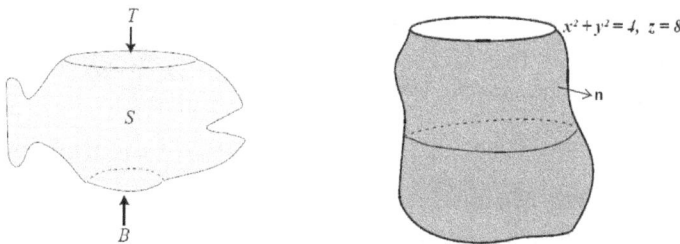

Figure 10.30: A solid region W lying between the planes z = a and z = b whose boundary consists of a surface S, a top T , and a bottom B (left) and a lumpy shopping bag of volume 75 (right)

5.7. A shopping bag is modeled as a lumpy surface S having an open circular top. Suppose that the bag is placed in $\mathbb{R}^3$ in such a way that the boundary of the opening is the circle $x^2 + y^2 = 4$ in the plane z = 8. See Figure 10.30, right. Assume that, when filled exactly to the brim, the bag has a volume of 75.

Let F(x, y, z) = (x, y, z). If S is oriented by the outward normal, use Gauss's theorem to find $\iint_S \mathbf{F} \cdot d\mathbf{S}$ (Note that S is not a closed surface.)

Let U be an open set in $\mathbb{R}^3$. A smooth real-valued function h : U → R is called harmonic on U if:

$$\frac{\partial^2 h}{\partial x^2} + \frac{\partial^2 h}{\partial y^2} + \frac{\partial^2 h}{\partial z^2} = 0 \quad \text{at all points of } U.$$

Exercises 5.8–5.10 concern harmonic functions. The results are the analogues of Exercises 4.12–4.15 in Chapter 9, which were about harmonic functions of two variables.

5.8. Show that h(x, y, z) = e^{-5x} sin(3y + 4z) is a harmonic function on $\mathbb{R}^3$.

5.9. Let W be a bounded subset of an open set U in $\mathbb{R}^3$ whose boundary ∂W consists of a finite number of piecewise smooth closed surfaces. Assume that every pair of points in W can be connected by a piecewise smooth curve in W . If h is harmonic on U and if h = 0 at all points of ∂W , prove that h = 0 on all of W . (Hint: Consider the vector field $\mathbf{F} = h\nabla h = h(\frac{\partial h}{\partial x}, \frac{\partial h}{\partial y}, \frac{\partial h}{\partial z})$.)

5.10. Let U and W be as in the preceding exercise. Show that, if h_1 and h_2 are harmonic on U and if $h_1 = h_2$ at all points of ∂W , then $h_1 = h_2$ on all of W . Thus a harmonic function is determined on W by its values on the boundary.

Section 6 The inverse square field

6.1. Let S_a denote the sphere $x^2 + y^2 + z^2 = a^2$ of radius a, oriented by the outward normal. If we did not already know the answer, it might be tempting to calculate the integral of the inverse square field $\mathbf{G}(x, y, z) = \left(-\frac{x}{(x^2+y^2+z^2)^{3/2}}, -\frac{y}{(x^2+y^2+z^2)^{3/2}}, -\frac{z}{(x^2+y^2+z^2)^{3/2}}\right)$ over S_a by filling in the sphere with the solid ball W_a given by $x^2 + y^2 + z^2 \leq a^2$ and using Gauss's theorem. Unfortunately, this is not valid (and would give an incorrect answer), because W_a contains the origin and the origin is not in the domain of G. On S_a, however, G agrees with the vector field $\mathbf{F}(x, y, z) = \left(-\frac{x}{a^3}, -\frac{y}{a^3}, -\frac{z}{a^3}\right) = -\frac{1}{a^3}(x, y, z)$, and F is defined on all of $\mathbb{R}^3$.

Find $\iint_{S_a} \mathbf{G} \cdot d\mathbf{S}$ by replacing G with F and applying Gauss's theorem.

6.2. Let F be a vector field defined on all of $\mathbb{R}^3$, except at the two points p = (2, 0, 0) and q = (-2, 0, 0). Let S_1, S_2, and S be the following spheres, centered at (2, 0, 0), (-2, 0, 0), and (0, 0, 0), respectively, each oriented by the outward normal.

- ⊙ S_1: $(x - 2)^2 + y^2 + z^2 = 1$
- ⊙ S_2: $(x + 2)^2 + y^2 + z^2 = 1$
- ⊙ S: $x^2 + y^2 + z^2 = 25$

Assume $\nabla \cdot \mathbf{F} = 0$. If $\iint_{S_1} \mathbf{F} \cdot d\mathbf{S} = 5$ and $\iint_{S_2} \mathbf{F} \cdot d\mathbf{S} = 6$, what is $\iint_S \mathbf{F} \cdot d\mathbf{S}$?

6.3. Give the details of the justification of the instance of Gauss's law cited in the text. That is, if S is a piecewise smooth closed surface in $\mathbb{R}^3 - \{p_1, p_2, \ldots, p_k\}$, oriented by the outward normal, $F(x) = \sum_{i=1}^{k} G(x - p_i)$ is the vector field of (10.18), show that:

$$-\frac{1}{4\pi} \iint_S F \cdot dS \text{ equals the number of the points } p_1, p_2, \ldots, p_k \text{ that lie in the interior of } S.$$

6.4. Suppose that objects of mass $m_1, m_2, \ldots, m_k$, not necessarily equal, are located at the points $p_1, p_2, \ldots, p_k$, respectively, in $\mathbb{R}^3$. The gravitational force that they exert on another object whose position is x is described by the vector field:

$$F(x) = \sum_{i=1}^{k} m_i G(x - p_i).$$

where G is our usual inverse square field (10.16). Let S be a piecewise smooth closed surface in $\mathbb{R}^3 - \{p_1, p_2, \ldots, p_k\}$, oriented by the outward normal. What can you say about $-\frac{1}{4\pi} \iint_S F \cdot dS$ in this case?

Section 7 A more substitution-friendly notation for surface integrals

7.1. Let S be the unit sphere $x^2 + y^2 + z^2 = 1$, oriented by the outward normal. Consider the surface integral:

$$\iint_S xy\, dy \wedge dz + yz\, dz \wedge dx + xz\, dx \wedge dy.$$

    a. What is the vector field that is being integrated?
    b. Evaluate the integral.

7.2. Suppose that a surface S in $\mathbb{R}^3$ is the graph $z = f(x, y)$ of a smooth real-valued function $f: D \to R$ whose domain D is a bounded subset of $\mathbb{R}^2$. Let S be oriented by the upward normal. Show that:

$$\iint_S 1\, dx \wedge dy = \text{Area}(D).$$

In other words, the integral is the area of the projection of S on the xy-plane. Similar interpretations hold for $\iint_S 1\, dy \wedge dz$ and $\iint_S 1\, dz \wedge dx$ under analogous hypotheses.

Section 8 Independence of parametrization

8.1. Let S be the portion of the cone $z = \sqrt{x^2 + y^2}$ in $\mathbb{R}^3$ where $0 \leq z \leq 2$. In Section 5.4, we showed that S could be parametrized in more than one way. One possibility is based on using polar/cylindrical coordinates as parameters (Example 5.14):

$$\sigma : [0,2] \times [0,2\pi] \to \mathbb{R}^3, \quad \sigma(s,t) = (s\cos t, \ s\sin t, \ s).$$

Another is based on spherical coordinates (Example 5.17):

$$\tau : [0, 2\sqrt{2}] \times [0, 2\pi] \to \mathbb{R}^3, \quad \tau(u,v) = (\frac{\sqrt{2}}{2}u\cos v, \ \frac{\sqrt{2}}{2}u\sin v, \ \frac{\sqrt{2}}{2}u).$$

Find a formula for a function $T : [0, 2\sqrt{2}] \times [0,2\pi] \to [0,2] \times [0,2\pi]$ that satisfies τ (u, v) = σ (T (u, v)), as in equation (10.25).

11 Chapter

WORKING WITH DIFFERENTIAL FORMS

This final chapter is a whirlwind survey of how to work with differential forms. We focus particularly on what differential forms have to say about some of the main theorems we have learned recently. In what follows, all functions, including vector fields, are assumed to be smooth.

Four theorems, version 1

1. **Conservative vector fields.** Let C be a piecewise smooth oriented curve in $\mathbb{R}^n$ from p to q. If $F = \nabla f$ is a conservative vector field on an open set containing C, then:

$$\int_C \nabla f \cdot d\mathbf{s} = f(\mathbf{q}) - f(\mathbf{p}). \tag{C}$$

2. **Green's theorem.** Let D be a bounded subset of $\mathbb{R}^2$. If $F = (F1, F2)$ is a vector field on an open set containing D, then:

$$\iint_D \left(\frac{\partial F_2}{\partial x} - \frac{\partial F_1}{\partial y} \right) dx\, dy = \int_{\partial D} \mathbf{F} \cdot d\mathbf{s}. \tag{Gr}$$

3. **Stokes's theorem.** Let S be a piecewise smooth oriented surface in $\mathbb{R}^3$. If $F = (F_1, F_2, F_3)$ is a vector field on an open set containing S, then:

$$\iint_S (\nabla \times \mathbf{F}) \cdot d\mathbf{S} = \int_{\partial S} \mathbf{F} \cdot d\mathbf{s}. \tag{S}$$

4. **Gauss's theorem.** Let W be a bounded subset of $\mathbb{R}^3$. If $F = (F_1, F_2, F_3)$ is a vector field on an open set containing W , then:

$$\iiint_W (\nabla \cdot \mathbf{F})\, dx\, dy\, dz = \iint_{\partial W} \mathbf{F} \cdot d\mathbf{S}. \tag{Ga}$$

Our goal is to present just enough of the properties of differential forms to indicate how forms enter the picture as far as these theorems are concerned. Further information is developed in the exercises.

11.1 INTEGRALS OF DIFFERENTIAL FORMS

Roughly speaking, a differential k-form is an expression that can be integrated over an oriented k-dimensional domain. For instance, in $\mathbb{R}^n$, here are the cases at opposite ends of the range of dimensions.

⊙ n-forms. In $\mathbb{R}^n$ with coordinates $x_1, x_2, \ldots, x_n$, an n-form is an expression:

$$f\,dx_1 \wedge dx_2 \wedge \cdots \wedge dx_n,$$

where f is a real-valued function of $x = (x_1, x_2, \ldots, x_n)$. It could also be written $f \wedge dx_1 \wedge dx_2 \wedge \cdots \wedge dx_n$, though f is treated commonly as a scalar factor. If W is an n-dimensional bounded subset of $\mathbb{R}^n$, which by default we say has the "positive" orientation, then the integral $\int_W f\,dx_1 \wedge dx_2 \wedge \cdots \wedge dx_n$ is defined to be the usual Riemann integral:

As indicated on the left of equation (11.1), we won't use multiple integral signs with differential forms. The dimension of the integral can be inferred from the type of form that is being integrated.

– +{p}

0-forms. A 0-form is simply another name for a real-valued function f. 0-forms are integrated over zero-dimensional domains, i.e., individual points, by evaluating the function at that point and where we orient a point by attaching a + or sign. So $f_{+\{p\}} f = f(p)$ and $f_{-\{p\}} f = -f(p)$. If C is an oriented curve from p to q, we define its oriented boundary to be the oriented points $\partial C = (+\{q\}) \cup (-\{p\})$.

We restrict our attention here to differential forms on subsets of $\mathbb{R}^3$ (or $\mathbb{R}$ or $\mathbb{R}^2$, depending on the context), real-valued functions f of three variables (or one or two), and vector fields on subsets of $\mathbb{R}^3$ (or $\mathbb{R}$ or $\mathbb{R}^2$), since these are the cases of interest in the four big theorems listed above.

Table 11.1 below gives the notation for 0, 1, 2, and 3-forms and their associated integrals. The integrals that the notation stands for, shown in the last column, are the kinds of integrals we have been studying: 1-forms integrate as line integrals, 2-forms as surface integrals, and 3-forms as triple integrals.

Table 11.1: Integrals of differential forms

Type of form	Domain of integration	Notation for integral using forms	Meaning of integral
0-form	oriented points	$f+\{p\}\ f$ or $f-\{p\}\ f$	$=f(p)$ or $-f(p)$
1-form	oriented curves	$f_C\ F_1\,dx + F_2\,dy + F_3\,dz$	$=f_C\ F.ds$

Type of form	Domain of integration	Notation for integral using forms	Meaning of integral
2-form	oriented surfaces	$\int_S F_1\,dy \wedge dz + F_2\,dz \wedge dx$ $+F_3\,dx \wedge dy$	$=\iint_S F.\,ds$
3-form	3-dim solids	$\int_W f\,ds \wedge dy \wedge dz$	$=\iint_W f\,dx\,dy\,dz$

In this notation, the four theorems (C), (Gr), (S), and (Ga) translate as integrals of forms as follows.

Four theorems, version 2

1. Conservative vector fields.

$$\int_C \frac{\partial f}{\partial x}\,dx + \frac{\partial f}{\partial y}\,dy + \frac{\partial f}{\partial z}\,dz = f(\mathbf{q}) - f(\mathbf{p}) = \int_{\partial C} f \tag{C}$$

2. Green's theorem.

$$\int_D \left(\frac{\partial F_2}{\partial x} - \frac{\partial F_1}{\partial y} \right) dx \wedge dy = \int_{\partial D} F_1\,dx + F_2\,dy \tag{Gr}$$

3. Stokes's theorem.

$$\int_S \left(\frac{\partial F_3}{\partial y} - \frac{\partial F_2}{\partial z} \right) dy \wedge dz + \left(\frac{\partial F_1}{\partial z} - \frac{\partial F_3}{\partial x} \right) dz \wedge dx + \left(\frac{\partial F_2}{\partial x} - \frac{\partial F_1}{\partial y} \right) dx \wedge dy$$
$$= \int_{\partial S} F_1\,dx + F_2\,dy + F_3\,dz \tag{S}$$

4. Gauss's theorem.

$$\int_W \left(\frac{\partial F_1}{\partial x} + \frac{\partial F_2}{\partial y} + \frac{\partial F_3}{\partial z} \right) dx \wedge dy \wedge dz = \int_{\partial W} F_1\,dy \wedge dz + F_2\,dz \wedge dx + F_3\,dx \wedge dy \tag{Ga}$$

11.2 DERIVATIVES OF DIFFERENTIAL FORMS

In this brief foray, we focus on how differential forms behave without worrying about exactly what they are. Here are some simple rules for working with forms.

1. **Arithmetic.** These are the rules we encountered earlier that reflect a connection between differential forms and determinants (see the end of Section 10.7):

  a. dy ∧ dx = -dx ∧ dy, dz ∧ dy = -dy ∧ dz, dx ∧ dz = -dz ∧ dx,
  b. dx ∧ dx = dy ∧ dy = dz ∧ dz = 0.

These rules have the effect of simplifying what differential forms can look like. For instance, a 2-form should be a generalized double integrand. In $\mathbb{R}^3$, such an expression would have the form:

$$f_1\, dx \wedge dx + f_2\, dx \wedge dy + f_3\, dx \wedge dz + f_4\, dy \wedge dx + \cdots + f_9\, dz \wedge dz,$$

where $f_1, f_2, \ldots, f_9$ are real-valued functions of x, y, z. According to the rules, however, three of the nine terms—the ones involving dx∧dx, dy∧dy, and dz∧dz—are zero and therefore can be omitted. Similarly, the remaining six terms can be combined in pairs: for instance, f_2 dx ∧ dy + f_4 dy ∧ dx = f_2 dx ∧ dy - f_4 dx ∧ dy = $(f_2 - f_4)$ dx ∧ dy. Hence we may assume that any 2-form on a subset of R3 is a sum of three terms having the form η = F_1 dy ∧ dz + F_2 dz ∧ dx + F_3 dx ∧ dy.

2. Differentiation. In general, the derivative of a k-form is a (k + 1)-form. Specifically, for differential forms on subsets of $\mathbb{R}^3$, we define:

For 0-forms:

$$df = \frac{\partial f}{\partial x}\, dx + \frac{\partial f}{\partial y}\, dy + \frac{\partial f}{\partial z}\, dz.$$

For 1-forms:

$$d(F_1\, dx + F_2\, dy + F_3\, dz) = dF_1 \wedge dx + dF_2 \wedge dy + dF_3 \wedge dz.$$

For 2-forms:

$$d(F_1\, dy \wedge dz + F_2\, dz \wedge dx + F_3\, dx \wedge dy) = dF_1 \wedge dy \wedge dz + dF_2 \wedge dz \wedge dx + dF_3 \wedge dx \wedge dy.$$

In (b) and (c), F_1, F_2, and F_3 are 0-forms, so their derivatives are as defined in (a).

Example 11.1. Let x = r cos θ and y = r sin θ. These are real-valued functions of r and θ—in other words, they are 0-forms—so by definition:

$$dx = \frac{\partial x}{\partial r}\, dr + \frac{\partial x}{\partial \theta}\, d\theta = \cos \theta\, dr - r \sin \theta\, d\theta$$
$$\text{and} \quad dy = \frac{\partial y}{\partial r}\, dr + \frac{\partial y}{\partial \theta}\, d\theta = \sin \theta\, dr + r \cos \theta\, d\theta.$$

Then, by the arithmetic rules:

$$\begin{aligned}
dx \wedge dy &= (\cos \theta\, dr - r \sin \theta\, d\theta) \wedge (\sin \theta\, dr + r \cos \theta\, d\theta) \\
&= \cos \theta \sin \theta\, dr \wedge dr + r \cos^2 \theta\, dr \wedge d\theta - r \sin^2 \theta\, d\theta \wedge dr - r^2 \sin \theta \cos \theta\, d\theta \wedge d\theta \\
&= \cos \theta \sin \theta \cdot 0 + r \cos^2 \theta\, dr \wedge d\theta - r \sin^2 \theta \cdot (-dr \wedge d\theta) - r^2 \sin \theta \cos \theta \cdot 0 \\
&= (r \cos^2 \theta + r \sin^2 \theta)\, dr \wedge d\theta \\
&= r\, dr \wedge d\theta.
\end{aligned}$$

Example 11.2. Let σ(s, t) = (x(s, t), y(s, t), z(s, t)) be a smooth parametrization of a surface S. Then x, y, and z are each real-valued functions of s and t, so, for example:

$$dy = \frac{\partial y}{\partial s}\,ds + \frac{\partial y}{\partial t}\,dt \qquad \text{and} \qquad dz = \frac{\partial z}{\partial s}\,ds + \frac{\partial z}{\partial t}\,dt.$$

It follows that:

$$
\begin{aligned}
dy \wedge dz &= (\frac{\partial y}{\partial s}\,ds + \frac{\partial y}{\partial t}\,dt) \wedge (\frac{\partial z}{\partial s}\,ds + \frac{\partial z}{\partial t}\,dt) \\
&= 0 + \frac{\partial y}{\partial s}\frac{\partial z}{\partial t}\,ds \wedge dt + \frac{\partial y}{\partial t}\frac{\partial z}{\partial s}\,dt \wedge ds + 0 \\
&= (\frac{\partial y}{\partial s}\frac{\partial z}{\partial t} - \frac{\partial y}{\partial t}\frac{\partial z}{\partial s})\,ds \wedge dt \\
&= \det \begin{bmatrix} \frac{\partial y}{\partial s} & \frac{\partial y}{\partial t} \\ \frac{\partial z}{\partial s} & \frac{\partial z}{\partial t} \end{bmatrix}\,ds \wedge dt.
\end{aligned}
$$

This is the same as the expression for dy ∧dz that we obtained when we discussed the substitutions used to compute surface integrals, as in equation (10.21) of Chapter 10. Perhaps the calculation given here is a more organic way of obtaining the substitution. Either way, the formula is consistent with what the change of variables theorem says to do to pull back an integral with respect to y and z to one with respect to s and t.

Next, we derive formulas for the derivatives of 1 and 2-forms.

Example 11.3. For a 1-form ω = F_1 dx + F_2 dy + F_3 dz, we find d_ω by calculating d(F_1 dx), d(F_2 dy), and d(F_3 dz) separately. For instance, by definition:

$$
\begin{aligned}
d(F_1\,dx) &= dF_1 \wedge dx \\
&= (\frac{\partial F_1}{\partial x}\,dx + \frac{\partial F_1}{\partial y}\,dy + \frac{\partial F_1}{\partial z}\,dz) \wedge dx \\
&= \frac{\partial F_1}{\partial x}\,dx \wedge dx + \frac{\partial F_1}{\partial y}\,dy \wedge dx + \frac{\partial F_1}{\partial z}\,dz \wedge dx \\
&= 0 - \frac{\partial F_1}{\partial y}\,dx \wedge dy + \frac{\partial F_1}{\partial z}\,dz \wedge dx \\
&= -\frac{\partial F_1}{\partial y}\,dx \wedge dy + \frac{\partial F_1}{\partial z}\,dz \wedge dx.
\end{aligned}
$$

Similarly:

$$d(F_2\,dy) = \frac{\partial F_2}{\partial x}\,dx \wedge dy - \frac{\partial F_2}{\partial z}\,dy \wedge dz$$

$$\text{and} \quad d(F_3\,dz) = -\frac{\partial F_3}{\partial x}\,dz \wedge dx + \frac{\partial F_3}{\partial y}\,dy \wedge dz.$$

Adding these calculations gives:

$$d(F_1\,dx + F_2\,dy + F_3\,dz) = \left(\frac{\partial F_3}{\partial y} - \frac{\partial F_2}{\partial z}\right) dy \wedge dz$$

$$+ \left(\frac{\partial F_1}{\partial z} - \frac{\partial F_3}{\partial x}\right) dz \wedge dx + \left(\frac{\partial F_2}{\partial x} - \frac{\partial F_1}{\partial y}\right) dx \wedge dy.$$

a formula that is reproduced in row 2(b) of Table 11.2 below.

Example 11.4. For the derivative of a 2-form $\eta = F_1\,dy \wedge dz + F_2\,dz \wedge dx + F_3\,dx \wedge dy$, we compute, for example:

$$d(F_2\,dz \wedge dx) = dF_2 \wedge dz \wedge dx$$

$$= \left(\frac{\partial F_2}{\partial x}\,dx + \frac{\partial F_2}{\partial y}\,dy + \frac{\partial F_2}{\partial z}\,dz\right) \wedge dz \wedge dx$$

$$= \frac{\partial F_2}{\partial x}\,dx \wedge dz \wedge dx + \frac{\partial F_2}{\partial y}\,dy \wedge dz \wedge dx + \frac{\partial F_2}{\partial z}\,dz \wedge dz \wedge dx$$

$$= -\frac{\partial F_2}{\partial x}\,dx \wedge dx \wedge dz - \frac{\partial F_2}{\partial y}\,dy \wedge dx \wedge dz + 0$$

$$= 0 + \frac{\partial F_2}{\partial y}\,dx \wedge dy \wedge dz$$

$$= \frac{\partial F_2}{\partial y}\,dx \wedge dy \wedge dz.$$

Likewise:

$$d(F_1\,dy \wedge dz) = \frac{\partial F_1}{\partial x}\,dx \wedge dy \wedge dz \quad \text{and} \quad d(F_3\,dx \wedge dy) = \frac{\partial F_3}{\partial z}\,dx \wedge dy \wedge dz.$$

Summing these calculations gives:

$$d(F_1\,dy \wedge dz + F_2\,dz \wedge dx + F_3\,dx \wedge dy) = \left(\frac{\partial F_1}{\partial x} + \frac{\partial F_2}{\partial y} + \frac{\partial F_3}{\partial z}\right) dx \wedge dy \wedge dz.$$

as in row 3 of Table 11.2.

We collect the results in a table. Below it, for future reference, we reproduce the most recent version of our four big theorems (version 2).

Table 11.2 Formulas for derivatives

1. 0-forms	
a. $f = f(x)$	$df = f'(x)\,dx$
b. $f = f(x, y)$	$df = \frac{\partial f}{\partial x}\,dx + \frac{\partial f}{\partial y}\,dy$
c. $f = f(x, y, z)$	$df = \frac{\partial f}{\partial x}\,dx + \frac{\partial f}{\partial y}\,dy + \frac{\partial f}{\partial z}\,dz$

2. 1-forms	
a. $\omega = F_1(x, y)\, dx + F_2(x, y)\, dy$	$d\omega = \left(\frac{\partial F_2}{\partial x} - \frac{\partial F_1}{\partial y}\right) dx \wedge dy$
b. $\omega = F_1(x, y, z)\, dx + F_2(x, y, z)\, dy + F_3(x, y, z)\, dz$	$d\omega = \left(\frac{\partial F_3}{\partial y} - \frac{\partial F_2}{\partial z}\right) dy \wedge dz$ $+\left(\frac{\partial F_1}{\partial z} - \frac{\partial F_3}{\partial x}\right) dz \wedge dx + \left(\frac{\partial F_2}{\partial x} - \frac{\partial F_1}{\partial y}\right) dx \wedge dy$
3. 2-forms	
$\eta = F_1\, dy \wedge dz + F_2\, dz \wedge dx + F_3\, dx \wedge dy$	$d\eta = \left(\frac{\partial F_1}{\partial x} + \frac{\partial F_2}{\partial y} + \frac{\partial F_3}{\partial z}\right) dx \wedge dy \wedge dz$

Four theorems, version 2 (again)

1. **Conservative vector fields.**

$$\int_C \frac{\partial f}{\partial x}\, dx + \frac{\partial f}{\partial y}\, dy + \frac{\partial f}{\partial z}\, dz = f(\mathbf{q}) - f(\mathbf{p}) = \int_{\partial C} f \qquad \text{(C)}$$

2. **Green's theorem.**

$$\int_D \left(\frac{\partial F_2}{\partial x} - \frac{\partial F_1}{\partial y}\right) dx \wedge dy = \int_{\partial D} F_1\, dx + F_2\, dy \qquad \text{(Gr)}$$

3. **Stokes's theorem.**

$$\int_S \left(\frac{\partial F_3}{\partial y} - \frac{\partial F_2}{\partial z}\right) dy \wedge dz + \left(\frac{\partial F_1}{\partial z} - \frac{\partial F_3}{\partial x}\right) dz \wedge dx + \left(\frac{\partial F_2}{\partial x} - \frac{\partial F_1}{\partial y}\right) dx \wedge dy$$
$$= \int_{\partial S} F_1\, dx + F_2\, dy + F_3\, dz \qquad \text{(S)}$$

4. **Gauss's theorem.**

$$\int_W \left(\frac{\partial F_1}{\partial x} + \frac{\partial F_2}{\partial y} + \frac{\partial F_3}{\partial z}\right) dx \wedge dy \wedge dz = \int_{\partial W} F_1\, dy \wedge dz + F_2\, dz \wedge dx + F_3\, dx \wedge dy \qquad \text{(Ga)}$$

11.3 A LOOK BACK AT THE THEOREMS OF MULTIVARIABLE CALCULUS

Take a moment to compare the formulas in Table 11.2 with the four theorems. It soon becomes apparent that, in each theorem, a differential form and its derivative appear. In this new framework, the theorems can be written in spectacularly streamlined form.

Four theorems, final version

1. Conservative vector fields. If f is a real-valued function, then:

$$\int_C df = \int_{\partial C} f. \qquad\qquad (C)$$

2. **Green's theorem.** If $\omega = F_1\, dx + F_2\, dy$, then:

$$\int_D d\omega = \int_{\partial D} \omega. \qquad\qquad (Gr)$$

3. **Stokes's theorem.** If $\omega = F_1\, dx + F_2\, dy + F_3\, dz$, then:

$$\int_S d\omega = \int_{\partial S} \omega. \qquad\qquad (S)$$

4. **Gauss's theorem.** If $\eta = F_1\, dy \wedge dz + F_2\, dz \wedge dx + F_3\, dx \wedge dy$, then:

$$\int_W d\eta = \int_{\partial W} \eta. \qquad\qquad (Ga)$$

They're all the same theorem! But there's even more to it, for suppose we go back even further, to first-year calculus and a real-valued function $f(x)$ of one variable. The fundamental theorem of calculus says that $\int_a^b f'(x)\, dx = f(b) - f(a)$. Denoting the interval [a, b] by I and converting to differential form notation, this becomes:

$$\int_I df = \int_{\partial I} f.$$

In other words, the four theorems (C), (Gr), (S), and (Ga) of multivariable calculus are generalizations of something that we've known for a long time.

The pattern continues in higher dimensions and with more variables. One studies subsets of $\mathbb{R}^n$ that are k-dimensional in the sense that they can be traced out using k independent parameters. In other words, they are the images of smooth functions of the form $\psi : D \to \mathbb{R}^n$, where $D \subset \mathbb{R}^k$. These objects are called **k-dimensional manifolds**. Differential k-forms can be integrated over them, or at least over the ones for which there is a notion of orientation.

A typical k-form in $\mathbb{R}^n$ is a k-dimensional integrand, that is, a sum of terms of the form $f\, dx_{i1} \wedge dx_{i2} \wedge \cdots \wedge dx_{ik}$, where f is a real-valued function of n variables x_1, $x_2, \ldots, x_n$. By the arithmetic rules for differential forms, we may assume that the subscripts $i_1, i_2, \ldots, i_k$ are distinct and in fact that they are arranged, say in increasing order $i_1 < i_2 < \cdots < $ ik. If ζ is a k-form and M is an oriented k-dimensional manifold in $\mathbb{R}^n$ with an orientation-preserving parametrization $\psi : D \to \mathbb{R}^n$ as above, then the integral $\int M \ \zeta$ is defined using substitution and the rules of forms to pull the integral back to an ordinary Riemann integral over the k-dimensional parameter domain D in $\mathbb{R}^k$. (See the remarks before Exercise 4.14 at the end of the chapter for more about the pullback process.) In the end, the relation between the integral and its pullback is summarized by the formula:

$$\int_M \zeta = \int_D \psi^*(\zeta). \tag{11.2}$$

That the value of the integral does not depend on the parametrization rests ultimately on the change of variables theorem and the chain rule. In fact, you have seen the proof, at least in outline form. It is (10.30).

Perhaps it is worth reinforcing the point that equation (11.2) agrees with the expressions for line integrals and surface integrals that we obtained in the previous two chapters. See (9.6) and (10.23). In particular, line and surface integrals fit in as two special cases of a general type of integral, something that may not have been clear from the original motivations in terms of tangent and normal components of vector fields.

The general theory has a striking coherence and consistency. The culmination may be a theorem that describes how information over the entire manifold is related to behavior along the boundary. The boundary of a k-dimensional manifold M , denoted by ∂M , is (k - 1)-dimensional, as was the case with curves (k = 1) and surfaces (k = 2). If ω is a differential (k - 1)-form, then this theorem, which is known as Stokes's theorem, states that:

$$\boxed{\int_M d\omega = \int_{\partial M} \omega.}$$

It's the fundamental theorem of calculus.

11.4 EXERCISES FOR CHAPTER 11

In Exercises 4.1–4.5, calculate the indicated product.

4.1. $(x\ dx + y\ dy) \wedge (-y\ dx + x\ dy)$

4.2. $(x\ dx + y\ dy) \wedge (x\ dx - y\ dy)$

4.3. $(x\ dx + y\ dy + z\ dz) \wedge (y\ dx + z\ dy + x\ dz)$

4.4. $(-y\ dx + x\ dy + z\ dz) \wedge (yz\ dy \wedge dz + xz\ dz \wedge dx + xy\ dx \wedge dy)$

4.5. $(x^2\ dy \wedge dz) \wedge (yz\ dy \wedge dz + xz\ dz \wedge dx + xy\ dx \wedge dy)$

4.6.   a. If $\omega = F_1\,dx_1 + F_2\,dx_2 + \cdots + F_n\,dx_n$ is a 1-form on $\mathbb{R}^n$, show that $\omega \wedge \omega = 0$.

b. Find an example of a 2-form η on $\mathbb{R}^4$ such that $\eta \wedge \eta \neq 0$.

In Exercises 4.7–4.11, find the derivative of the given differential form.

4.7. $f = x^2 + y^2$

4.8. $f = x \sin y \cos z$

4.9. $\omega = (x^2 - y^2)\,dx + 2xy\,dy$

4.10. $\omega = 2xy^3z^4\,dx + 3x^2y^2z^4\,dy + 4x^2y^3z^3\,dz$

4.11. $\eta = (x + 2y + 3z)\,dy \wedge dz + e^{xyz}\,dz \wedge dx + x^4y^5\,dx \wedge dy$

4.12. Let U be an open set in $\mathbb{R}^3$.

a. Let $f = f(x, y, z)$ be a 0-form (= real-valued function) on U . Find $d(df)$.

b. Let $\omega = F_1\,dx + F_2\,dy + F_3\,dz$ be a 1-form on U . Find $d(d\omega)$.

c. To what facts about vector fields do the results of parts (a) and (b) correspond?

4.13. A typical 3-form on $\mathbb{R}^4$ has the form:

$$\zeta = F_1\,dx_2 \wedge dx_3 \wedge dx_4 + F_2\,dx_1 \wedge dx_3 \wedge dx_4 + F_3\,dx_1 \wedge dx_2 \wedge dx_4 + F_4\,dx_1 \wedge dx_2 \wedge dx_3,$$

where F_1, F_2, F_3, and F_4 are real-valued functions of x_1, x_2, x_3, x_4. The notation is arranged so that, in the term with leading factor F_i, the differential dxi is omitted.

As a 4-form on $\mathbb{R}^4$, $d\zeta$ has the form $d\zeta = f\,dx1 \wedge dx2 \wedge dx3 \wedge dx4$ for some real-valued function f. Find a formula for f in terms of F_1, F_2, F_3, and F_4.

We lay out more formally some of the rules that govern substitutions and pullbacks. In Rm with coordinates x_1, x_2, , x_m, a typical differential k-form is a sum of terms of the form:

$$\zeta = f\,dx_{i_1} \wedge dx_{i_2} \wedge \cdots \wedge dx_{i_k}, \tag{11.3}$$

where f is a real-valued function of $x_1, x_2, \ldots, x_m$ and $1 \le i_j \le m$ for $j = 1, 2, , k$. Suppose that each of the coordinates xi can be expressed as a smooth function of n other variables $u_1, u_2, \ldots, u_n$, say $x_i = \gamma_i(u_1, u_2, \ldots, u_n)$ for $i = 1, 2, \ldots, m$. This is equivalent to having a smooth function $\gamma : \mathbb{R}^n \to Rm$ from $(u_1, u_2, \ldots, u_n)$-space to $(x_1, x_2, \ldots, x_m)$-space, where, in terms of component functions, $\gamma = (\gamma_1, \gamma_2, \ldots, \gamma_m)$.

Using γ to substitute for $x_1, x_2, \ldots, x_m$ converts the k-form ζ into a k-form in the variables $u_1, u_2, \ldots, u_n$. This k-form is what we have been calling the pullback of ζ, denoted by $\gamma*(\zeta)$. For instance, substituting $x = \gamma(u)$ converts $f(x)$ into $f(\gamma(u)) = (f \circ \gamma)(u)$. In other words, f pulls back to the composition $f \circ \gamma : \mathbb{R}^n \to Rm \to R$. We write this as:

$$\gamma^*(f) = f \circ \gamma.$$

Similarly, after substitution, dxi becomes $d\gamma_i$, i.e.:

$$\gamma^*(dx_i) = d\gamma_i,$$

where the expression on the right is the derivative of the 0-form γ_i. For instance, if $(x_1, x_2) = \gamma(u_1, u_2) = (u_1 \cos u_2, u_1 \sin u_2)$, then $\gamma^*(dx_1) = d\gamma 1 = d(u_1 \cos u_2) = \cos u_2 \, du_1 - u_1 \sin u_2 \, du_2$. Applying these substitutions en masse, the k-form $\zeta = f \, dx_{i_1} \wedge dx_{i_2} \wedge \cdots \wedge dx_{i_k}$ of equation (11.3) pulls back as:

$$\gamma^*(\zeta) = (f \circ \gamma) \, d\gamma_{i_1} \wedge d\gamma_{i_2} \wedge \cdots \wedge d\gamma_{i_k}.$$

Perhaps a slicker way of writing this is:

$$\gamma^*(f \, dx_{i_1} \wedge dx_{i_2} \wedge \cdots \wedge dx_{i_k}) = \gamma^*(f) \, \gamma^*(dx_{i_1}) \wedge \gamma^*(dx_{i_2}) \wedge \cdots \wedge \gamma^*(dx_{i_k}).$$

If $\zeta = \zeta_1 + \zeta_2$ is a sum of k-forms, then carrying out the substitutions in both summands gives the rule $\gamma^*(\zeta_1 + \zeta_2) = \gamma^*(\zeta_1) + \gamma^*(\zeta_2)$.

Exercises 4.14–4.18 are intended to provide some practice working with pullbacks.

4.14. Let $\alpha : \mathbb{R} \to \mathbb{R}^3$ be given by $\alpha(t) = (\cos t, \sin t, t)$, where we think of $\mathbb{R}^3$ as xyz-space.

   a. Find $\alpha^*(x^2 + y^2 + z^2)$.
   b. Show that $\alpha^*(dx) = -\sin t \, dt$.
   c. Find $\alpha^*(dy)$ and $\alpha^*(dz)$.
   d. Let ω be the 1-form $\omega = -y \, dx + x \, dy + z \, dz$ on $\mathbb{R}^3$. Find $\alpha*(\omega)$.

4.15. Let $T : \mathbb{R}^2 \to \mathbb{R}^2$ be the function from the uv-plane to the xy-plane given by T$(u, v) = (u^2 - v^2, 2uv)$.

   a. Show that $T^*(dx) = 2u \, du - 2v \, dv$.
   b. Find $T^*(dy)$.
   c. Let $\omega = -y \, dx + x \, dy$. Find $T^*(\omega)$.
   d. Let $\eta = xy \, dx \wedge dy$. Find $T^*(\eta)$.

4.16. Let $\sigma : \mathbb{R}^2 \to \mathbb{R}^3$ be the function from the uv-plane to xyz-space given by $\sigma(u, v) = (u \cos v, u \sin v, u)$.

   a. Let $\omega = x \, dx + y \, dy - z \, dz$. Show that $\sigma^*(\omega) = 0$.
   b. Let $\eta = z \, dx \wedge dy$. Find $\sigma^*(\eta)$.

4.17. Let $\alpha : \mathbb{R} \to \mathbb{R}^n$ be a smooth function, where $\alpha(t) = (x_1(t), x_2(t), \ldots, x_n(t))$. If $f : \mathbb{R}^n \to \mathbb{R}$ is a smooth real-valued function, $f = f(x_1, x_2, \ldots, x_n)$, then df is a 1-form on $\mathbb{R}^n$. Show that:

$$d(\alpha^*(f)) = \alpha^*(df).$$

(Hint: Little Chain Rule.)

4.18. Let $T : \mathbb{R}^2 \to \mathbb{R}^2$ be a smooth function, regarded as a map from the uv-plane to the xy-plane, so $(x, y) = T(u, v) = (T_1(u, v), T_2(u, v))$.

    a. Find $T^*(dx)$ and $T^*(dy)$ in terms of the partial derivatives of T_1 and T_2.

    b. Let η be a 2-form on the xy-plane: $\eta = f\, dx \wedge dy$, where f is a real-valued function of x and y. Show that:
$$T^*(\eta) = (f \circ T) \cdot \det DT\, du \wedge dv.$$

Let $S : \mathbb{R}^2 \to \mathbb{R}^2$ be a smooth function, regarded as a map from the st-plane to the uv-plane. Then the composition $T \circ S : \mathbb{R}^2 \to \mathbb{R}^2$ goes from the st-plane to the xy-plane. Let $\eta = f\, dx \wedge dy$, as in part (b). Show that:
$$(T \circ S)^*(\eta) = S^*(T^*(\eta)),$$

i.e., $(T \circ S)^* = S^* \circ T^*$ for 2-forms on $\mathbb{R}^2$. (Actually, the same relation holds for compositions and k-forms in general.)

We next formulate a version of the change of variables theorem for double integrals that is adapted for differential forms, something we have alluded to before. We are not trying to give another proof of the theorem. Rather, we take the theorem as known and describe how to restate it in the language of differential forms.

Differential forms are integrated over oriented domains, so, given a bounded subset D of $\mathbb{R}^2$, we first make a choice whether to assign it the positive orientation or the negative orientation. If this seems too haphazard, think of it as analogous to choosing whether, as a subset of $\mathbb{R}^3$, D is oriented by the upward normal n = k or the downward normal n = -k.

A typical 2-form on D has the form $f\, dx \wedge dy$, where f is a real-valued function on D. We define its integral over D by:
$$\int_D f\, dx \wedge dy = \begin{cases} \iint_D f(x, y)\, dx\, dy & \text{if } D \text{ has the positive orientation,} \\ -\iint_D f(x, y)\, dx\, dy & \text{if } D \text{ has the negative orientation,} \end{cases}$$

where, in each case, the integral on the right is the ordinary Riemann double integral. So, for instance, if D has the negative orientation, then we could also say that, as an integral of a differential form, $\int_D f\, dy \wedge dx$ is equal to the Riemann integral $\iint_D f(x, y)\, dx\, dy$.

Exercises 4.19–4.20 concern how to express the change of variables theorem in terms of integrals of differential forms.

4.19. We say that a basis (v, w) of $\mathbb{R}^2$ has the positive orientation if det $[v\ w] > 0$ and the negative orientation if det $[v\ w] < 0$. Here, we use $[v\ w]$ to denote the 2 by 2 matrix whose columns are v and w. In Section 2.5 of Chapter 2, we used the terms "right-handed" and "left-handed" to describe the same concepts, though there we put the vectors into the rows of the matrix, i.e., we worked with the transpose, which doesn't affect the determinant.

Let $L : \mathbb{R}^2 \to \mathbb{R}^2$ be a linear transformation represented by a matrix A with respect to the standard bases. Thus $L(x) = Ax$. In addition, assume that det $A \neq 0$.

    a. Show that $(L(e_1), L(e_2))$ has the positive orientation if det A > 0 and the negative orientation if det A < 0.

    b. More generally, let (v, w) be any basis of $\mathbb{R}^2$. Show that (v, w) and (L(v), L(w)) have the same orientation if det A > 0 and opposite orientations if det A < 0.

4.20. Let D^* and D be bounded subsets of $\mathbb{R}^2$, and let $T : D^* \to D$ be a smooth function that maps D^* onto D and that is one-to-one, except possibly on the boundary of D^*. As in the change of variables theorem, we think of T as a function from the uv-plane to the xy-plane. Moreover, assume that det DT (u, v) has the same sign at all points (u, v) of D^*, except possibly on the boundary of D^*, where det DT $(u, v) =$ 0 is allowed.

Let D^* be given an orientation, and let D have the orientation imposed on it by T . That is, let T (D^*) denote the set D with the same orientation, positive or negative, as D^* if det DT > 0 and the opposite orientation from D^* if det DT < 0. In the first case, we say that T is orientation-preserving and, in the second, that it is orientation-reversing. In other words, we categorize how T acts on orientation based on the behavior of its first-order approximation. Let $\eta = f\, dx \wedge dy$ be a 2-form on D. Use the change of variables theorem to prove that:

$$\int_{T(D^*)} \eta = \int_{D^*} T^*(\eta).$$

This equation has appeared in various guises throughout the book. You may assume in your proof that D^* has the positive orientation. The arguments are similar if it has the negative orientation. (Hint: See Exercise 4.18.)

The remaining exercises call for speculation rather than proofs. You are asked to propose reasonable solutions consistent with patterns that have come before . It is considered a bonus if what you say is true. The correct answers are known and are covered in more advanced courses that study manifolds, for instance, advanced real analysis or perhaps differential geometry or differential topology.

4.21. A surface S in $\mathbb{R}^4$ is a two-dimensional subset, something that can be parametrized by a function $\sigma : D \to \mathbb{R}^4$, where D is a subset of $\mathbb{R}^2$. Let us call the parameters s and t so that $\sigma(s, t) = (x_1(s, t), x_2(s, t), x_3(s, t), x_4(s, t))$. Assume that it is known how to define an orientation of S and what it means for a parametrization σ to be orientation-preserving.

    a. Write down the general form of a differential form η in the variables x_1, x_2, x_3, x_4 that could be integrated over S. (Hint: Six terms.)

b. Let $\sigma : D \to \mathbb{R}^4$ be an orientation-preserving parametrization of S. The integral $\int_S \eta$ should be defined via the parametrization as a double integral $\iint_D f(s,t)\, ds\, dt$ over the parameter domain D. Propose a formula for the function f.

4.22. Continuing with the preceding problem, assume that it is also known how to orient the boundary ∂S of an oriented surface S in $\mathbb{R}^4$ in a reasonable way. Propose an expression μ that makes the following statement true:

$$\int_{\partial S} F_1\, dx_1 + F_2\, dx_2 + F_3\, dx_3 + F_4\, dx_4 = \int_S \mu.$$

4.23. Let W be a bounded four-dimensional region in $\mathbb{R}^4$ whose boundary ∂W consists of a finite number of piecewise smooth three-dimensional subsets. Assuming that it is possible to define an orientation on ∂W in a reasonable way, propose a 3-form ζ on $\mathbb{R}^4$ such that:

$$\mathrm{Vol}\,(W) = \int_{\partial W} \zeta,$$

where the volume on the left refers to four-dimensional volume. More than one answer for ζ is possible. Try to find one with as much symmetry in the variables x_1, x_2, x_3, x_4 as you can.

ANSWERS TO SELECTED EXERCISES

Section 1.1 Vector arithmetic

1.1. $x + y = (5, -3, 9)$, $2x = (2, 4, 6)$, $2x - 3y = (-10, 19, -12)$

1.3. $(-5, 3, -9)$

1.5. (b) $\frac{1}{3}x + \frac{2}{3}y$

(c) $(-1, 1, 2)$

Section 1.2 Linear transformation

2.2. $T(x + y) = T(x_1 + x_2, y_1 + y_2) = (x_1 + y_1, x_2 + y_2, 0) = (x_1, x_2, 0) + (y_1, y_2, 0) = T(x) + T(y)$ $T(cx) = T(cx_1, cx_2) = (cx_1, cx_2, 0) = c(x_1, x_2, 0) = c\,T(x)$

Section 1.3 The matrix of a linear transformation

3.1. (a) $\begin{bmatrix} 1 & 3 \\ 2 & 4 \end{bmatrix}$

(b) $(23, 34)$

(c) $(x_1 + 3x_2, 2x_1 + 4x_2)$

3.3. $\begin{bmatrix} 0 & -1 \\ 1 & 0 \end{bmatrix}$

3.5. $\begin{bmatrix} 1 & 0 \\ 0 & 1 \end{bmatrix}$

3.9. (a) $T(e_1) = (1, 2)$, $T(e_2) = (2, 4)$

(b) all scalar multiples of $(1, 2)$

3.11. $\begin{bmatrix} 1 & 0 \\ 0 & 1 \\ 0 & 0 \end{bmatrix}$

3.13. $\begin{bmatrix} 0 & 1 & 0 \\ 1 & 0 & 0 \\ 0 & 0 & -1 \end{bmatrix}$

Section 1.4 Matrix multiplication

4.1. $AB = \begin{bmatrix} 3 & 1 \\ 1 & 7 \end{bmatrix}$, $BA = \begin{bmatrix} 6 & 2 \\ 2 & 4 \end{bmatrix}$

4.3. $AB = \begin{bmatrix} 1 & 0 & 0 \\ 0 & 1 & 0 \\ 0 & 0 & 1 \end{bmatrix}$, $BA = \begin{bmatrix} 1 & 0 & 0 \\ 0 & 1 & 0 \\ 0 & 0 & 1 \end{bmatrix}$

4.5. $AB = \begin{bmatrix} 21 & 0 \\ 0 & 14 \end{bmatrix}$, $BA = \begin{bmatrix} 13 & 4 & -5 \\ 4 & 5 & 6 \\ -5 & 6 & 17 \end{bmatrix}$

Section 1.5 The geometry of the dot product

5.1. (a) -3

(b) $\|x\| = \sqrt{6}$, $\|y\| = 3\sqrt{2}$

(c) $\arccos(-\frac{1}{2\sqrt{3}})$

5.3. $\pm \frac{1}{\sqrt{5}}(-2, 1)$

Section 1.6 Determinants

6.1. 23

6.3. 0

6.5. 12

6.9. (a) $T(x_1, x_2, x_3) = -3x_1 + 6x_2 - 3x_3$

(b) $T(x + y) = T(x_1 + y_1, x_2 + y_2, x_3 + y_3) = -3(x_1 + y_1) + 6(x_2 + y_2) - 3(x_3 + y_3) = -3x_1 + 6x_2 - 3x_3 - 3y_1 + 6y_2 - 3y_3 = T(x) + T(y)$

$T(cx) = T(cx_1, cx_2, cx_3) = -3(cx_1) + 6(cx_2) - 3(cx_3) = c(-3x_1 + 6x_2 - 3x_3) = c\,T(x)$

(c) $[-3\ 6\ -3]$

(d) $(-3, 6, -3) \cdot (1, 2, 3) = -3 + 12 - 9 = 0$

$(-3, 6, -3) \cdot (4, 5, 6) = -12 + 30 - 18 = 0$

Section 2.1 Parametrizations

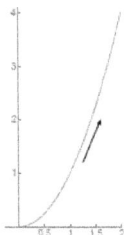

1.1.

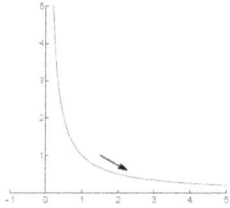

1.3.

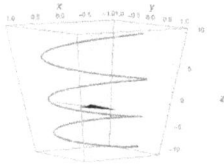

1.5.

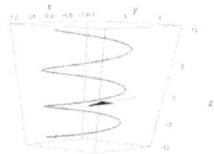

1.7.

1.9. $\alpha(x) = (x, x^2), -1 \leq x \leq 1$

1.11. $\alpha(t) = (a \cos t, -a \sin t)$, $0 \leq t \leq 2\pi$

1.13. $\alpha(t) = (1 + 4t, 2 + 5t, 3 + 6t)$

1.15. $\alpha(t) = (1 - t, 0, t)$

1.17. (a) $\dfrac{\mathbf{v} \cdot (\mathbf{p} - \mathbf{a})}{\mathbf{v} \cdot \mathbf{v}}$

(b) $\mathbf{a} + \dfrac{\mathbf{v} \cdot (\mathbf{p} - \mathbf{a})}{\mathbf{v} \cdot \mathbf{v}} \mathbf{v}$

1.19. (b) No intersection

(c) Intersection at $(2, 1, 3)$

Section 2.2 Velocity, acceleration, speed, arclength

2.1. (a) $v(t) = (3t^2, 6t, 6)$

$v(t) = 3(t^2 + 2)$

$a(t) = (6t, 6, 0)$

(b) 88

2.3. (a) $v(t) = (2 \sin 2t, 2 \cos 2t, 6t^{1/2})$

$v(t) = 2\sqrt{1 + 9t}$

(b) $\frac{4}{27}\left(19\sqrt{19} - 10\sqrt{10}\right)$

2.5. $\alpha(t) = (1 + \frac{1}{\sqrt{14}}t, 1 + \frac{2}{\sqrt{14}}t, 1 + \frac{3}{\sqrt{14}}t)$

Section 2.3 Integrals with respect to arclength

3.1. $\frac{150}{7}$

3.3. (a) $5\sqrt{2}$

(b) $\sqrt{2}$

Section 2.4 The geometry of curves: tangent and normal vectors

4.1. $\mathbf{T}(t) = \frac{1}{5}(4\cos 4t, -4\sin 4t, 3)$

$\mathbf{N}(t) = (-\sin 4t, -\cos 4t, 0)$

$\mathbf{T}(t) \cdot \mathbf{N}(t) = \frac{1}{5}(-4\cos 4t \sin 4t + 4\sin 4t \cos 4t + 0) = 0$

Section 2.5 The cross product

$$$$

5.1. (a) (2, -4, 2)

(b) $\mathbf{v} \cdot (\mathbf{v} \times \mathbf{w}) = 2 - 4 + 2 = 0$, $\mathbf{w} \cdot (\mathbf{v} \times \mathbf{w}) = 2 - 12 + 10 = 0$

5.3. (a) (-10, -11, -3)

(b) $\mathbf{v} \cdot (\mathbf{v} \times \mathbf{w}) = -10 + 22 - 12 = 0$, $\mathbf{w} \cdot (\mathbf{v} \times \mathbf{w}) = 20 - 11 - 9 = 0$

5.5. (-4, 5, -1)

5.7. (a) (-3, -9, -3)

(b) $3\sqrt{11}$

(c) $\mathbf{u} \cdot (\mathbf{v} \times \mathbf{w}) = -12$, $\mathbf{v} \cdot (\mathbf{u} \times \mathbf{w}) = 12$, $(\mathbf{u} \times \mathbf{v}) \cdot \mathbf{w} = -12$

(d) 12

5.11. (b) $\frac{1}{3}\sqrt{41}$

5.15. (a) $\mathbf{u} \cdot (\mathbf{u} \times \mathbf{v} \times \mathbf{w}) = \det \mathbf{u}\,\mathbf{u}\,\mathbf{v}\,\mathbf{w} = 0$ (two equal rows) and so on

(b) Left-handed

(c) (-4, -4, -4, -4)

Section 2.6 The geometry of space curves: Frenet vectors

6.1. $(\frac{3}{5}\cos 4t, -\frac{3}{5}\sin 4t, -\frac{4}{5})$

Section 2.7 Curvature and torsion

7.1. $\alpha(t) = (2 \cos t, 2 \sin t, 0)$ is one possibility.

Section 2.9 The classification of space curves

9.1. (a) $\sqrt{2}$

(b) $\frac{1}{\sqrt{2}}(-\sin t, -\cos t, 1)$

(c) $(-\cos t, \sin t, 0)$

(d) $-\frac{1}{\sqrt{2}}(\sin t, \cos t, 1)$

(e) $\frac{1}{2}$

(f) $-\frac{1}{2}$

(g) $a = 1$, $b = -1$. Both α and β trace out the same helix but in opposite directions. The rotation $F(x, y, z) = (x, -y, -z)$ about the x-axis by π transforms one para-metrization into the other.

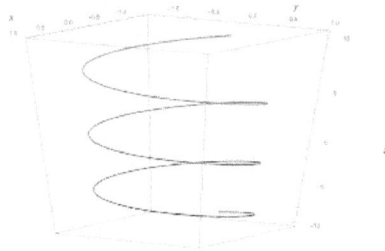

9.3. (a) $2\sqrt{2}$

(b) $\frac{1}{2\sqrt{2}}(1 + \sqrt{3}\cos t, \sqrt{3} - \cos t, -2\sin t)$

(c) $\frac{1}{2}(-\sqrt{3}\sin t, \sin t, -2\cos t)$

(d) $\frac{1}{2\sqrt{2}}(1 - \sqrt{3}\cos t, \sqrt{3} + \cos t, 2\sin t)$

(e) $\frac{1}{4}$

(f) $\frac{1}{4}$

(g) It's congruent to the helix parametrized by $\beta(t) = (2 \cos t, 2 \sin t, 2t)$.

9.7. $\frac{1}{7}\sqrt{\frac{19}{14}}$

9.9. (a) $\frac{\sqrt{2}}{2(1+\sin^2 t)^{3/2}}$

(b) $\tau(t) = 0$

$$\mathbf{T}(t) = -\frac{\sqrt{2}}{2}\left(1, \sin 2t, \cos 2t\right)$$
$$\mathbf{N}(t) = (0, -\cos 2t, \sin 2t)$$
$$\kappa(t) = \sqrt{2}$$

9.13. (a) $\tau(t) = \sqrt{2}$

(b)
$$\alpha(t) = \left(-\frac{\sqrt{2}}{2}t, -\frac{\sqrt{2}}{4} + \frac{\sqrt{2}}{4}\cos 2t, -\frac{\sqrt{2}}{4}\sin 2t\right)$$

Section 3.1 Graphs and level sets

1.1. (b)

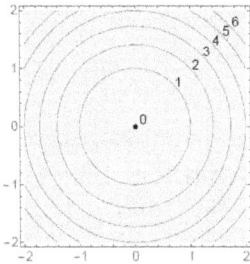

(c)

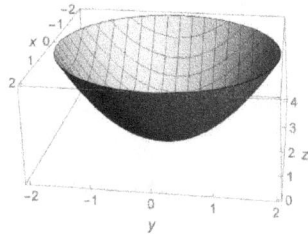

1.3. (b)

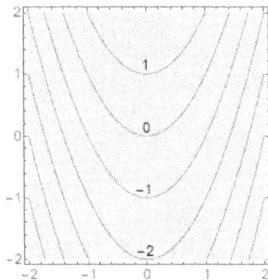

(c)

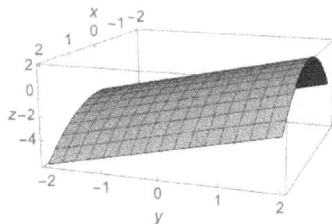

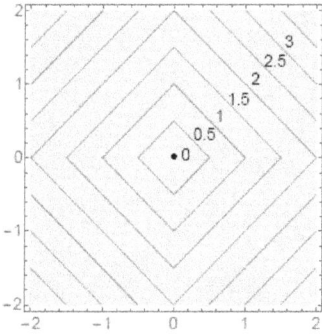

1.5. (b)

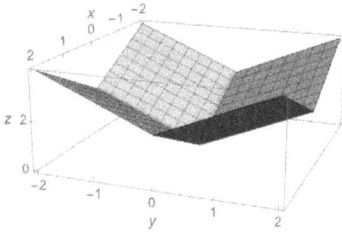

(c)

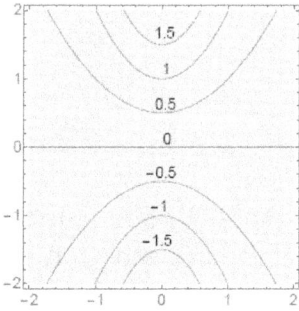

1.7. (b)

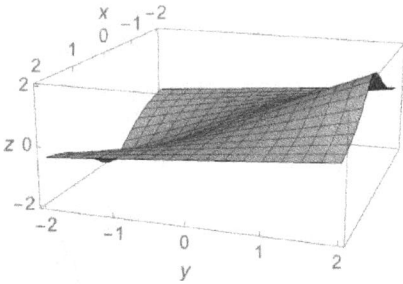

(c)

Section 3.2 More surfaces in $\mathbb{R}^3$

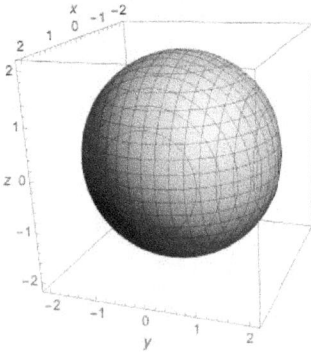

2.1.

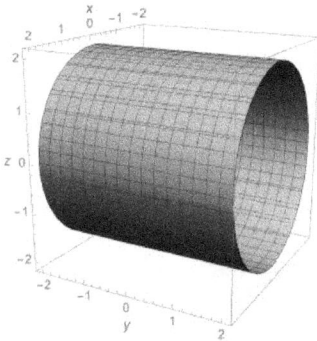

2.3.

2.5. The level set corresponding to $c = 0$ consists of the single point $(0, 0, 0)$. Those for $c = 1$ and $c = 2$ are spheres centered at the origin of radius 1 and $\sqrt{2}$, respectively.

2.7. $c = -1, 0, 1$ (left, center, right)

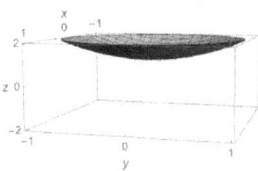 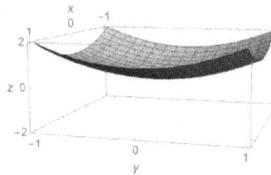 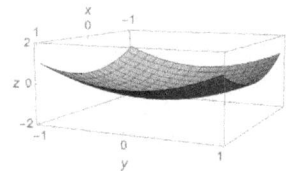

2.9. $c = -1, 0, 1$ (left, center, right)

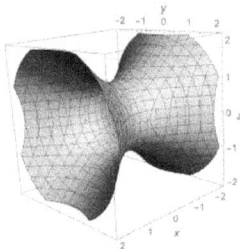 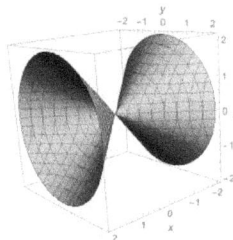 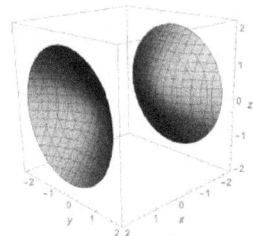

Section 3.3 The equation of a plane

3.1. $4x + 5y + 6z = 32$

3.3. For example, $p = (9, 0, 0)$, $n = (1, 3, 5)$

3.5. Perpendicular

3.7. $6x + 3y + 2z = 6$

3.9. $2x + z = 3$

3.11. $\left(\frac{8}{3}, \frac{1}{3}, \frac{1}{3}\right)$

3.13. (b) It's the triangular region with vertices $(1, 0, 0)$, $(0, 1, 0)$, and $(0, 0, 1)$.

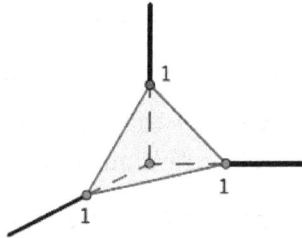

3.15. (c) $\dfrac{4}{\sqrt{3}}$

Section 3.4 Open sets

4.1. Let $a = (c, d)$ be a point of U, and let $r = \min\{c, 1 - c, d, 1 - d\}$. Then $B(a, r) \subset U$, so U is an open set.

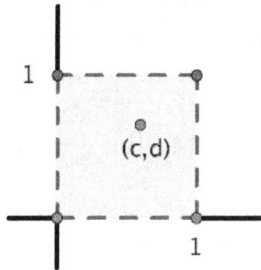

4.3. The point $a = (1, 0)$ is in U, but no open ball centered at a stays within U. Thus U is not an open set.

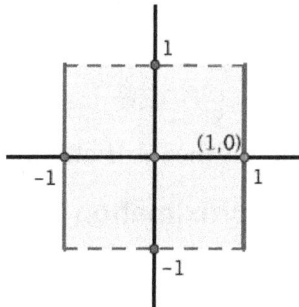

Section 3.5 Continuity

5.1. (a) If $(x, y) \neq (0, 0)$, then, in polar coordinates,

$$f(x,y) = f(r\cos\theta, r\sin\theta) = \frac{(r\cos\theta)^3(r\sin\theta)}{r^2} = r^2\cos^3\theta\sin\theta,$$

so $|f(x, y)| = |\mathbb{R}^2\cos 3\,\theta\sin\theta| \leq \mathbb{R}^2 = \|(x, y)\|2$. The inequality also holds if $(x, y) = (0, 0)$ (both sides equal 0). Thus, if $\|(x, y)\| < a$, then $|f(x, y)| \leq \|(x, y)\|^2 < a^2$.

(b) Let $\epsilon > 0$ be given, and let $\delta = \sqrt{\epsilon}$. By part (a), if $\|(x, y) - (0, 0)\| = \|(x, y)\| < \delta$, then $|f(x, y) - f(0, 0)| = |f(x, y)| < \delta^2 = \epsilon$. Therefore f is continuous at $(0, 0)$.

Section 3.6 Some properties of continuous functions

6.1. $f(x, y) = (1, 2) \cdot (x, y) = x + 2y$. We know that the projections x and y are continuous, so, as an algebraic combination of continuous functions, f is continuous as well.

Section 3.7 The Cauchy-Schwarz and triangle inequalities

7.1. (a) By the triangle inequality, $\|v\| = \|(v-w)+w\| \leq \|v-w\|+\|w\|$, so $\|v\|-\|w\| \leq \|v-w\|$.

(b) Reversing the roles of v and w gives $\|w\|-\|v\| \leq 1\|w-v\| = 1\|v-w\|$, or $-(\|v\|-\|w\|) \leq \|v - w\|$. Hence $\pm(\|v\| - \|w\|) \leq \|v - w\|$, so $\|v\| - \|w\| \leq \|v - w\|$.

7.4. First, note that $|cf(x) - cf(a)| = |c|\,|f(x) - f(a)| \leq (|c| + 1)\,|f(x) - f(a)|$. Now, let $\epsilon > 0$ be given. Since f is continuous at a, there exists a $\delta > 0$ such that, if $\|x - a\| < \delta$, then $|f(x) - f(a)| < \frac{\epsilon}{|c|+1}$. Then, for this choice of δ, if $\|x - a\| < \delta$, $|cf(x) - cf(a)| \leq (|c| + 1)\,|f(x) - f(a)| < (|c| + 1)\cdot \frac{\epsilon}{|c|+1} = \epsilon$. Hence cf is continuous at a.

Section 3.8 Limits

8.1. For sums: Let $L = \lim_{x\to a} f(x)$ and $M = \lim_{x\to a} g(x)$, and consider the functions:

$$\tilde{f}(x) = \begin{cases} f(x) & \text{if } x \neq a, \\ L & \text{if } x = a \end{cases} \quad \text{and} \quad \tilde{g}(x) = \begin{cases} g(x) & \text{if } x \neq a, \\ M & \text{if } x = a. \end{cases}$$

By definition of limit, $\tilde{f}$ and $\tilde{g}$ are continuous at a, hence so is their sum. In other words, the function given by

$$\tilde{f}(x) + \tilde{g}(x) = \begin{cases} f(x) + g(x) & \text{if } x \neq a, \\ L + M & \text{if } x = a \end{cases}$$

is continuous at a. By definition, this means that $\lim_{x\to a}(f(x) + g(x)) = L + M$.

Section 4.1 The first-order approximation

1.1. (a) $\frac{\partial f}{\partial x}(a) = 4,\ \frac{\partial f}{\partial y}(a) = -2$

(b) $Df(a) = 4\ -2$, $\nabla f(a) = (4, -2)$

(c) $\ell(x, y) = -3 + 4x - 2y$

(d) $f(2.01, 1.01) = 3.02$, $\ell(2.01, 1.01) = 3.02$

1.3. (a) $\dfrac{\partial f}{\partial x} = 3x^2 - 4xy + 3y^2$

$\dfrac{\partial f}{\partial y} = -2x^2 + 6xy - 12y^2$

(b) $[3x^2 - 4xy + 3y^2 \ -2x^2 + 6xy - 12y^2]$

1.5. (a), $\dfrac{\partial f}{\partial x} = (1 + xy)e^{xy}$

$\dfrac{\partial f}{\partial y} = x^2 e^{xy}$

(b) $(1 + xy)e^{xy} \ x^2 e^{xy}]$

1.7. (a) $\dfrac{\partial f}{\partial x} = -\dfrac{2x}{(x^2+y^2)^2}$

$\dfrac{\partial f}{\partial y} = -\dfrac{2y}{(x^2+y^2)^2}$

(b) $\left[-\dfrac{2x}{(x^2+y^2)^2} \quad -\dfrac{2y}{(x^2+y^2)^2} \right]$

1.9. (a) $\dfrac{\partial f}{\partial x} = 1, \ \dfrac{\partial f}{\partial y} = 2, \ \dfrac{\partial f}{\partial z} = 3$

(b) $(1, 2, 3)$

1.11. (a) $\dfrac{\partial f}{\partial x} = \big(\cos(x + 2y) - 2x \sin(x + 2y)\big)e^{-x^2-y^2-z^2}$

$\dfrac{\partial f}{\partial y} = 2\big(\cos(x + 2y) - y \sin(x + 2y)\big)e^{-x^2-y^2-z^2}$

$\dfrac{\partial f}{\partial z} = -2z \sin(x + 2y)e^{-x^2-y^2-z^2}$

(b) $((\cos(x + 2y) \ 2x \sin(x + 2y) \ e^{-x2-y2-z2}$, $2 (\cos(x + 2y) -y \sin(x + 2y))\ e^{-x2-y2-z2}$, $-2z \sin(x + 2y)e^{-x2-y2-z2}$)

1.13. (a) $\dfrac{\partial f}{\partial x} = \dfrac{2x}{x^2+y^2}$

$\dfrac{\partial f}{\partial y} = \dfrac{2y}{x^2+y^2}$

$\dfrac{\partial f}{\partial z} = 0$

(b) $\left(\dfrac{2x}{x^2+y^2}, \dfrac{2y}{x^2+y^2}, 0 \right)$

1.17. (a) $\dfrac{\partial f}{\partial x} = \dfrac{2x^3}{\sqrt{x^4+y^4}}$

$\dfrac{\partial f}{\partial y} = \dfrac{2y^3}{\sqrt{x^4+y^4}}$

(b) $\dfrac{\partial f}{\partial x}(0,0) = 0, \ \dfrac{\partial f}{\partial y}(0,0) = 0$

Section 4.2 Conditions for differentiability

2.1. $\lim_{(x,y)\to(0,0)} f(x, y) = \lim_{(x,y)\to(0,0)} (x^2 - y^2) = 0$, while $f(0, 0) = \pi$, so f is not continuous at $(0, 0)$, hence not differentiable there either.

Section 4.4 The C¹ test

4.1. For $(x, y) \ne (0, 0)$, the partial derivatives are given by $\frac{\partial f}{\partial x} = \frac{x}{\sqrt{x^2+y^2}}$ and $\frac{\partial f}{\partial y} = \frac{y}{\sqrt{x^2+y^2}}$. As algebraic combinations and compositions of continuous functions, both of these are continuous at all points other than the origin. Thus, by the C¹ test, f is differentiable at all points other than the origin.

4.2. Differentiable at all (x, y) in $\mathbb{R}^2$

4.4. Differentiable at all (x, y) where $x \ne 0$ and $y \ne 0$

4.6. Differentiable at all (w, x, y, z) in R4

Section 4.5 The Little Chain Rule

5.1. (b) $(f \circ \alpha)(t) = f(t^2, t^3) = t^4 + t^6$, so $(f \circ \alpha)'(t) = 4t^3 + 6t^5$ and $(f \circ \alpha)'(1) = 10$.

5.3. $\frac{\partial f}{\partial x}\frac{dx}{dt} + \frac{\partial f}{\partial y}\frac{dy}{dt} + \frac{\partial f}{\partial z}\frac{dz}{dt}$ where $\frac{\partial f}{\partial x}, \frac{\partial f}{\partial y},$ and $\frac{\partial f}{\partial z}$ are evaluated at $\alpha(t)$

Section 4.6 Directional derivatives

6.1. $-\frac{5}{\sqrt{17}}$

6.3. $\frac{2}{\sqrt{6}}$

6.5. $\frac{6}{\sqrt{17}}(4, 1)$

6.7. $r = \frac{5}{\sqrt{2}}, s = \frac{5}{2\sqrt{2}}$

6.9. (a) $\frac{1}{\sqrt{38}}(-1, 1, -6)$

(b) $\frac{1}{\sqrt{38}}(1, -1, 6)$

(c) $800\sqrt{38}\, e^{-42}\, {}^\circ C/\sec$

6.11. (a) Direction: $u = \frac{1}{\sqrt{13}}(3, 2)$

Value: $\sqrt{13}$

(b) Direction: $u = \frac{1}{\sqrt{13}}(-3, -2)$

Value: $-\sqrt{13}$

(c) $y = \frac{1}{3}x^2 - \frac{1}{3}$

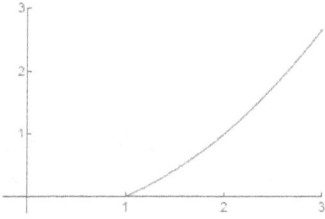

(d) $y = -\frac{3}{2}\ln x$

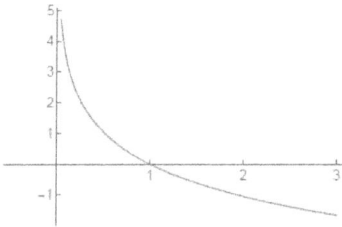

Section 4.7 ∇f as normal vector

7.1. 2x - 4y - z = -3
7.3. 4y + 3z = 7
7.5. α(t) = (1 + 14t, 2 + 5t, 3 - 8t)

Section 4.8 Higher-order partial derivatives

8.1.
$$\frac{\partial^2 f}{\partial x^2} = 12x^2 - 12xy + 6y^2$$
$$\frac{\partial^2 f}{\partial y\,\partial x} = -6x^2 + 12xy - 12y^2$$
$$\frac{\partial^2 f}{\partial x\,\partial y} = -6x^2 + 12xy - 12y^2$$
$$\frac{\partial^2 f}{\partial y^2} = 6x^2 - 24xy + 60y^2$$

8.3.
$$\frac{\partial^2 f}{\partial x^2} = (4x^2 - 2)e^{-x^2-y^2}$$
$$\frac{\partial^2 f}{\partial y\,\partial x} = 4xye^{-x^2-y^2}$$
$$\frac{\partial^2 f}{\partial x\,\partial y} = 4xye^{-x^2-y^2}$$
$$\frac{\partial^2 f}{\partial y^2} = (4y^2 - 2)e^{-x^2-y^2}$$

8.5. $i = 4, j = 3, \frac{\partial^7 f}{\partial x^4 \partial y^3}(0,0) = 144$

Section 4.10 Max/min: critical points

10.1. $(1, 2)$ saddle point

10.3. $(0, 0)$ local minimum

10.5. $(0, 0)$ saddle point, $(3, 9)$ local minimum

10.7. $k \neq 0$: $(0, 0)$ saddle point; $\left(\frac{k}{3}, \frac{k}{3}\right)$ local minimum if k > 0, local maximum if k < 0 k = 0: $(0, 0)$ neither local maximum nor minimum (and, in addition, degenerate)

10.11. $y = x + \frac{1}{3}$

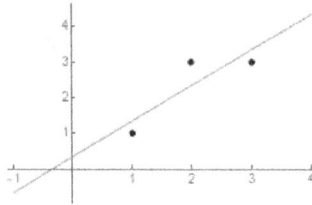

Section 4.11 Classifying nondegenerate critical points

11.1. $1 - \frac{1}{2}(x+y)^2$

11.3. (a) $h^3 \frac{\partial^3 f}{\partial x^3} + 3h^2 k \frac{\partial^3 f}{\partial x^2 \partial y} + 3hk^2 \frac{\partial^3 f}{\partial x \partial y^2} + k^3 \frac{\partial^3 f}{\partial y^3}$

(b) $f(\mathbf{a}) + \frac{\partial f}{\partial x}(\mathbf{a}) h + \frac{\partial f}{\partial y}(\mathbf{a}) k + \frac{1}{2}\left(\frac{\partial^2 f}{\partial x^2}(\mathbf{a}) h^2 + 2\frac{\partial^2 f}{\partial x \partial y}(\mathbf{a}) hk + \frac{\partial^2 f}{\partial y^2}(\mathbf{a}) k^2\right) + \frac{1}{6}\left(\frac{\partial^3 f}{\partial x^3}(\mathbf{a}) h^3 + 3\frac{\partial^3 f}{\partial x^2 \partial y}(\mathbf{a}) h^2 k + 3\frac{\partial^3 f}{\partial x \partial y^2}(\mathbf{a}) hk^2 + \frac{\partial^3 f}{\partial y^3}(\mathbf{a}) k^3\right)$

Section 4.12 Max/min: Lagrange multipliers

12.1. (a) x can be chosen large and positive and y large and negative so that x + 2y = 1. Thus $f(x, y)$ = xy can be made as large and negative as you like.

(b) Maximum value $\frac{1}{8}$ at $\left(\frac{1}{2}, \frac{1}{4}\right)$

12.3. Maximum at $\frac{1}{\sqrt{29}}(2, -3, 4)$ and minimum at $\frac{1}{\sqrt{29}}(-2, 3, -4)$

12.5. $x = \frac{2}{3}, y = 1, z = \frac{6}{5}$

12.9. x = 400, y = 150, z = 100

Section 5.1 Volume and iterated integrals

1.1. $\frac{49}{6}$

1.3. $\frac{7}{6}$

1.5. $\frac{26}{3} \ln 2$

1.7. $e^2 - 2e + 1$

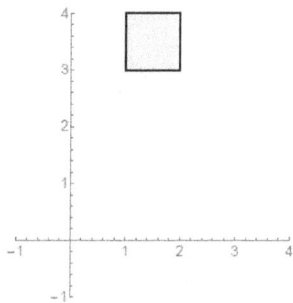

1.9. (a)

(b)
$$\int_3^4 \left(\int_1^2 f(x, y)\, dx \right) dy$$

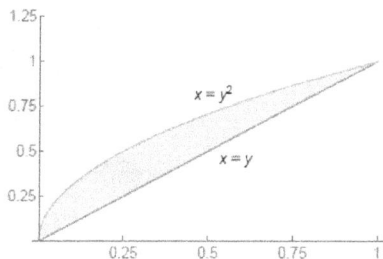

1.11. (a)

(b),
$$\int_0^1 \left(\int_x^{\sqrt{x}} f(x, y)\, dy \right) dx$$

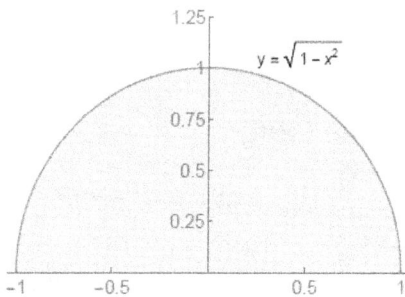

1.13. (a)

(b)
$$\int_0^1 \left(\int_{-\sqrt{1-y^2}}^{\sqrt{1-y^2}} f(x, y)\, dx \right) dy$$

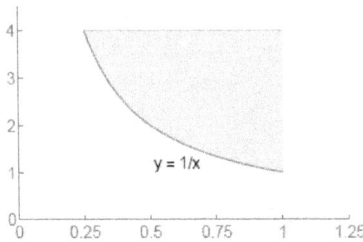

1.15. (a)

$$\int_1^4 \left(\int_{\frac{1}{y}}^1 y e^{xy}\, dx \right) dy$$

(b)

(c) $e^4 - 4e$

1.17. $\frac{2}{3}$

1.19. $\frac{2}{3}$

1.21. $\frac{1}{3}$

Section 5.2 The double integral

2.1. (a) $S = 0$ or $\frac{1}{9}$

(b) $S = 0, \frac{1}{16}, \frac{1}{8}, \frac{3}{16},$ or $\frac{1}{4}$

(c) If n is odd, $S = 0$ or $\frac{1}{n^2}$.. If n is even, $S = 0, \frac{1}{n^2}, \frac{2}{n^2}, \frac{3}{n^2},$ or $\frac{4}{n^2}$.

(e) Let $\epsilon > 0$ be given, and let $\delta = \frac{\sqrt{\epsilon}}{2}$. If R is subdivided into subrectangles of dimensions Δx_i by Δy_j, where $\Delta x_i < \delta$ and $\Delta y_j < \delta$ for all i, j, then, by part (d), all Riemann sums based on the subdivision satisfy $|\sum_{i,j} f(\mathbf{p}_{ij}) \Delta x_i \Delta y_j - 0| < 4\delta^2 = \epsilon$. Therefore f is integrable over R, and $\iint_R f(x, y)\, dA = 0$.

Section 5.3 Interpretations of the double integral

3.1. 18 hundred birds

3.3. (a) 80 miles per hour

(b) All points of R that lie on the circle $x^2 + y^2 = \frac{8}{3}$, a quarter-circular arc

3.5. $\left(\frac{2}{3}, \frac{5}{3} \right)$

3.7. $\sqrt{\frac{2}{3}}$

Section 5.4 Parametrization of surfaces

4.1. (a) $\sigma : D \; \mathbb{R}^3$, $\sigma(x, y) = (x, y, \sqrt{a^2 - x^2 - y^2})$, where D is the disk $x^2 + y^2$ a^2

(b) $\sigma : D \to \mathbb{R}^3$, $\sigma(r, \theta) = (r\cos\theta, r\sin\theta, \sqrt{a^2 - r^2})$, where $D = [0, a] \times [0, 2\pi]$

(c) $\sigma : D \to \mathbb{R}^3$, $\sigma(\phi, \theta) = (a\sin\phi\cos\theta, a\sin\phi\sin\theta, a\cos\phi)$, where $D = [0, \frac{\pi}{2}] \times [0, 2\pi]$

Section 5.5 Integrals with respect to surface area

5.1. (a) $\sigma : D \to \mathbb{R}^3$, $\sigma(x, y) = (x, y, 1 - x - y)$, where D is the triangular region in the xy-plane with vertices $(0, 0), (1, 0), (0, 1)$

(b) $\dfrac{\sqrt{3}}{2}$

(c) $\dfrac{7\sqrt{3}}{2}$

5.2. 36π

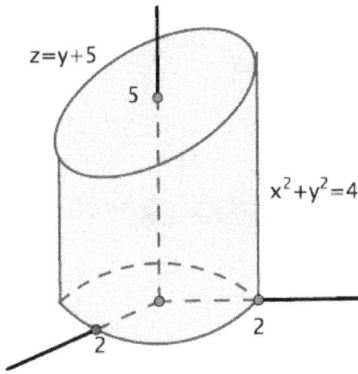

5.5. (a)

(b) $4(6 + \sqrt{2})\pi$

(c) $2(27 + 10\sqrt{2})\pi$

5.7. (a) 8

(b) 4

Section 5.6 Triple integrals and beyond

6.1. $\dfrac{1}{12}$

6.3. $\dfrac{7}{48}$

6.5. $\dfrac{64}{15}$

6.7. a^2

6.8. (a)

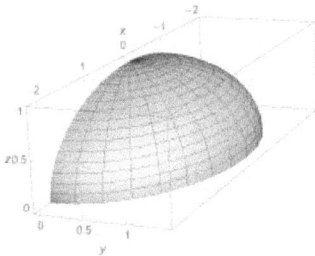

(b)
$$\int_0^{\sqrt{2}}\left(\int_0^{\sqrt{1-\frac{y^2}{2}}}\left(\int_{-2\sqrt{1-\frac{y^2}{2}-z^2}}^{2\sqrt{1-\frac{y^2}{2}-z^2}} f(x,y,z)\,dx\right)dz\right)dy$$

(c)
$$\int_{-2}^{2}\left(\int_0^{\sqrt{1-\frac{x^2}{4}}}\left(\int_0^{\sqrt{2-\frac{x^2}{2}-2z^2}} f(x,y,z)\,dy\right)dz\right)dx$$

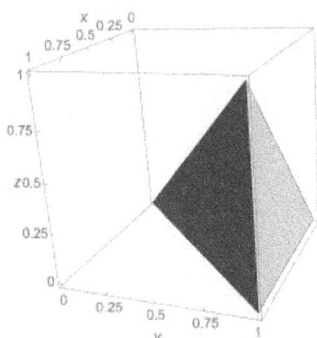

6.10. (a)

(b)
$$\int_0^1\left(\int_0^y\left(\int_z^y f(x,y,z)\,dx\right)dz\right)dy$$

(c)
$$\int_0^1\left(\int_0^x\left(\int_x^1 f(x,y,z)\,dy\right)dz\right)dx$$

Section 6.1 Continuity revisited

1.2. Given any $\epsilon > 0$, there exists a $\delta > 0$ such that $\|f(x) - L\| < \epsilon$ whenever $\|x - a\| < \delta$, except possibly when x = a.

Section 6.2 Differentiability revisited

2.1. (a) $\begin{bmatrix} 2x & -2y \\ 2y & 2x \end{bmatrix}$

(b) $\begin{bmatrix} 2 & -4 \\ 4 & 2 \end{bmatrix}$

2.3. $[1\ 2y\ 9z^2]$

2.5. $\begin{bmatrix} 1 & 0 \\ 0 & 1 \\ 0 & 0 \end{bmatrix}$

2.7. (a) $\begin{bmatrix} e^{x+y} & e^{x+y} \\ -e^{-x}\cos y & -e^{-x}\sin y \\ -e^{-x}\sin y & e^{-x}\cos y \end{bmatrix}$

(b) All entries of $Df(x, y)$ are continuous on $\mathbb{R}^2$, so, by the C^1 test, f is differentiable at every point of $\mathbb{R}^2$.

2.9. (a)
$$\begin{bmatrix} a & b & c \\ d & e & f \\ g & h & i \end{bmatrix} \quad (= A)$$

Section 6.3 The chain rule: a conceptual approach

3.1. (a)
$$Df(s,t) = \begin{bmatrix} 1 & -1 \\ 2e^{2s+3t} & 3e^{2s+3t} \\ \cos t & -s \sin t \end{bmatrix}, \quad Dg(x, y, z) = \begin{bmatrix} yz & xz & xy \\ 2x & 0 & 3z^2 \end{bmatrix}$$

(b)
$$\begin{bmatrix} 2 & 0 \\ 4 & -4 \end{bmatrix}$$

(c)
$$\begin{bmatrix} 0 & 0 & 0 \\ 0 & 0 & 0 \\ 0 & 0 & 0 \end{bmatrix}$$

3.3. (a)
$$\begin{bmatrix} 1 & 0 \\ 0 & 1 \end{bmatrix}$$

(b)
$$\begin{bmatrix} 1 & 0 \\ 0 & 1 \end{bmatrix}$$

Section 6.4 The chain rule: a computational approach

4.1.
$$\frac{\partial w}{\partial s} = \frac{\partial f}{\partial x} + 3\frac{\partial f}{\partial y} + 5\frac{\partial f}{\partial z}$$
$$\frac{\partial w}{\partial t} = 2\frac{\partial f}{\partial x} + 4\frac{\partial f}{\partial y} + 6\frac{\partial f}{\partial z}$$

4.3. (a)
$$\frac{\partial w}{\partial \rho} = \frac{\partial f}{\partial x} \sin \phi \cos \theta + \frac{\partial f}{\partial y} \sin \phi \sin \theta + \frac{\partial f}{\partial z} \cos \phi$$
$$\frac{\partial w}{\partial \phi} = \frac{\partial f}{\partial x}\rho \cos \phi \cos \theta + \frac{\partial f}{\partial y}\rho \cos \phi \sin \theta - \frac{\partial f}{\partial z}\rho \sin \phi$$
$$\frac{\partial w}{\partial \theta} = -\frac{\partial f}{\partial x}\rho \sin \phi \sin \theta + \frac{\partial f}{\partial y}\rho \sin \phi \cos \theta$$

(b) 0

4.5. $\frac{\partial f}{\partial x} = \frac{5}{2}, \frac{\partial f}{\partial y} = \frac{3}{4}$

4.9. $\frac{\partial z}{\partial x} = -\frac{1}{3}, \frac{\partial z}{\partial y} = -\frac{2}{3}$

4.11. (b) $\frac{\partial f}{\partial x}\frac{d^2 x}{dt^2} + \frac{\partial f}{\partial y}\frac{d^2 y}{dt^2} + \frac{\partial^2 f}{\partial x^2}\left(\frac{dx}{dt}\right)^2 + 2\frac{\partial^2 f}{\partial x \partial y}\frac{dx}{dt}\frac{dy}{dt} + \frac{\partial^2 f}{\partial y^2}\left(\frac{dy}{dt}\right)^2$

Section 7.1 Change of variables for double integrals

1.2. $\frac{\pi}{8}$

1.4. $(\sin 16 - \sin 4)\pi$

1.6. 2π

1.8. (b) $(1 - e^{-a^2})\pi$

Section 7.3 Examples: linear changes of variables, symmetry

3.1. (a) D is the disk of radius 2 centered at (3, 2).

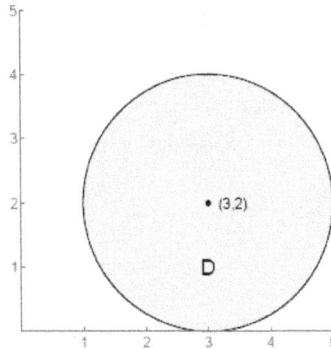

(b) D* is the disk $u^2 + v^2 \leq 4$ of radius 2 centered at the origin.

(c) $\iint_{D^*} (u + v + 5)\, du\, dv = 20\pi$

3.4. (a) 150

(b) 100

3.7. (a) $T(u, v) = \left(\sqrt{\frac{u}{v}}, \sqrt{uv} \right)$

(b) D* = $[1, 2] \times [1, 4]$

(c) $\frac{3}{2}e(e - 1)$

3.9. $\frac{8}{9}$

Section 7.4 Change of variables for n-fold integrals

4.1. $\frac{48}{5}\pi$

4.3. $\frac{\pi}{6}\left(1 + 3\ln\frac{3}{4}\right)$

4.5. $\left(0, 0, \frac{3}{8}a\right)$

4.7. $4(2 - \sqrt{2})$

4.9. $\frac{8\pi^2}{15}$

Section 8.2 Exercises for Chapter 8 (Vector fields)

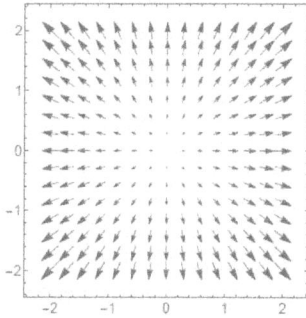

1.1.

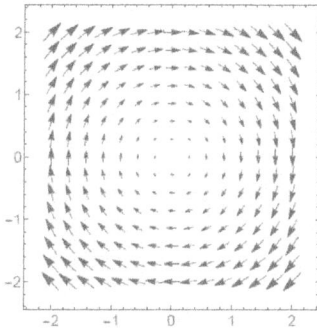

1.3.

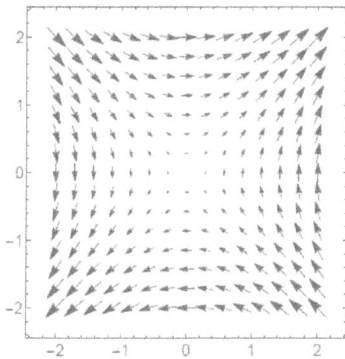

1.5.

1.7. (a) $\mathbf{F}(x, y) = \frac{1}{\sqrt{4x^2+1}}(-2x, 1)$ or its negative

(b) $\mathbf{F}(x, y) = \frac{1}{\sqrt{4x^2+1}}(1, 2x)$ or its negative

1.9. (a)

$\alpha'(t) = (-2a\sin(\sqrt{2}t), \sqrt{2}a\cos(\sqrt{2}t))$ and $\mathbf{F}(\alpha(t)) = \mathbf{F}(\sqrt{2}a\cos(\sqrt{2}t), a\sin(\sqrt{2}t)) = (-2a\sin(\sqrt{2}t), \sqrt{2}a\cos(\sqrt{2}t))$, so $\alpha'(t) = \mathbf{F}(\alpha(t))$.

$\frac{x^2}{2} + y^2 = \frac{2a^2\cos^2(\sqrt{2}t)}{2} + a^2\sin^2(\sqrt{2}t) = a^2\cos^2(\sqrt{2}t) + a^2\sin^2(\sqrt{2}t) = a^2$

(b)

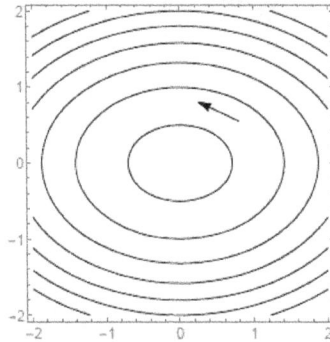

1.12. (a) $\alpha(t) = (x0e^t, y0e^t) = et(x_0, y_0)$

(b) The integral curves are rays emanating out from the origin.

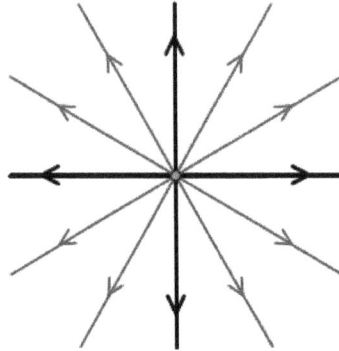

Section 9.1 Definitions and examples

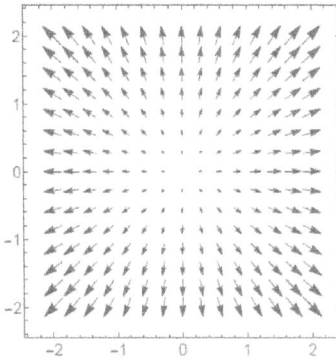

1.1. (a)

(b) For instance , any line segment radiating directly away from the origin. If $\alpha(t) =$ (t, t), $0 \le t \le 1$, then $\int_C x\,dx + y\,dy = 1$.

(c) For instance, any circle centered at the origin. If $\alpha(t) = $ (cos t, sin t), $0 \le t \le 2\pi$, then $\int_C x\,dx + y\,dy = 0$.

1.3. $\frac{1}{2}(\pi + \sin^2(\pi^2))$

1.4. $\frac{179}{60}$

1.6. (a) F(x, y) = (x + y, y - x)

(b) 1

(c) 0

(d) 2

1.8. $\frac{69}{2}$

1.10. (a)

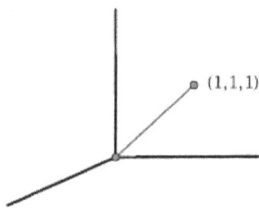

(b) e⁴ + 2

Section 9.3 Conservative fields

3.1. (a) It's conservative with potential function f (x, y) = xy.

(b) 2

3.3. (a) (-6x²y - 8x³z, 6y⁴z + 6xy², 12x²z² - 12y³z²)

(b) Not conservative

3.5. Not conservative

3.7. Conservative with potential function $f(x, y, z) = \frac{1}{2}x^2 + \frac{1}{2}y^2 + \frac{1}{2}z^2 + xyz$

3.9. (a) q = 1 - 2p

(b) All positive p

3.13. 3

3.17. The vector field F = ∇(fg) is conservative with potential function fg, so $\int_C \nabla(fg) \cdot ds = 0$ for all piecewise smooth oriented closed curves C. By Exercise 1.19 in Chapter 4, ∇(fg) =f ∇g + g ∇f , so $\int_C (f \nabla g + g \nabla f) \cdot ds = 0$. Hence $\int_C (f \nabla g) \cdot ds = - \int_C (g \nabla f) \cdot ds$.

Section 9.4 Green's theorem

4.1. 2

4.3. $\frac{9\pi^3}{2}$

4.6. 4

4.8. $\frac{1}{2}$

4.10. 12

Section 9.5 The vector field W

5.1. (a) 27

Section 9.6 The converse of the mixed partials theorem

6.1. (a) C traverses the unit circle once clockwise. Winding number = -1.

(b) Cn goes around the unit circle |n| times, counterclockwise if n > 0 and clockwise if n < 0. It stays stuck at the point (1, 0) if n = 0. Winding number = n.

6.2. Winding number = 0

Section 10.1 What the surface integral measures

1.1. (a) $2\pi a^3$

1.2. $\frac{1}{2}$

Section 10.2 The definition of the surface integral

2.1. $\frac{7}{2}$

2.3. $2\pi + \frac{32\pi^3}{3}$

2.5. $\frac{8}{3}$

Section 10.3 Stokes's theorem

3.1. (a) For example, G(x, y, z) = (-y cos z, x, y²z)

(b) 2π

(c) 2π

3.3. (a) (y - 2x, x + y, z)

(b) $\frac{3}{2}\pi$

Section 10.4 Curl fields

4.1. $\nabla \cdot F = 3$, not 0, so F is not a curl field.

4.2. It's a curl field with $G(x, y, z) = (\frac{1}{2}z^2, \frac{1}{2}x^2, \frac{1}{2}y^2)$, for example.

Section 10.5 Gauss's theorem

5.1. $\frac{384}{5}\pi$

5.3. $\frac{104}{3}\pi$

5.7. 225 - 32π

Section 10.6 The inverse square field

6.2. 11

6.4. It's the total mass contained in the interior of S, i.e., the sum of those mi for which pi is in the interior of S.

Section 10.7 A more substitution-friendly notation for surface integrals

7.1. (a) F(x, y, z) = (xy, yz, xz)

(b) 0

Section 11.4 Exercises for Chapter 11 (Working with differential forms)

4.1. $(x^2 + y^2)\, dx \wedge dy$

4.3. $(xy - z^2)\, dy \wedge dz + (yz - x^2)\, dz \wedge dx + (xz - y^2)\, dx \wedge dy$

4.5. 0

4.7. 2x dx + 2y dy

4.9. 4y dx ∧ dy

4.11. $(1 + xze^{xyz})\, dx \wedge dy \wedge dz$

4.13. $\frac{\partial F_1}{\partial x_1} - \frac{\partial F_2}{\partial x_2} + \frac{\partial F_3}{\partial x_3} - \frac{\partial F_4}{\partial x_4}$

4.14. (a) $1 + t^2$

(c) $\alpha^*(dy) = \cos t\, dt,\ \alpha^*(dz) = dt$

(d) (1 + t) dt

4.16. (b) $u^2\, du \wedge dv$

4.18. (a)
$$T^*(dx) = \frac{\partial T_1}{\partial u}\, du + \frac{\partial T_1}{\partial v}\, dv$$
$$T^*(dy) = \frac{\partial T_2}{\partial u}\, du + \frac{\partial T_2}{\partial v}\, dv$$

4.21. (a) $F_1\, dx_1 \wedge dx_2 + F_2\, dx_1 \wedge dx_3 + F_3\, dx_1 \wedge dx_4 + F_4\, dx_2 \wedge dx_3 + F_5\, dx_2 \wedge dx_4 + F_6\, dx_3 \wedge dx_4$

www.ingramcontent.com/pod-product-compliance
Lightning Source LLC
Chambersburg PA
CBHW082003190326
41458CB00010B/3048